The Handbook of Enviro

Volume 89

Founding Editor: Otto Hutzinger

Series Editors: Damià Barceló • Andrey G. Kostianoy

In over three decades, *The Handbook of Environmental Chemistry* has established itself as the premier reference source, providing sound and solid knowledge about environmental topics from a chemical perspective. Written by leading experts with practical experience in the field, the series continues to be essential reading for environmental scientists as well as for environmental managers and decision-makers in industry, government, agencies and public-interest groups.

Two distinguished Series Editors, internationally renowned volume editors as well as a prestigious Editorial Board safeguard publication of volumes according to high scientific standards.

Presenting a wide spectrum of viewpoints and approaches in topical volumes, the scope of the series covers topics such as

- local and global changes of natural environment and climate
- anthropogenic impact on the environment
- water, air and soil pollution
- remediation and waste characterization
- environmental contaminants
- biogeochemistry and geoecology
- chemical reactions and processes
- chemical and biological transformations as well as physical transport of chemicals in the environment
- environmental modeling

A particular focus of the series lies on methodological advances in environmental analytical chemistry.

The Handbook of Envir onmental Chemistry is available both in print and online via http://link.springer.com/bookseries/698. Articles are published online as soon as they have been reviewed and approved for publication.

Meeting the needs of the scientific community, publication of volumes in subseries has been discontinued to achieve a broader scope for the series as a whole.

Volatile Methylsiloxanes in the Environment

Volume Editors: Vera Homem · Nuno Ratola

With contributions by

M. Alaee · S. Augusto · A. Cincinelli · J. L. Cortina · M. Crest ·
N. de Arespacochaga · M. Farré · K. Gaj · E. Gallego · C. Guerranti ·
V. Homem · Y. Horii · K. Kannan · A. Katsoyiannis · I. S. Krogseth ·
T. Martellini · M. S. McLachlan · J. F. Perales · J. Raich-Montiu ·
N. Ratola · F. J. Roca · J. Sanchís · R. Scodellini · C. Scopetani ·
P. Teixidor · D.-G. Wang · N. A. Warner

 Springer

Editors
Vera Homem
LEPABE – Laboratory for Process
Engineering, Environment, Biotechnology
and Energy
Faculty of Engineering
University of Porto
Porto, Portugal

Nuno Ratola
LEPABE – Laboratory for Process
Engineering, Environment, Biotechnology
and Energy
Faculty of Engineering
University of Porto
Porto, Portugal

ISSN 1867-979X   ISSN 1616-864X   (electronic)
The Handbook of Environmental Chemistry
ISBN 978-3-030-50137-2   ISBN 978-3-030-50135-8   (eBook)
https://doi.org/10.1007/978-3-030-50135-8

This Springer imprint is published by the registered company Springer Nature Switzerland AG.
The registered company address is: Gewerbestrasse 11, 6330 Cham, Switzerland

Series Preface

With remarkable vision, Prof. Otto Hutzinger initiated *The Handbook of Environmental Chemistry* in 1980 and became the founding Editor-in-Chief. At that time, environmental chemistry was an emerging field, aiming at a complete description of the Earth's environment, encompassing the physical, chemical, biological, and geological transformations of chemical substances occurring on a local as well as a global scale. Environmental chemistry was intended to provide an account of the impact of man's activities on the natural environment by describing observed changes.

While a considerable amount of knowledge has been accumulated over the last four decades, as reflected in the more than 150 volumes of *The Handbook of Environmental Chemistry*, there are still many scientific and policy challenges ahead due to the complexity and interdisciplinary nature of the field. The series will therefore continue to provide compilations of current knowledge. Contributions are written by leading experts with practical experience in their fields. *The Handbook of Environmental Chemistry* grows with the increases in our scientific understanding, and provides a valuable source not only for scientists but also for environmental managers and decision-makers. Today, the series covers a broad range of environmental topics from a chemical perspective, including methodological advances in environmental analytical chemistry.

In recent years, there has been a growing tendency to include subject matter of societal relevance in the broad view of environmental chemistry. Topics include life cycle analysis, environmental management, sustainable development, and socio-economic, legal and even political problems, among others. While these topics are of great importance for the development and acceptance of *The Handbook of Environmental Chemistry*, the publisher and Editors-in-Chief have decided to keep the handbook essentially a source of information on "hard sciences" with a particular emphasis on chemistry, but also covering biology, geology, hydrology and engineering as applied to environmental sciences.

The volumes of the series are written at an advanced level, addressing the needs of both researchers and graduate students, as well as of people outside the field of "pure" chemistry, including those in industry, business, government, research establishments, and public interest groups. It would be very satisfying to see these volumes used as a basis for graduate courses in environmental chemistry. With its high standards of scientific quality and clarity, *The Handbook of Environmental Chemistry* provides a solid basis from which scientists can share their knowledge on the different aspects of environmental problems, presenting a wide spectrum of viewpoints and approaches.

The Handbook of Environmental Chemistry is available both in print and online via www.springerlink.com/content/110354/. Articles are published online as soon as they have been approved for publication. Authors, Volume Editors and Editors-in-Chief are rewarded by the broad acceptance of *The Handbook of Environmental Chemistry* by the scientific community, from whom suggestions for new topics to the Editors-in-Chief are always very welcome.

Damià Barceló
Andrey G. Kostianoy
Series Editors

Preface

The challenges of progress and sustainable evolution of life on our planet are becoming increasingly difficult to meet but at the same time have been fuelling unprecedented technological advances at an exponential rate unforeseen a few years ago. While for many centuries the natural resources were more than enough for the needs of human life and all ecosystems, their depletion and exponential growth of the population in the last decades have been leading us through a complicated path. Thus, the urge for the mass production of consumer goods triggers a quest for synthetic materials that can join an easy accessibility to a wide range of applications. With the awareness of scientists on an initial stage and eventually of the general population later on of the potential hazardous effects for life and the environment of some of these materials, a considerable number of chemicals that were employed profusely had to be replaced due to restrictions or ultimately a ban on their use. With the help of high-performance analytical protocols and equipment, the scrutiny of each new substance is currently much stronger (and potentially more accurate), which can reduce their lifespan. One class of chemicals that has been used for some time in numerous products and industrial applications is siloxanes. These silicon–oxygen-based structures possess an array of unique properties that combined make them extremely attractive for sealing, coating, insulating, stabilizing, wetting, moulding and waterproofing, among other applications. Not surprisingly, their global manufacture is increasing exponentially and, naturally, concerns of possible impacts of their use due to the release into the environment and the exposure of humans and biota were recently activated among the scientific community. This emerging effort has been engaging both academia and industry and has produced already several studies that in some cases point towards contradictory views. However, considerable gaps are still identified in the knowledge of the levels, trends, behaviour and effects of siloxanes. But nevertheless, some governing bodies (namely the EU) are already considering recommendations on the restriction of these chemicals in some personal care products. Therefore, we believe this is the appropriate moment to put forward this book, which intends to shed light on the state of the art of siloxanes research and promote the widening of the discussion among the different stakeholders.

With a focus on volatile methylsiloxanes (VMSs), the shorter-chained organosiloxanes that have been the object of most works so far, this book comprises 12 chapters that shed light on the main areas and environmental compartments where they have been found and studied. The first one is an overall description of the structural and functional properties, toxic risks and possible transformations of VMSs in the environment, while the second focuses on their main uses in numerous activities and products as well as the identification of the major sources of emission. The next chapter deals with the analytical strategies and protocols that have been used to address the quantification of VMSs, including the issue of possible cross-contaminations, and the following two are devoted to the presence of VMSs in wastewater treatment plants (WWTPs) in two ways: first the complete balance of their presence in all components of the treatment facilities and second the implications and effects on the production of biogas, both in WWTPs and in landfills. There is also one chapter dedicated to present levels, trends, fate and exposure of VMS in water bodies, while three others discuss one of the main receivers of VMS emissions – the atmosphere – focussing on indoor air, outdoor air and the general atmospheric fate, including modelling approaches. The following two chapters are dedicated to the levels in biota and the potential for biomagnification in the atmosphere and to the findings of VMSs in remote areas of the world, respectively. Finally, some conclusions and future directions for the upcoming studies are presented as closing chapter. This book is not by any means intended as a finishing line but rather an important step to improve the general state of the art of the knowledge of VMSs, to fuel new collaborations between research groups and/or with the industry and finally to convince more researchers to join the winding path that unravels the behaviour and effects of these ubiquitous but still understudied chemicals.

Porto, Portugal   Vera Homem
Porto, Portugal   Nuno Ratola

Contents

Properties, Potential Toxicity, and Transformations of VMSs in the Environment

Kazimierz Gaj

Contents

Abstract The main applications of siloxanes, sources of their release, their production volume, and the reasons for the growing interest in them are presented. The premises for which they are currently considered as potential environmental pollutants are explained and commented on. The physical and chemical properties of selected volatile methylsiloxanes are characterized with regard to both the increasing number of their applications and the impact on ecosystems and humans. On the basis of the available scientific literature and official risk assessments, their toxic potential, durability, circulations and transformations in the environment, long-range transport

K. Gaj (✉)
Faculty of Environmental Engineering, Wrocław University of Science and Technology, Wrocław, Poland
e-mail: kazimierz.gaj@pwr.edu.pl

V. Homem and N. Ratola (eds.), *Volatile Methylsiloxanes in the Environment*,
Hdb Env Chem (2020) 89: 1–32, DOI 10.1007/698_2018_360,
© Springer Nature Switzerland AG 2018, Published online: 21 August 2018

potential, degradation mechanisms, and fate in some environmental compartments are discussed. It is argued that they can interfere with the female reproductive system, the liver, and the lungs and irritate the skin, the eyes, and the respiratory system, considering that their bioaccumulation potential is relatively high. Due to their relative durability, they can be accumulated in various environmental matrices and transported to regions distant from the release sites. It is shown that the main mechanisms of their degradation are hydrolysis and demethylation with OH radicals in, respectively, the water and soil environment and in the air (referred to as the final sink in which the fundamental process of their degradation and decay take place). In conclusion, the need for further research on their impact on human health, their bioaccumulation and biodegradation, their life time and long-term impact on eco-systems, and the development of methods of removing VMSs from wastewater, sludge, and biogas is highlighted.

Keywords Emission, Environmental risk, Mechanisms of degradation, Persistence, Siloxanes

1 Introduction

The term "siloxanes" refers to a group of oligomeric spaces containing repeated Si–O–Si chemical linkages with one or several hydrocarbyl derivatives bound to Si, including alkyl, vinyl, phenyl polyester, aminopropyl, trifluoropropyl, and other functional groups. The alkyl group is a typical side and end group of siloxanes. Both due to common applications and potential nuisances, the most important alkyl siloxanes are volatile methylsiloxanes (VMSs) with the methyl substituent group.

Siloxanes have been considered as potential environmental contaminants for less than two decades, despite the fact that they have been present in our daily lives for almost 70 years, mainly as personal care products (PCPs).[1] The main reasons for the increasing concern about the rapid environmental burden of these unnatural substances are:

- Relative durability, which promotes their spreading over a large area, and accumulation in various compartments of the environment
- Lipophilicity, which can be the reason for their bioaccumulation and bioconcentration
- Low biodegradability
- Toxic effects (demonstrated in animal studies for some VMSs)

Owing to their unique properties, siloxanes are used in many industries, and their production is growing rapidly. Potentially, the greatest threat to the environment is their use for the production of cosmetics, cleaning agents, and solvents, whereby, due to their high volatility, they escape into the atmosphere and through wastewater

---

[1]The first attempts to use silicones in cosmetics production took place in the late 1940s.

treatment plants (WWTPs) and landfills penetrate into surface waters and soil. But most siloxanes are used as monomers for the production of silicone polymers in which they can occur as both intentional components and unbound impurities. Generally, they are widely used in personal care, in pharmaceutical and household products, as well as in many industrial applications. The PCPs in which they are most often used are shampoos, shower gels, creams and lotions, body moisturizers, hair conditioners, stick or cream deodorants, antiperspirants, solid and gel soaps, shaving foams and gels, aftershaves, dentifrice products, sun protection products, etc. Owing to their distinctive physicochemical properties (e.g., low surface tension, hydrophobic nature, transparency, lack of smell), VMSs are used mainly as softeners and carrier solvents. The content of siloxanes in the above products ranges from a few per cent to over 90%. As the ingredients of cosmetics, cyclomethylsiloxanes (cVMSs), in particular octamethylcyclotetrasiloxane (D4), decamethylcyclopentasiloxane (D5), and dodecamethylcyclohexasiloxane (D6), are mainly used. The highest cVMS concentrations in cosmetics are observed for D5 – from 0.01% in hand creams to more than 35% in deodorants [1] – and for D4, from 0.1 to 54% [2]. Typically, a mixture of cVMSs is used. A study of the cosmetics available on the market in Portugal showed the presence of VMSs in 96% of the analyzed samples [3]. Among them, cVMSs – mainly D4 and D6 – were most commonly detected (94%). Linear VMSs were detected in 52% of the samples (hexamethyldisiloxane (L2) was dominated). It is estimated that over 50% of the cosmetics launched in the last 10 years contained siloxanes [4]. Environment Canada found cVMSs to be present in about 6,000 cosmetics [5]. However, with the increase in the number of reports on the potential harmfulness of VMSs, the upward trend in their use in cosmetics has declined.

VMSs (and silicone polymers made from them) are also used in many industrial and home applications, e.g., as washing, cleaning, and impregnating agents, paints and coatings, thinners, textile dyes, polishes and wax blends, medical and pharmaceutical preparations, pH regulators, precipitants, flocculants, adhesives, sealants, heat transfer and dielectric fluids, defoamers and degreasers, lubricant oils, surfactants, etc. They are used in the optical, electronic, textile, leather, and building materials industry (e.g., as sealants and water repellent coatings). As defoamers, VMSs are used in the manufacture of pulp and paper, food, and petrochemicals as well as in water treatment. Thanks to their good adhesion to the skin and durability, silicone-based materials are used in blood-handling equipment, catheters, contact lenses, implants and prostheses, drains, heart bypasses and valves, respirators, bandages and slices, as protective barriers and lubricants for medical devices, and so on. Their use as an alternative to chlorofluorocarbons is important for reducing the depletion of the ozone layer and global warming.

As the volume of production of siloxanes and their polymers increases, so does the scale of the potential exposure to them. The emission of VMSs to the individual matrices of the environment increases, which raises concerns about their environmental fate and potential impact on humans and ecosystems. The total worldwide production of silicones in 2002 amounted to about 2 million Mg [6–8]. The dominant countries were the USA (~0.8 million Mg cVMSs) and China (~0.5 million Mg cVMSs) [9]. In the EU this production was about 0.5 million Mg in 2006, being construction the largest-using sector [10]. One should note that the above production figures can be underrated due to the confidentiality of data claimed by some companies and countries, especially in the EU. The largest increase has been observed in China, where in the years 1993–2002, the production tripled. Another increase (about twofold) took place in the period 2002–2008. China has actually become the largest producer and consumer of silicones in the world [11, 12]. Detailed data on the production of siloxanes and siloxane polymers, broken down into various industry sectors, can be found in [4]. They show that in 2013, the worldwide production of silicones exceeded 3 million Mg. Mojsiewicz-Pieńkowska and Krenczkowska [9] estimate that the global production currently amounts to 10 million Mg and that the global market has a value of around 19 billion $. More cautious data are provided by Rücker and Kümmerer [13], who estimate the total annual production of silicone polymers in 2017 at 6.5 million Mg and its annual increase since 2011 at 6.5%. Arespacochaga et al. [11] predict an increase of similar size (6%) by 2022. Most data on the production volume are for cVMSs, especially the ones potentially toxic and widely used in PCPs – D4 and D5. OECD has identified D4 and D5 as high-production volume chemicals [14]. According to the European Chemical Agency (ECHA), the recorded production volume is in the range of 100,000–1,000,000 Mg/year for D4 and 10,000–100,000 Mg/year for D5 [15]. The quantity of D5 used in PCPs in the EU–27 in 2012 amounted to about 20,000 Mg. The precise figure for D4 is unknown, but it is much lower (below 375 Mg/year) than for D5. The use of D4 in PCPs has been rapidly decreasing since 2002 owing to its substitution with potentially less harmful D5 [15].

Because of their special properties (volatility, relative durability, poor biodegradability, hydrophobicity), high-production volume, and widening range of applications, VMSs migrate into the air, surface waters, sediments, and soils, and through them they get to the biota. Considering the potential harm to humans and the environment, and the rapidly growing scale of applications and production volume, the environmental fate of VMSs is of critical importance.

Because of the highest exposure, the potentially adverse health effects and the most frequent detection in different environmental matrices and biogases, the analysis of VMSs was limited to eight compounds (Table 1), particularly focusing on cVMSs – potentially the most onerous.

2 General Characteristic of VMSs

Volatile methylsiloxanes differ significantly in their polymerization degree and spatial configuration, forming linear and cyclic structures. VMSs having a chain structure are most commonly designated with the letter L or, rather rarely, with the letter M, while the ones with a ring structure are designated with the letter D (Table 1). The number following the letter L or D refers to the number of silicon atoms in the compound.

Table 1 Nomenclature and structure of selected VMSs

Chemical names	Abbreviations	CAS/EC No.	Chemical formula/molar mass, g/mol	Structural pattern
Hexamethyldisiloxane	L2, MM	107-46-0 203-492-7	$Si_2-O-(CH_3)_6$ 162.4	
Octamethyltrisiloxane	L3, MDM	107-51-7 203-497-4	$Si_3-O_2-(CH_3)_8$ 236.5	
Decamethyltetrasiloxane	L4, MD$_2$M	141-62-8 205-491-7	$Si_4-O_3-(CH_3)_{10}$ 310.7	
Dodecamethylpentasiloxane	L5, MD$_3$M	141-63-9 205-492-2	$Si_5-O_4-(CH_3)_{12}$ 384.8	
Hexamethylcyclotrisiloxane, cyclotrisiloxane	D3	541-05-9 208-765-4	$Si_3-O_3-(CH_3)_6$ 222.5	
Octamethylcyclotetrasiloxane, cyclotetrasiloxane	D4	556-67-2 209-136-7	$Si_4-O_4-(CH_3)_8$ 296.6	
Decamethylcyclopentasiloxane, cyclopentasiloxane	D5	541-02-6 208-764-9	$Si_5-O_5-(CH_3)_{10}$ 370.8	
Dodecamethylcyclohexa-siloxane, cyclohexasiloxane	D6	540-97-6 208-762-8	$Si_6-O_6-(CH_3)_{12}$ 444.9	

According to INCI (International Nomenclature for Cosmetic Ingredients), instead of the chemical names of the individual compounds, the trade names of siloxane groups – dimethicone and cyclomethicone – are usually used on PCP labels. The former group comprises linear VMSs (as monomer or polydimethylsiloxane – PDMSs) and the latter group – cyclic VMSs. Both groups can contain VMS mixtures in varying proportions.

3 Physical and Chemical Properties of Major VMSs

Generally, siloxanes are characterized by high thermal stability, hydrophobicity, high resistance to oxidation, UV radiation and biodegradation, low surface tension and viscosity, high compressibility, and low chemical reactivity. Siloxane polymers retain their properties even at temperatures in the order of 300–400°C, provided that they are free from impurities in the form of volatile monomers [16]. Moreover, they are characterized by dispersion, antifoam, and damping properties, low crystallization temperature, and high spreadability. At ambient conditions most of them are colorless and odorless fluids well-permeable to UV radiation and visible light. These useful properties largely depend on the length of the Si–O backbone and the type of substituted side and end groups. Some of the properties (e.g., low viscosity) stem from the unique flexibility of the polysiloxane chain, greater than in the case of the C–C chains in standard organic polymers. This is mainly due to the larger distances between the Si–O atoms in the chain [13].

Besides the desirable functional properties, siloxanes have undesirable characteristics, such as relative persistence in the environment, lipophilicity (posing a bioaccumulation risk), high vapor pressure, and volatility at ambient conditions (posing a threat to the air quality). The fact that siloxanes are durable means that they potentially can undergo long-range transport (LRT). The group of analyzed VMSs contains many physically different compounds. Cyclic are more stable than the linear ones capable of self-disintegrating to L2. Among cVMSs, D6 is characterized by a relatively low vapor pressure, whereby it does not volatilize readily, whereas chemically unstable D3 evaporates rapidly due to its high vapor pressure, whereby its concentration in sewage and sludge (and therefore in biogas) is low. The high vapor pressure and poor solubility in water of VMSs result in high air/water partition coefficients (K_{AW}) and octanol/water partition coefficients (K_{OW}) (Table 2). The high value of log K_{OW} indicates a potential for biomagnification. Because of the low water solubility, the Henry's law constants are high, whereby VMSs have a strong tendency to partition from water and wet soil into the air.

Due to their high K_{OW} coefficients, VMSs exhibit lower solubility in water than many organic compounds and have a unique ability to adsorb on numerous materials, which promotes their accumulation in sediment and sewage sludge. The latter feature is associated with the relatively high organic carbon/water partition coefficient (K_{OC}) of some cVMSs. A higher log K_{OC} value was measured for D5 than for D4, but both adsorb strongly to organic particles [15]. Along with sewage sludge,

Table 2 Environmentally important properties of selected VMSs

	Physical state at 20°C, 101.3 kPa [17]	Water solubility at 25°C µg/dm³ [18]	Henry's law constant at 25°C kPa m³/mol [18]	Vapor pressure at 25°C Pa [17]	Boiling point at 101.3 kPa °C [17]	Melting point °C [17]	Flash point °C [17]	BCF dm³/kg [17, 19]	Log K_{OW} at 25°C [18]	Log K_{OC} at 25°C [20]	Log K_{OA} at 25°C [21]
L2	Liquid	930	486	5,613	101	−66	1	1,300	5.24	3.75	2.98
L3	Liquid	34	2,027	445	152	−80	39	10,000	6.6	4.37	3.77
L4	Liquid	6.7	11,146	50	195	−76	63	6,910	8.21	4.80	4.64
L5	Liquid	0.3	27,358	13	232	−80	117	4,260[a]	9.61	5.54	NA
D3	Solid	1,560	152	1,160	135	64	35	2,500	5.64	3.55	NA
D4	Liquid	56.2	1,210	141	176	17.5	56	14,900[a]	6.74	4.17	4.31
D5	Liquid	17	3,354	27	210	−38	76	16,200[a]	8.03	4.60	4.95
D6	Liquid	5.1	4,945	3	245	−3	105	2,860[a]	9.06	5.08	5.77

NA not available

[a]European Chemical Agency [19]

siloxanes can enter fermentation chambers at WWTPs and volatilize into biogas. Then, as a result of biogas combustion, they form hardly removable deposits blocking biogas energy-utilizing devices (boilers, turbines, internal combustion engines, heat exchangers, exhaust catalysts, fuel cells, and so on). Such coating is mainly composed of silica and silicates [22]. In the case of piston engines, it covers the inside walls of their chambers, the cylinder heads, the valves, and the spark plugs and accumulates in the engine oil. This results in worse heat exchange, an increased flue gas temperature, reduced engine performance, the increased emission of air pollutants, and the deterioration of the oil lubricating properties. Therefore, it is necessary to remove siloxanes from biogas, which is neither easy nor cheap, the more so that comprehensive methods of biogas purification are still in the research phase [23]. An additional environmental hazard is the emission of fine crystalline silica into the atmospheric air [24]. The strong tendency of siloxanes to adsorb on various matrices makes their sampling and chemical analysis very difficult, which can lead to serious analytical errors.

The octanol/air partition coefficient (K_{OA}) is primarily used to estimate the partition of substances from the air to aerosol particles and to the surfaces of solids, such as soil and vegetation. It can also be used to estimate the biomagnification potential. The value of log K_{OA} is relatively low for VMSs in comparison with that for, e.g., PCBs. Hence, the potential of VMSs for depositing from the air onto surface media and for biomagnification is low.

An overview of the main physical and chemical properties of VMSs, based on a literature survey and the available chemical databases, is presented in Table 2. Because of the mentioned above unfavorable environmental properties of some VMSs, numerous studies have been carried out to check whether they should be classified as persistent organic pollutants (POPs) and as substances undergoing LRT according to the criteria of the Stockholm Convention [25]. It is also examined whether they meet the PBT (persistence, bioaccumulation, toxicity) criteria according to REACH [26].

More detailed information on the physicochemical properties of organosiloxane compounds can be found in several recent review papers [13, 27–31].

4 VMSs as Potentially Toxic Pollutants

4.1 Background

The increasing production of siloxanes and their wider application result in higher levels of VMS emission and greater human exposure. Therefore, there is a need for more detailed studies of the potential adverse or harmful effects of VMSs on organisms.

The toxicity of silicone compounds, in particular PDMS, has been thoroughly investigated due to their widespread use in medical technology and cosmetics. A considerable number of test results have indicated that these polymers are neutral to warm-blooded organisms. Until recently, individual siloxanes were also believed to be nontoxic. However, the studies carried out on animals in recent years have shown that some of them may directly or indirectly affect the human body. The greatest number of data has been published on the toxicity of cVMSs, especially D4 and D5, which may pose a special threat to people due to their widespread use in cosmetics. The toxic properties of D3, D6, and linear VMSs are poorly understood.

Environmental siloxane hazards have been analyzed by several government agencies in various countries. Risk assessments of cVMSs have been carried out in Denmark [32] and the other Nordic countries [33], Canada [5, 34, 35], the UK [6–8], and the EU [2, 36, 37]. In the USA, at the request of the US EPA, 2-year studies on rats have been conducted. In its fact sheet on D5 in dry-cleaning applications, EPA announced risk assessments for D5 [38]. So far no report on their results has been published.

4.2   Main Routes of Human Exposure

People are exposed to VMSs by breathing indoor and outdoor air (including dust on which VMSs can be adsorbed or condensed), through food (VMSs are soluble in fat), by drinking water (this applies mainly to the VMSs more soluble in water, such as L2 and D3), and through direct contact with the products containing them in their formulations (PCPs applied to the skin, the hair, and the mouth). The routes of most serious exposure are inhalation and absorption through the skin. Due to the high vapor pressure and elevated concentrations of VMSs in indoor air, breathing is considered the primary absorption path [39]. Most of the cVMSs contained in PCPs evaporate during their application. It is estimated that less than 1% gets into the sewerage system [13].

In studies [39], the daily inhalation exposure to the sum of VMSs was estimated to be in the range of 0.27–3.18 µg/kg bw.,[2] depending on the age of the examined individual. There are few publications on dermal absorption of VMSs, but this absorption was found to be poor – below 1% [40, 41]. It decreases with decreasing molar mass and increasing volatility. Capela and co-authors [3] studied the daily exposure to VMSs for 12 of the most popular types of cosmetics and toiletries. Their modelling shows that daily dermal exposure for adults is within the range of 25.04–89.25 µg/kg bw. Of the PCPs studied, body moisturizers were the main contributors (86%), followed by face creams (9%) and aftershaves (3%). The cVMSs (mainly D5) content was found to be the highest. Of the linear VMSs, the highest level of exposure was found for face creams. In general, the highest dermal

---

[2]Body weight.

and inhalation exposure, especially to D5, has been linked to the use of body moisturizers. However, considering the low permeability of D5 through the human skin and its rapid evaporation, it can be expected that systemic toxicity is unlikely to occur via this route. A major route is the respiratory tract since D5 from lungs can be distributed to fatty tissues and bioaccumulate there [42]. A relevant opinion has been issued by the Scientific Committee on Consumer Safety (SCCS) [43]. It states that the use of D5 in cosmetic products is safe for human health, with the exception of some cosmetics in the form of aerosols and pressurized sprays. In these cases, excessive exposure to the respiratory tract may occur. This opinion is based on the maximum concentrations in the finished products. An earlier opinion from the SCCS [44] also states that D4 and D5 used in cosmetics do not pose a risk to human health. Other uses were not considered.

4.3   Toxicity of VMSs According to Official Risk Assessments and Animal Studies

4.3.1   Criteria for Assessment of Toxicity and Bioaccumulation

According to REACH [26], a given substance fulfils the toxicity criteria in each of the following cases:

- The long-term no-observed effect concentration (NOEC) for marine or freshwater organisms is less than 10 $\mu g/dm^3$.
- The substance is classified as carcinogenic (category 1A or 1B), mutagenic (category 1A or 1B), or toxic for reproduction (category 1A, 1B, or 2) according to [45].
- There is other evidence of chronic toxicity, as defined by the classifications STOT RE (specific target organ toxicity after repeated exposure), category 1 (oral, dermal, inhalation of gases/vapors, inhalation of dust/mist/fume), or category 2 (oral, dermal, inhalation of gases/vapors, inhalation of dust/mist/fume, according to [45]).

A substance is considered as bioaccumulative (B) when the bioconcentration factor (BCF[3]) in aquatic species is higher than 2,000 dm^3/kg and very bioaccumulative (vB) when BCF is above 5,000 dm^3/kg or, in the absence of this factor, that the log K_{OW} is greater than 5.

---

[3]BCF (dm^3/kg bw.) is the mass ratio of the accumulated substance and body weight (kg/kg bw.) divided by the concentration of this substance in the surroundings (kg/dm^3).

4.3.2 Toxicity of Cyclic VMSs

D3

The data on the cVMSs analyzed show D3 to be the least toxic. The few studies presented by ECHA [19] indicate that it has no potential for bioaccumulation and does not irritate the skin or the eyes. Mutagenicity tests also proved negative. There are no data on its long-term toxicity to aquatic organisms. Only short-term toxicity toward fish (*Rainbow trout*) and protozoa (*Daphnia magna*) was investigated, and no toxic effects were observed.

D4

According to [45], D4 belongs to the environmental hazard class and category "Aquatic Chronic 4" and hazard statement "H413," which means that it may cause long-lasting harmful effects to aquatic life. Due to the impact on human health, its hazard class and category is "Repr.2" (toxic for reproduction) and hazard statement "H361f" (suspected of damaging fertility).

The last classification is based on studies of rats subjected to inhalation exposure. The studies are expertly presented and summarized in detail in [6, 15, 46], and in them, all authors dispute the claim that the mechanism responsible for D4's adverse reproductive effects in animals is also relevant to human health. The more so because the adverse effects occurred at concentrations much higher (in the order of several hundred ppm_v) than the ones recorded in the atmospheric and indoor air. The inhalation of D4 can also cause liver enlargement and respiratory tract irritation in rats. The observed hepatomegaly appears to be associated with the effect of D4 on liver metabolic enzymes.

The carcinogenicity of D4 was discussed by Brooke et al. [6]. The studies were carried out on rats exposed daily to concentrations of $10-700$ ppm_v for up to 2 years. Uterine adenomas were found in 11% of the females at the concentration of 700 ppm_v. There were no cases of uterine adenomas in the parallel control groups. It has been shown that endometrial cancers arise because D4 acts as a dopamine antagonist. Studies of the toxicity of D4 (up to 1 g/kg/day) absorbed via the oral route were carried out on rats, but no evidence of any adverse developmental effects caused by D4 was found. The reproductive toxicity of D4 in rats can be associated with the inhibition of the release of the luteinizing hormone. It is considered unlikely that such effects will occur in humans because of the significant differences in the cycle regulation mechanisms. Fertility impairment (the estrogenic effect) as a critical effect of D4 was confirmed by the Danish EPA on the basis of rat inhalation studies [32], which also showed the effect of D4 on the liver (an increase in liver weight) and the lung (chronic interstitial inflammation of the lung). Similar results are reported by Zhang et al. [47]. Assessing D4 toxicity from a lethal dose for mice (LD_{50} ~6–7 g/kg [48]), it was concluded that it is similar to carbon tetrachloride or trichlorethylene. The effect of D4 on the estrous cycle of rats and mice has been repeatedly confirmed [49–56]. Quinn et al. [52] showed ovulation suppression in female rats, resulting from the oral dosing of D4. Also the effect on the sexual cycle was observed in rats subjected to the inhalation of D4 vapor. The reduction in the number of pregnancies occurred under the influence of the D4 concentration of 700 ppm$_v$ for 6 h prior to conception [55].

In study [57] rats were exposed in inhalation chambers to 0, 10, 30, 150, or 700 ppm$_v$ of D4 fumes for 6 h/day, 5 days/week, and 104 weeks. As a result of 2-year exposure, there were an enlargement of the liver, the kidneys, and the testes, an increase in uterus weight, hypertrophy of the interstitial cells, endometrial hyperplasia, and upper respiratory tract irritation. The impact of D4 on the respiratory system, the liver, and the kidneys did not result in carcinogenic changes in these organs. However, an increased incidence of uterine adenoma was observed. The mechanisms of the abovementioned interactions after inhalation and dermal and oral exposure are uncertain.

The harmful effects of D4 on aquatic organisms have been experimentally demonstrated [5]. The authors concluded that D4 "has or may have immediate or long-term harmful effect on the environment or its biological diversity." On the other hand, they stated that D4 is not introduced into the environment in an amount or concentration, which can pose danger to human life or health. According to literature review [31], the NOEC of D4 for some aquatic organisms is below 10 µg/dm^3.

The conclusion is that D4 meets the toxicity criteria of the REACH [26]. Environment Canada [5] stated that D4 has the potential to cause ecological harm, particularly as a result of long-term exposures, and classified D4 among substances with some potential for bioaccumulation in aquatic organisms. Furthermore, according to Brooke et al. and the European Chemical Agency [6, 15], D4 should be classified as "very persistent and very bioaccumulative" (vPvB) and as a "persistent, bioaccumulative, and toxic" (PBT) substance in the environment. The most important adverse effect of D4 exposure, confirmed by the European Commission, is the impairment of fertility by estrogenic activity.

D5

In contrast to D4, D5 is nontoxic to aquatic organisms. At concentrations up to water solubility, there were no observable effects in both short- and long-term studies [37]. D5 does not meet the mentioned in Sect. 4.3.1 criteria for substances cancerogenic, mutagenic, or toxic to specific organs. The available data show no evidence of chronic D5 toxicity. Several studies have shown no toxic effects of D5 on the reproductive process [52–54]. No negative effects of D5 skin dosing have been noted [15]. Therefore, D5 is not classified among human health hazards. On the other hand, its adverse effects on animals, e.g., liver enlargement [58–60] and increased incidence of uterine endometrial adenomas and adenocarcinomas [32], have been observed. Despite this, the ECHA [15] and the UK Environment Agency [6] have classified D5 among vPvB substances under the REACH criteria, mainly because of its relatively long hydrolysis half-life in freshwater and low biodegradability. However, according to the Environment Canada, D5 cannot be classified as a substance meeting the criterion for bioaccumulation, due to the contradictory evidence between laboratory studies and model calculations [34].

The statistically significant increase in uterine adenocarcinomas in female rats in inhalation study [58] was probably due to the antagonistic effect of D5 on dopamine. Considering the differences in the reproductive process between rats and humans, this mechanism will presumably not occur in humans. However, it can occur in other mammals and birds. Moreover, it was found that the enlargement of the liver is not accompanied by any functional or histopathological changes [37]. D5 is considered to have no mutagenic potential.

Disturbing research results, which may indicate the carcinogenicity of D5, were presented in 2005 by US EPA [38]. The study was conducted on 120 individuals in the population of laboratory rats, which were exposed to D5 with vapor concentrations of 0, 10, 40, and 160 ppm_v for 6 h/day, 5 days/week, and 24 months. The results showed a statistically significant increase of uterine cancer among female rats exposed to D5 highest concentration. The potential carcinogenicity of D5 (the increase in uterine adenocarcinoma) was also found by Dekant and Klaunig [42] and Jean et al. [61]. Therefore, the reproductive toxicity of D5 can raise concerns. However, there are no data to draw unambiguous conclusions about the risk to humans. The risk does not seem to be significant because the concentrations used in the rat studies were usually much higher than the ones detected in the atmospheric and indoor air.

According to the SCCS opinion regarding D5 in cosmetic products [43], the acute toxicity of D5 by inhalation, dermal, and oral routes is relatively low, and it can be considered to have no genotoxic potential. The document states that D5 is slightly irritating to the skin and the eyes and that the liver, the lungs, and the uterus are potential target organs after repeated-dose inhalation exposure. The reported conclusion is that the use of D5 in cosmetic products is safe at the reported concentrations, except for hair styling aerosols and sun care spray products, for which, at the maximum concentrations declared by the applicant, exposure to D5 may lead to locally toxic air concentrations. The authors of the opinion note that it does not apply to the use of D5 in oral care products. They recommend that the purity of D5 in cosmetic products should exceed 99%, considering the risk of exposure to D4 – a trace contaminant of D5.

D6

Because of the larger size of the D6 molecule, and so its lower penetrability through cell membranes, it can be expected that its toxicity is lower than that of D4 and D5. The toxicity of D6 has been studied only for four aquatic organisms [30], and the authors conclude that the typical concentrations of D6 in the environment do not have any toxic effects on those organisms.

There are no data on the carcinogenicity and chronic toxicity of D6, and there is only limited information about the effects of D6 on reproduction and genotoxicity. In addition, the available data are mainly based on oral exposure (there are no data for inhalation and dermal exposure). The results presented in [35], which refers to an unpublished piece of research, indicate that D6 does not adversely affect fish (*fathead minnow*) and *daphnia* at concentrations close to its water solubility. But D6 toxicity forecasts made using the model [62] indicate that at these concentrations, chronic adverse effects can develop in aquatic organisms. The reliability of the results is, however, questioned. The lower bioavailability of D6 relative to D4 and D5, resulting mainly from its lower solubility in water and lower bioaccumulation potential, indicates, in the above authors' opinion, that no toxic threshold for the adverse effect should be expected for D6 at its water solubility limit. The liver was indicated as the human target organ for oral and inhalation exposure. The conclusion was that D6 is not introduced into the environment in amounts and concentrations and under conditions, which can adversely affect the environment or its biological diversity. Such conclusions can be found in a similar screening environmental risk assessment of D6 [6].

According to the PBT Profiler chemical database [18], D6 is classified as persistent in the environment because its half-life in sediment (540 days) exceeds the US EPA criteria (6 months). D6 is also classified as a substance with potential for bioconcentration in fish and aquatic organisms (it may accumulate in the food chain). However, according to [45], D6 (similarly as D5) is not classified among dangerous substances.

Therefore, due to the limited number of studies at present, it is not possible to draw unambiguous conclusions about D6 toxicity.

4.3.3 Toxicity of Linear VMSs

L2

There are few data on the toxicity of linear VMSs. Their systemic toxicity after oral, dermal, or inhalation exposure is considered low. However, they have potential for dermal irritation [33, 63]. Due to its possible carcinogenic and toxic effects, L2 was put on the OSPAR list of Chemicals for Priority Action in 2000 [64]. However, in 2007, OSPAR agreed to deselect L2 from this list because foregoing research had not confirmed its toxic or carcinogenic properties.

The analyses presented in [64] indicate that L2 seems to have no carcinogenic potential and does not affect the one-generation reproductive assay in rats and there is no evidence of its mutagenic effects in vitro or in vivo. An analysis of compliance with the EU PBT criteria shows that only criterion P may be exceeded. Additionally, L2 can be considered a weak antiestrogen compound with no measurable effect on uterine weight [56].

According to [45], the available oral, dermal, and inhalation toxicity data show that L2 does not meet the criteria for classification for acute toxicity. However, in the

ECHA database [19], there is information that L2 is toxic to aquatic life with long-lasting effects and it is suspected of causing cancer. Similar information can be found in the PBT Profiler database [18] – L2 is chronically toxic to fish, and in the PubChem database [65], L2 is very toxic to aquatic life (hazard statement *H400*) and toxic to aquatic life with long-lasting effects (*H411*), which means that it is hazardous to the aquatic environment in both the short and long term. There is also a warning that L2 can irritate the eyes.

The absorption of L2 through the human skin is very low (0.023%), and the bioaccumulation of L2, despite its lipophilicity, is unlikely due to its effective removal through metabolism and exhalation [66]. It was not excluded that L2 may have properties that threaten human health at high concentrations, especially target organs such as kidneys and testicles. It was clearly stated that L2 is not expected to be genotoxic or oncogenic. However, considering the above facts, this does not seem so obvious.

According to [45], L2 does not meet the criteria for classification as irritant to the skin and the eyes, and it is not classified with regard to skin sensitization, mutagenicity, and immunotoxic effects. The available data suggest that L2 does not need to be classified for carcinogenicity and do not imply that L2 should be classified for adverse effects on fertility or offspring development [19].

L3

Detailed toxicokinetics and acute toxicological data are not available for L3. According to the European Chemical Agency [19], no short- and long-term exposure hazards have been identified. Due to its relatively high K_{OW} and low water solubility (Table 2), dermal absorption is unlikely, especially since L3 is very volatile. However, there are no dermal toxicity studies to verify this. On the basis of the tests of the structurally similar L2 molecule, it can be predicted that the absorption of L3 by inhalation is also low.

According to [44], L3 is neither classified for mutagenicity and reproductive or developmental toxicity nor as a skin- or eye-irritating substance. However, this compound is potentially toxic to aquatic organisms with long-lasting effects (hazard statement *H410*) and can temporarily irritate the eyes [65]. On the other hand, L3 is not irritating to the skin of rabbits and is not a sensitizer in the human patch test, but it has potential to bioaccumulate and is not readily biodegradable [67]. In accordance with [18], L3 is chronically toxic to fish and can be toxic also to other aquatic organisms. No data are available on the carcinogenicity of L3.

Considering that the bioavailability of L3 is lower, its solubility in water almost 30 times lower, and its molecule size larger than that of L2, one can assume that the adverse effect of L3 is much weaker.

L4

There are only three reliable inhalation studies of L4 [19]. According to one of them (the longest duration study on rats), L4 does not cause any toxicologically significant adverse effects via this route of exposure. In the rat uterotrophic assay, only a weak estrogenic response was noted. There are no studies on acute toxicity via the inhalation route, respiratory irritation, repeated-dose toxicity for systemic and local dermal exposures, genetic toxicity in vivo, the effect on fertility via the dermal route, and the effect on developmental toxicity via oral and dermal routes. Adverse effects were observed after the repeated dosing of L4 via the oral route only [19].

According to [45], based on the analyses contained in [19], L4 does not require classification for acute toxicity, adverse effects of repeated-dose toxicity, skin or eye irritation and sensitization, mutagenicity, and effects on reproduction.

On the basis of the chemical structure of L4, the PBT Profiler [18] predicts that L4 cannot be classified as a chemical raising concerns about human health.

L5

Considering the similar structure and physicochemical properties of L4 and L5, it can be expected that the toxicological profiles of the substances should also be similar. A few available studies indicate that the substances are neither acutely toxic nor irritating [19]. The high hydrophobicity of L5 means that it is unlikely to be absorbed through the skin and enter the bloodstream. The inhalation of L5 fumes is also very limited as its vapor pressure is the lowest of the VMSs analyzed. Only the inhalation of L5 aerosols could take place. But absorption via this route is also unlikely due to the relatively high log K_{OW} (Table 2). However, there has been no research in this area so far. There are no studies on L5 metabolism, its toxic effects on reproduction and development, respiratory sensitization and irritation, fertility, and genetic toxicity.

According to [45], L5 does not require the same classification as L4.

4.4   Summary Comments on VMS Toxicity

On the basis of the available literature data, it is difficult to unambiguously assess the toxicity of this group of compounds. Only D4 and D5 have been thoroughly tested on animals for their environmental and health impact. The prevailing view in the scientific literature is that VMSs are nontoxic to humans and the environment or their toxicity is very limited. For example, the authors of recently published studies and reviews [42, 46, 68, 69] concluded that the changes in the reproductive cycle and uterine tumors caused by D4 and D5 in rats are not relevant to the human risk. They claim this on the basis of the differences between the species in the regulation of their reproductive systems and the inadequately high levels of exposure (up to several hundred ppm$_v$) in the tests on rats. But since the cancer-inducing mechanism has not been established so far, such unambiguous conclusions seem to be premature. There is also no full explanation of the mechanisms of the influence of VMSs on various organs. Undoubtedly, the short-term impact of VMSs, especially at low realistic air and water concentrations, is well tolerated by humans. However, their long-term effect on health and the environment needs further research. Because of the high lipophilicity of VMSs, particular attention should be given to the risk of their bioaccumulation and bioconcentration. All the analyzed VMSs, except L2, can be considered as bioaccumulative (BCF > 2,000 dm^3/kg) while L3, L4, D4, and D5 as highly bioaccumulative (BCF > 2,000 dm^3/kg) (Table 2).

Many studies on animals indicate that cVMSs can damage the liver and the lungs and cause endocrine and immunological disorders. In addition, oral exposure to D4 may cause estrogenic effects, including infertility. However, there are no representative long-term studies on humans, which would confirm these adverse effects. An additional threat posed by cVMSs is their relative durability and the risk that they can impact a large area.

Another problem is the omission of cumulative exposure to VMSs in risk assessments. Because VMSs are able to interconvert via hydrolysis and condensation, they should be considered both separately and jointly (synergism). The more so that their mixtures are usually used in cosmetics and PCPs. On the other hand, the large number of elements in this group, their varied properties, and the lack of uniform classification criteria can pose a significant difficulty.

For the above reasons, it is certainly worth gradually reducing their use (mainly D4 and D5) by replacing them with less stable and less potentially toxic substitutes and limiting their content in PCPs. The ECHA [15] has issued an opinion recommending restrictions on D4 and D5 in PCPs. It reads as follows: "D4 and D5 shall not be placed on the market or used in concentrations equal to or greater than 0.1% by weight of each in personal care products that are washed off in normal use conditions." The opinion refers to PCPs, which are washed away with water and discharged to sewage within a few minutes of application. One of the reasons for the restriction is that D5 is vPvB, while D4 is both a vPvB and PBT substance in accordance with the REACH [26]. Similar restrictions were previously proposed by the UK Competent Authority [70].

A separate problem relating to the toxicity of VMSs is the final product of their oxidation. Both in the atmosphere and as a result of the combustion of biogas and waste or sludge containing VMSs, the final product of their transformation is (besides CO_2 and H_2O) crystalline submicron silica, which, apart from being able to penetrate, owing to its small particle size (below 100 nm), directly into the bloodstream, is classified as a potential carcinogen – because of its fibrous form [71]. According to the classification [72], silica dust has carcinogenic and mutagenic properties and can induce asthma and affect reproductive system. An additional threat results from the large specific surface area of these particles, on which other dangerous air pollutants can adsorb and accumulate together with SiO_2, e.g., in the liver or other organs.

5 Persistence, Circulation, and Transformation of VMSs in the Environment

5.1 Distribution Routes of VMSs in Main Compartments of the Environment

Waste VMSs, from both the production of semifinished products containing them (oligomers and siloxane polymers) and the production and use of the end products (mainly PCPs), are mostly emitted directly to the atmosphere via ventilation systems. More than 90% of the cVMSs used in PCPs enter the atmosphere [73]. The main source of their emission to the air are body moisturizers [3]. Some of the less volatile and heavier VMSs, especially the cyclic ones, get (e.g., from rinsed cosmetics) into the sewerage system and WWTPs. The rest is transferred to waste disposal. The D5 content in municipal sewage amounts to over 90% of the total VMSs [74]. Siloxane polymers (mainly PDMS) and the spent packaging with their residues are also sent to WWTPs and to landfills. Then they undergo hydrolytic depolymerization, transforming into VMSs and their decomposition products – silanes and silanols. From the landfills, VMSs get into the air and biogas, or, along with the leachate, they return to WWTPs or leak into the surrounding soils. About a few per cent of D4 reaches WWTPs [75]. The larger sources of VMSs in the sewerage systems are wash-off products like shower gels, shampoos, soaps, and hair conditioners [15]. An additional VMS source at WWTPs is their local use as antifoaming agents.

The relative durability and poor biodegradability of VMSs do not allow their effective removal in the conventional process of biological wastewater treatment. In water, VMSs undergo hydrolysis, whose efficiency depends mainly on the congener, the pH, the temperature, and the presence of catalysts in the form of acids or bases. Their $\tau_{1/2}$ in this medium ranges from several hours to several months. The main ways to remove and transfer VMSs from wastewater to other environmental compartments are adsorption on flocs of activated sludge and sediments (owing to their high log K_{OC} values) and volatilization. The latter method especially applies to smaller siloxane molecules with a high vapor pressure (D3, L2), which are released into the air during sewage treatment, mainly in the aeration chambers or already in the sewerage system. The VMSs that have not evaporated and are more soluble in water (e.g., L3) get into the receivers of treated wastewater. The typical removal efficiency of D4 is about 96% (48% to sludge and 48% to the air). For D5 it is c.a. 95% (~73% to sludge and 22% to the air) [6, 15]. Higher removal efficiencies for D5 and D6 (above 98%), based on their influent and effluent concentration tests, were obtained by Van Egmond et al. [76].

Sewage sludge, along with the adsorbed VMSs, can be deposited on landfills, burned after drying, composted and used for fertilizing and soil enrichment, or fermented as part of the sludge stabilization processes. Depending on the adopted procedure, VMSs will get into different compartments of the environment. As a result of the fermentative decomposition of the sludge and the elevated temperature

(35–55°C) in the fermentation chamber, they are released into the biogas and flow with it to the air or to the combustion plant. This applies in particular to D4 and D5, which are mainly detected in biogas from WWTPs. During biogas combustion, VMSs partially precipitate in the form of silica and silicate deposits on the internal surfaces of the apparatus. The remaining part is emitted into the air as fine crystalline silica or in the unchanged form. It is estimated that only 0.5–1% of VMS content in biogas converts to SiO_2, while the rest is emitted into the air with exhaust gases [22]. VMSs with a lower vapor pressure and a higher molecular weight (D6, L5) mostly remain associated with post-fermentation sludge. This can be a problem in the case of their energy use. The combustion of biogas and sewage sludge, the storage of unfermented sewage sludge, or its use for soil enrichment can contribute to the release of D4 to other environmental media. Additionally, as a result of its contact with soil, PDMS can be decomposed to D4 and D5 (as transient degradation products), and their increased release into the air from agricultural areas and landfills can occur. About 71% of the D5 content in sewage sludge can partition to the air after contact with moist soil [34]. D6 behaves completely differently. Because of its high K_{OC}, it is expected to adsorb strongly to soil, and therefore its evaporation potential is much lower. In the case of dry soil, VMSs are quickly hydrolyzed by clay minerals to form dimethylsilanediol (DMSD) as the final breakdown compound.

5.2   Biodegradability of VMSs

Biochemical processes leading to the breaking of bonds between silicon atoms and the methyl groups were described by Dewil et al. [77]. Some researchers [6, 78–81] also described the biodegradation of PDMS in contact with dry soil. Its first step is hydrolysis to simplest siloxanes (including D3, D4, and D5) and silanols (mainly DMSD). Then, under anaerobic conditions, DMSD is biodegraded in soil or released into the air. The biological degradation of DMSD is carried out by bacteria *Schlechtendahl* and *Arthrobacter* and fungi – *Fusarium oxysporum* [82].

The anaerobic decomposition of the cVMSs contained in activated sludge was investigated by Xu et al. [83]. After 60 h of the process, they found that the degree of degradation was in the range of 44.4–62.8% for D4 and D5 and in the range of 3.0–1.8% for D3 and D6. Thus, D3 and D6 have little potential to biodegrade in aqueous environments at anaerobic conditions. VMSs can be decomposed also by *Pseudomonas* bacteria in aerobic conditions [84]. However, this is a very slow process. For example, the $\tau_{1/2}$ of D5 in aerobic conditions exceeds 1,000 days [50]. A flow diagram of a feasible D4 biodegradation process was presented by Accettola and Haberbauer [85]. The degradation of cVMSs is described as a multi-step hydrolysis process beginning with the ring-opening hydrolysis, proceeding through the formation of linear oligomeric siloxane diols, and ending with DMSD.

VMSs are not readily biodegradable in water at aerobic conditions (Table 3).

Table 3 Biodegradation of VMSs in water (aerobic conditions, 20°C) [19]

Compound	% Degradation/time
L2	2/28 days
L3, L4, L5	0/28 days
D3	0.06/28 days
D4	3.7/29 days
D5	0.14/28 days
D6	4.5/28 days

Table 4 REACH persistence criteria [26]

Compartment	$\tau_{1/2}$, days	
	Persistent (P)	Very persistent (vP)
Marine water	>60	>60
Fresh or estuarine water	>40	>60
Marine sediment	>180	>180
Fresh or estuarine water sediment	>120	>180
Soil	>120	>180
Air	>2	–

5.3   Durability and Mechanisms of VMS Decline in the Environment

5.3.1   Persistence Criteria and Degradation Processes of VMSs

A substance is considered to be persistent (P) or very persistent (vP) if its $\tau_{1/2}$ meets the criteria listed in Table 4.

If $\tau_{1/2}$ in the air exceeds 2 days, it is presumed that a substance has the potential to be transported over long distances in the atmosphere. LRT potential is one of the key parameters used to assess hazards caused by persistent organic pollutants (POPs). This means the ability to transfer substances via the atmospheric air, water, and migratory species to regions distant from the source of their release. In addition, the substance should cause adverse human health or environmental effects in the remote regions.

5.3.2   Persistence in Water

Because of their high values of vapor pressure and Henry's law constants, VMSs, which enter the water phase, escape relatively quickly to the atmospheric air. In the case of D5, its volatilization half-life from river water is estimated to be about 2 h [15]. The ones that do not evaporate undergo hydrolysis – a major degradation process for VMSs in water. The rate of hydrolysis depends on the pH and the temperature. It increases with both the growth of hydronium and hydroxyl ions

Table 5 Hydrolysis half-life of VMSs in water [19]

Compound	Temperature, °C	pH	$\tau_{1/2}$, days
L2	24.8	5, 7, 9	0.06, 4.83, 0.52
L3	25	5, 7, 9	0.21, 13.71, 0.41
L4, L5[a]	25	5, 7, 9	0.58, 30.33, 0.88
D3	25	4, 7, 9	2 min, 23 min, 0.4 min
D4	10	4, 7, 9	0.20, 22.5, 0.25
	25	4, 5, 7, 9	4.5 min, 1.38, 2.88–6, 1.4–60 min
	35	4, 7, 9	2.25 min, 1.04, 0.5 min
D5	10	4, 7, 9	1.46, 416.67, 5.83
	25	4, 5.5, 7, 8, 9	0.39, 14.63, 66.25–73.4, 8.92, 1.03–1.32
	35	4, 7, 9	0.18, 24.58, 0.27
D6	25	7	>365

[a]No hydrolysis study is available for L5; data were taken as for structurally similar L4

and the increase in temperature. At a neutral pH, it is relatively slow. The rate of hydrolysis can be slowed down due to the strong tendency of VMSs to adsorb on the surface of the particles suspended in water or contained in sediments. Both D4 and D5 undergo rapid hydrolysis in acidic (pH 4) and alkaline (pH 9) conditions, and their $\tau_{1/2}$ at the temperature of 10–35°C ranges from a few minutes to several hours for D4 and from a few hours to less than 6 days for D5 (Table 5). However, it is significantly longer at a neutral pH in the above temperature interval, ranging from 1 to 22.5 days for D4 and from 24.6 to 417 days for D5. The longest $\tau_{1/2}$ at these conditions is for D6 – above 1 year. The most unstable in water is D3, for which $\tau_{1/2}$ is in the order of minutes at 25°C and the pH of 4–9. As a result of the hydrolysis, VMSs disintegrate into oligomeric diols as intermediates, and the final product is DMSD.

Based on the data from Table 5, it can be suspected that D5 and D6 meet the assessment criteria for P or vP in water (although not in all water reservoirs).

5.3.3 Persistence in Soil

VMSs may enter soil mainly as a result of its intentional enrichment and/or fertilization with sewage sludge. They can also reach soils via direct infiltration from poorly sealed landfills and, to a negligible extent, through the deposition of particles with VMSs adsorbed from the air. After application, most of them (88.5% of D4 – [5]) partition to the air via volatilization. The remaining part is bound to solid particles and undergoes hydrolytic degradation. Its effectiveness depends mainly on the congener, the soil type, the humidity, the temperature, and the organic and mineral matter (acting as a catalyst) content. The volatilization of VMSs is fastest in wet soil, unlike hydrolysis, which is most effective in dry soil. The degradation rate constant increases by 3–5 orders of magnitude when the water content in soil decreases from saturation to air-dry [28]. For example, the degradation of PDMS occurs faster in soil with a lower moisture content (50% in a few days at humidity below 3%) than in soil with a higher moisture content (3% in 6 months at a moisture content of 12%) [79]. Furthermore, the degradation rate of VMSs decreases as the molecule size increases, which is probably associated with a decrease in the diffusion rate. In the case of air-dry soil (RH = 32%), at 22°C, the $\tau_{1/2}$ of cVMSs ranges from ~50 min to ~3 days and that of linear VMSs – from ~6 h to ~4 days. For wet soils (RH = 90–100%), at 22°C, it can be >400 days for L2 and >200 days for D6 (Table 6).

As in the water phase, the degradation of VMSs in soil is a multi-step process leading to the formation of linear oligomeric siloxane diols, which are unstable and further hydrolyzed to the DMSD monomer.

According to REACH criteria, only D6 may be considered persistent in soil.

5.3.4   Persistence in Sediment

The degradation of VMSs in sediments is also based on the hydrolysis process. However, it is much slower than in water and soil. There are no available data on the $\tau_{1/2}$ of linear VMSs in sediments. The $\tau_{1/2}$ of D4 in freshwater sediment at 24°C is about 242 days in aerobic and ~365 days in anaerobic conditions [15] and is expected to be longer at lower temperatures (e.g., at 10 and 5°C, it is 294 and 588 days, respectively [5]). In the case of D5, due to its larger molecule size, higher hydrophobicity, and adsorption capacity (a higher log K_{OC}), $\tau_{1/2}$ at 24°C is extended to 1,200 and 3,100 days, depending on the access of oxygen [19]. The relatively long $\tau_{1/2}$ for D6 is estimated at ~540 days by PBT Profiler [18].

Generally, VMSs are more persistent in sediments than in other environmental matrices. All the analyzed cVMSs meet the REACH criteria for both a persistent (P) and very persistent (vP) substance in sediment.

5.3.5   Persistence in Air

VMSs can be removed from the atmosphere by wet and dry deposition and chemical transformations. Both deposition mechanisms, due to poor VMS water solubility and high vapor pressure, are negligible. Condensation of VMSs in air is also practically impossible, because their concentration is several orders of magnitude smaller than the one corresponding to the saturation pressure. The main degradation process of VMSs in the air is indirect photolysis by gas-phase oxidation with OH radicals, formed by UVB radiation. For the direct photolysis to occur, the decomposed substance should have the ability to absorb radiation in the range of 290–800 nm. However, VMSs do not absorb radiation longer than 190 nm [86]. There was also no observed influence of the concentration of other photooxidative species in the atmosphere (such as O_3 and NO_3 radicals) on the decline of VMS level in the air. On the other hand, VMSs react with OH radicals, which are formed in the troposphere mainly by O_3 photolysis (at 290–320 nm radiation) and reaction of atomic

Table 6 Hydrolysis half-life of VMSs in soils [19]

Compound	The origin of the sample	Temperature (°C)	RH (%)	$\tau_{1/2}$ (days)	Key value of $\tau_{1/2}$ for chemical safety assessment at 20°C (days)
L2	Michigan Londo (USA)[a]	22	32	1.8	NA
		22	100	407.6	
L3	Michigan Londo (USA)[a]	22.5	32	1.48	10
		22.5	100	119.5	
	UK[b]	22.5	32	0.26	
L4	Michigan Londo (USA)[a]	22	32	3.7	10
		22	100	106.6	
L5	Michigan Londo (USA)[a]	22	32	3.7	10
		22	100	106.6	
D3	NA	NA	NA	NA	NA
D4	Wahiawa (Hawaii)[c]	22	32	0.04	5.25
		22	100	0.89	
	Michigan Londo (USA)[a]	22	32	3.54	
		22	92	5.25	
D5	Wahiawa (Hawaii)[c]	22	50	0.11	12.5
		22	90	0.19	
	Michigan Londo (USA)[a]	22	50	9.7	
		22	90	12.5	
D6	Wahiawa (Hawaii)[c]	22	32	1.38	202 (22°C)
	Temperate soil[d]	22	50	158	
		22	90	202	
	Tropical soil[d]	22	50	1.8	
		22	90	3	

[a]Temperate soil
[b]Loamy silt soil
[c]Tropical soil, 55% clay content
[d]Estimated by modelling

oxygen with H_2O. VMS lifetime calculations for the three abovementioned oxidation mechanisms led to the following results: 150 days in the case of NO_3, 1.5 year for O_3, and 8 (D5) to 23 days (D3) for OH radicals [86]. The lifetime of VMSs resulting from the reaction with OH radicals ranges from 10 days in the case of D5 to 30 days in the case of D3 [87]. Similar values were obtained in model tests [88] for D4 and D5, respectively, 11.5 and 7.5 days, assuming the average concentration of OH radical representative of the Earth's north–central latitudes. D6 reacts with OH radicals the fastest, with 1.6 days as the calculated $\tau_{1/2}$ [89]. Hobson et al. [86] have found VMS lifetime inversely proportional to the OH radical concentration in the troposphere and seasonal variations in VMS concentrations as a result of UV radiation variability, which affects the formation of OH radicals. They can be degraded faster in city centers due to the automotive pollution, mainly NO_x and VOC, contributing to the formation of the precursor of OH radicals – O_3.

The products of siloxane oxidation with OH radicals are mainly silanols – compounds with much higher solubility in water, lower vapor pressure, and lesser potential for bioaccumulation and toxicity. They are further removed from the atmosphere by wet deposition and hydrolysis after dissolve in water. Silanols can also condense and adsorb on the particles of suspended dust and be subject to dry and wet deposition. VMS reactions with mineral aerosols (e.g., kaolinite, illite, mica, hematite) can significantly accelerate their removal from the atmosphere, especially under dry conditions [90, 91]. However, in the case of L4 and D4, 99% are removed in the form of silanols by wet deposition [92].

The $\tau_{1/2}$ of VMSs in the air varies from a few hours to several days depending on the temperature, the sun exposure, the concentration of OH radicals and its precursors, the congener and concentration of VMSs, the type and concentration of mineral aerosols, and the air humidity. In the case of L2 and L3, at the atmospheric concentration of OH radicals of 7.7×10^5 mol/cm^3 at 24°C, over a 24 h period, $\tau_{1/2}$ is estimated at 7.5 [66] and 8.8 days [67], respectively. The atmospheric $\tau_{1/2}$ of D4 is estimated at 12.7–15.8 days, assuming the constant rate of the reaction with OH radicals in the range of 1.01×10^{-12} to 1.26×10^{-12} cm^3/mol/s and the average OH radical concentration of 5×10^5 mol/cm^3 [36]. Taking into consideration the reactions with mineral aerosols, according to Navea et al. [91], the $\tau_{1/2}$ can be shortened to 6.9 days. In the case of D5, assuming the reaction rate of 1.55×10^{-12} cm^3/mol/s, at 24°C, $\tau_{1/2}$ is estimated at 10.4 days for the concentration of OH radicals of 5×10^5 mol/cm^3 [19]. On the basis of model calculations, atmospheric oxidation $\tau_{1/2}$ for D6 is predicted to be in the range of 2.6–12.8 [35]. Therefore, it can be assumed that all the analyzed siloxanes have $\tau_{1/2}$ above 2 days and so they are potentially subject to LRT.

5.4 Long-Range Transport Potential

After reaching the environment, VMSs move between its compartments, undergoing physical, chemical, and biological transformations. Because of their high vapor pressure and volatility, the major portion of the emitted VMSs reside in the air compartment as their main final sink. On the other hand, they undergo faster degradation in air than in the other matrices. According to the model tests [50], the range of this transport can exceed 5,000 km. The daily variability of VMS concentrations was tested by Yucuis et al. [93], proving its clear correlation with the height of the mixing layer, especially in the case of D5. The reason for such a relationship, in addition to the obvious effect of nocturnal temperature inversions, could be daily fluctuations in the concentrations of OH radicals, associated with UV radiation variability. Carrying out measurements in the cities with different population densities, they found yet another trend. The D5/D4 concentration ratio was positively correlated with the population density, and it ranged from 4.5 for Chicago to 2.1 for West Branch in the USA. This may result from a longer D4 lifetime comparing to D5, making the share of D4 increase in locations more distant from the emission sources.

The ability to LRT is often expressed as a characteristic travel distance (CTD), which is the distance at which 63% of the chemical is removed from the atmosphere. The eligibility criterion for substances with a high LRT potential, proposed by Beyer et al. [94], is met by most of the analyzed cVMSs at CDT > 2,000 km. The CDT for D4, D5, and D6, estimated using the OECD Screening Tool [95], amounts to, respectively, 5,254, 3,438, and 2,963 km [5, 34, 35]. This means that the VMSs can be transported to remote regions, such as the Arctic, which has been confirmed by Genualdi et al. [96]. The detected concentrations in the air were in the order of ng/m^3 and their seasonal variability corresponded to the fluctuation in OH radicals. However, their presence in the Arctic regions can be due to reasons other than LRT, such as releases from local sources [27]. Also, the samples can become contaminated, which is difficult to avoid due to the widespread use of siloxanes in PCPs. The more so that the measured concentrations are often at the threshold of detection and standardized VMS measurement methods are still missing. Assuming that VMSs are capable of LRT, the key issue from the point of view of environmental exposure is the possibility of their deposition on surface media in remote areas. Model calculations carried out by Xu and Wania [97] and Xu [98] showed that despite the ability of D4 and D5 to travel long distances in the atmosphere, their deposition potential is irrelevant due to the relatively high K_{AW} and low K_{OA} values.

In conclusion, D4, D5, and D6 have the potential for LRT in the atmosphere, but at their present concentrations, it seems that they do not have the potential to deposit in the waters and soils of remote regions, which could cause significant adverse health or environmental effects. Thus, according to the Stockholm Convention [25], the second condition of the potential for LRT is not met.

6   Summary and Conclusions

Due to the wide range of their applications and their ever-increasing production volume, VMSs are being emitted in increasing quantities – mainly directly to the air – which is referred to as the final sink. The remaining less volatile VMSs with a higher molecular weight get to WWTPs and landfills, from where they can migrate to the atmosphere as well as to soils and surface waters. In recent years, a large number of scientific reports have been published about VMS detection in the various matrices of the environment, also the ones far away from the places of release. Despite the rapid increase in knowledge about the physicochemical properties and environmental transformations of VMSs in recent years, the information concerning their toxicity and decay mechanisms is still insufficient. Further tests need to be carried out to determine also their biodegradation and bioaccumulation capabilities. Moreover, there are still many uncertainties as to their assessment in terms of PBT properties.

Until recently, it was commonly believed that VMSs are neutral to human health. Today it is claimed that they may be directly or indirectly toxic to various organs. They are relatively persistent, with high potential for bioaccumulation. Some of them are suspected of carcinogenic potential (D5); harmful effects on the female reproductive system, the liver, and the lungs (D4); and the irritating effect on the skin, eyes, and respiratory system (L2). But so far no legal restrictions have been introduced to control the use of VMSs and their emissions. There is no limit on concentrations acceptable in the indoor and ambient air. Therefore, no control is performed. They are not covered by the restrictions on VOC as well. Only D4 and D5 are registered under the REACH by European Commission Regulation.

The main processes of the degradation of VMSs in the environment are demethylation with OH radicals in the atmospheric air and hydrolysis in water and soil. The values of the $\tau_{1/2}$ of cVMSs, measured in various environmental compartments, vary from a few minutes in the case of decomposition in water (D3 and D4) to several years in the case of sediments (D5 and D6). But the literature reports on $\tau_{1/2}$ are not consistent and further research is needed.

cVMSs can undergo LRT in the atmosphere. However, their redeposition does not seem relevant. Due to their high potential for long-term exposure, they pose a greater threat in the aquatic environment and in sediments.

Therefore, considering that the toxic effect of D4 and D5 on rats and mice and their ability to bioaccumulate in aquatic organisms have been confirmed, it is necessary to reduce their content in wash-off PCPs, which are the largest source of D4 and D5 release to wastewater. The ECHA [15] proposes to restrict D4 and D5 content in wash-off PCPs to 0.1%. The easiest way to reduce exposure to VMSs is to use substitutes, based on, e.g., glycols or paraffins, which have less potential for bioaccumulation and toxicity. It is also worth undertaking research to increase the efficiency of removing VMSs from wastewater, sludge, and biogas, e.g., by developing biological methods and methods of their oxidation in the aqueous phase. For now, volatilization to the air by simply increasing sewage aeration seems to be the fastest and simplest way to degrade VMSs.

The question about the long-term impact of these relatively new chemicals on the environment and humans remains open.

References

1. Dudzina T, von Goetz N, Bogdal C, Biesterbos JWH, Hungerbühler K (2014) Concentrations of cyclic volatile methylsiloxanes in European cosmetics and personal care products: prerequisite for human and environmental exposure assessment. Environ Int 62:86–94
2. European Commission (2010) Opinion on cyclomethicone. Octamethylcyclotetrasiloxane (cyclotetrasiloxane, D4) and decamethylcyclopentasiloxane (cyclopentasiloxane, D5). Scientific Committee on Consumer Safety. SCCS/1241/10
3. Capela D, Alves A, Homem V, Santos L (2016) From the shop to the drain – volatile methylsiloxanes in cosmetics and personal care products. Environ Int 92–93:50–62
4. Global Silicones Council (2016) Socio-economic evaluation of the global silicones industry. Final report. London. https://sehsc.americanchemistry.com/Socio-Economic-Evaluation-of-the-Global-Silicones-Industry-Final-Report.pdf. Accessed 26 Jan 2018
5. Environment Canada (2008) Screening assessment for the challenge octamethylcyclotetrasiloxane. http://www.ec.gc.ca/ese-ees/default.asp?lang=en&n=2481b508-1. Accessed 26 Jan 2018
6. Brooke DN, Crookes MJ, Gray D, Robertson S (2009) Environmental risk assessment report: octamethylcyclotetrasiloxane. Environment Agency, Bristol
7. Brooke DN, Crookes MJ, Gray D, Robertson S (2009) Environmental risk assessment report: decamethylcyclopentasiloxane. Environment Agency, Bristol
8. Brooke DN, Crookes MJ, Gray D, Robertson S (2009) Environmental risk assessment report: dodecamethylcyclohexasiloxane. Environment Agency, Bristol
9. Mojsiewicz-Pieńkowska K, Krenczkowska D (2018) Evolution of consciousness of exposure to siloxanes – review of publications. Chemosphere 191:204–217
10. Centre European des Silicones (2008) A socio-economic study on silicones in Europe. Cambre Associates, Brussels
11. Arespacochaga N, Valderrama C, Raich-Montiu J, Crest M, Mehta S, Cortina JL (2015) Understanding the effects of the origin, occurrence, monitoring, control, fate and removal of siloxanes on the energetic valorization of sewage biogas – a review. Renew Sust Energ Rev 52:366–381
12. Jia H, Zhang Z, Wang C, Hong W-J, Sun Y, Li Y-F (2015) Trophic transfer of methyl siloxanes in the marine food web from coastal area of Northern China. Environ Sci Technol 49(5):2833–2840
13. Rücker C, Kümmerer K (2015) Environmental chemistry of organosiloxanes. Chem Rev 115:466–524
14. OECD (2009) The 2007 OECD list of high production volume chemicals. OECD environment, health and safety publications series No 112. ENV/JM/MONO(2009)40, Paris
15. European Chemical Agency (2016) Background document to the opinion on the Annex XV dossier proposing restrictions on octamethylcyclotetrasiloxane (D4) and decamethylcyclopentasiloxane (D5). Committee for Risk Assessment. ECHA/RAC/RES-O-0000001412-86-97/D
16. Su K, Jon V, DeGroot Jr JV, Norris AW, Lo PY (2005) Siloxane materials for optical applications. In: Lu W, Young JF (eds) ICO20: materials and nanotechnologies. Proc. of SPIE, vol 6029. Dow Corning Corporation, Auburn
17. ChemSpider – the free chemical database. Royal Society of Chemistry, UK, Cambridge. http://www.chemspider.com/. Accessed 15 Mar 2018
18. PBT Profiler (ver 2.001 upd Sept 28, 2016) Developed by the Environmental Health Analysis Center under contract to the Office of Chemical Safety and Pollution Prevention and US EPA. http://www.pbtprofiler.net. Accessed 26 Jan 2018
19. European Chemical Agency (2018) Information on chemicals. Last updated 10 Mar 2018. https://echa.europa.eu/information-on-chemicals/registered-substances. Accessed 10 Mar 2018
20. Mazzoni SM, Roy S, Grigoras S (1997) Eco-relevant properties of selected organosilicon materials. In: Chandra G (ed) Organosilicon materials. The handbook of environmental chemistry, vol 3. Springer, Berlin, pp 52–81

21. Xu S, Kropscott B (2013) Octanol/air partition coefficients of volatile methylsiloxanes and their temperature dependence. J Chem Eng Data 58:136–142
22. Tower P (2003) New technology for removal of siloxanes in digester gas results in lower maintenance costs and air quality benefits in power generation equipment. In: WEFTEC 03-78th annual technical exhibition and conference, pp 2–8
23. Gaj K (2017) Applicability of selected methods and sorbents to simultaneous removal of siloxanes and other impurities from biogas. Clean Techn Environ Policy 19(9):2181–2189
24. Gaj K, Pakuluk A (2015) Volatile methyl siloxanes as a potential hazardous air pollutants. Pol J Environ Stud 24(3):937–943
25. Stockholm Convention on Persistent Organic Pollutants (2010) Annex D, amended in 2009. The Secretariat of the Stockholm Convention, UNEP
26. Registration, Evaluation, Authorisation and Restriction of Chemicals (2011) Annex XIII criteria for the identification of persistent, bioaccumulative and toxic substances, and very persistent and very bioaccumulative substances. Official Journal of the European Union, Brussels
27. Xu S, Kozerski G, Mackay D (2014) Critical review and Interpretation of environmental data for volatile methylsiloxanes: partition properties. Environ Sci Technol 48(20):11748–11759
28. Mackay D, Cowan-Ellsberry CE, Powel DE, Woodburn KB, Xu S, Kozerski GE, Kim J (2015) Decamethylcyclopentasiloxane (D5) environmental sources, fate, transport, and routes of exposure. Environ Toxicol Chem 34(12):2689–2702
29. Fairbrother A, Burton GA, Klaine SJ, Powell DE, Staples CA, Mihaich EM, Woodburn KB, Gobas FAPC (2015) Characterization of ecological risks from environmental releases of decamethylcyclopentasiloxane (D5). Environ Toxicol Chem 34(12):2715–2722
30. Bridges J, Solomon KR (2016) Quantitative weight-of-evidence analysis of the persistence, bioaccumulation, toxicity, and potential for long-range transport of the cyclic volatile methyl siloxanes. J Toxicol Environ Health B 19(8):345–379
31. Homem V, Capela D, Silva JA, Cincinelli A, Santos L, Alves A, Ratola N (2017) An approach to the environmental prioritisation of volatile methylsiloxanes in several matrices. Sci Total Environ 579:506–513 Appendix A
32. Lassen C, Hansen CL, Mikkelsen SH, Maag J (2005) Siloxanes – consumption, toxicity and alternatives. Danish Environmental Protection Agency, Environmental Project No. 1031. http://www2.mst.dk/Udgiv/publications/2005/87-7614-756-8/pdf/87-7614-757-6.pdf. Accessed 26 Jan 2018
33. Kaj L, Schlabach M, Andersson J, Cousins AP, Remberger M, Broström-Lundén E, Cato I (2005) Siloxanes in the Nordic environment. Norden TemaNord 593. Nordic Council of Ministers, Copenhagen. http://www.norden.org/pub/miljo/miljo/uk/TN2005593.pdf. Accessed 7 March 2018
34. Environment Canada, Health Canada (2008) Screening assessment for the challenge decamethylcyclopentasiloxane. https://www.ec.gc.ca/ese-ees/default.asp?lang=En&n=13CC261E-1. Accessed 26 Jan 2018
35. Environment Canada, Health Canada (2008) Screening assessment for the challenge dodecamethylcyclohexasiloxane. https://www.ec.gc.ca/ese-ees/FC0D11E7-DB34-41AA-B1B3-E66EFD8813F1/batch2_540-97-6_en.pdf. Accessed 26 Jan 2018
36. European Chemical Agency (2012) Identification of PBT and vPvB substance. Results of evaluation of PBT/vPvB properties for octamethylcyclotetrasiloxane. Helsinki. https://echa.europa.eu/documents/10162/13628/octamethyl_pbtsheet_en.pdf. Accessed 26 Jan 2018
37. European Chemical Agency (2012) Identification of PBT and vPvB substance. Results of evaluation of PBT/vPvB properties for decamethylcyclopentasiloxane. Helsinki. https://echa.europa.eu/documents/10162/13628/decamethyl_pbtsheet_en.pdf. Accessed 26 Jan 2018
38. United States Environmental Protection Agency (2009) Siloxane D5 in drycleaning applications. Fact Sheet. Office of Pollution Prevention and Toxics (7404) 744-F-03-004. https://nepis.epa.gov/Exe/ZyPURL.cgi?Dockey=P1004S1S.txt. Accessed 6 Feb 2018
39. Tran TM, Kannan K (2015) Occurrence of cyclic and linear siloxanes in indoor air from Albany, New York, USA, and its implications for inhalation exposure. Sci Total Environ 511:138–144

40. Scientific Committee on Consumer Products (2005) Opinion on octamethylcyclotetrasiloxane (D4) cyclomethicone (INCI name). Adopted by the SCCP during the 6th plenary meeting of 13 Dec 2005. EC, Health & Consumer Protection Directorate–General, SCCP/0893/05

41. Jovanovic ML, McMahon JM, McNett DA, Tobin JM, Plotzke KP (2008) In vitro and in vivo percutaneous absorption of ^{14}C-octamethylcyclotetrasiloxane (^{14}C-D4) and ^{14}C-decamethylcyclopentasiloxane (^{14}C-D5). Regul Toxicol Pharmacol 50:239–248

42. Dekant W, Klaunig JE (2016) Toxicology of decamethylcyclopentasiloxane (D5). Regul Toxicol Pharmacol 74:S67–S76

43. Scientific Committee on Consumer Safety (2016) Opinion on decamethylcyclopentasiloxane (cyclopentasiloxane, D5) in cosmetic products. Final ver of 29 July 2016. EC, SCCS/1549/15. https://ec.europa.eu/health/scientific_committees/consumer_safety/docs/sccs_o_174.pdf. Accessed 11 Feb 2018

44. Scientific Committee on Consumer Safety (2010) Opinion on cyclomethicone – octamethylcyclotetrasiloxane (cyclotetrasiloxane, D4) and decamethylcyclopentasiloxane (cyclopentasiloxane, D5). SCCS/1241/10. http://ec.europa.eu/health/scientific_committees/consumer_safety/docs/sccs_o_029.pdf. Accessed 11 Feb 2018

45. Regulation of the European Parliament and of the Council (EC) No 1272/2008 of 16 Dec 2008, on the classification, labeling and packaging of substances and mixtures amending and repealing. Directives 67/548/EEC and 1999/45/EC, and amending Regulation (EC) No 1907/2006

46. Franzen A, Greene T, Van Landingham C, Gentry R (2017) Toxicology of octamethylcyclotetrasiloxane (D4). Toxicol Lett 279:2–22

47. Zhang J, Falany JL, Xie X, Falany CN (2000) Induction of rat hepatic drug metabolizing enzymes by dimethylcyclosiloxanes. Chem Biol Interact 124(2):133–147

48. Lieberman MW, Lykissa ED, Barrios R, Ou CN, Kala G, Kala SV (1999) Cyclosiloxanes produce fatal liver and lung damage in mice. Environ Health Perspect 107(2):161–165

49. Jean PA, Sloter ED, Plotzke KP (2017) Effects of chronic exposure to octamethylcyclotetrasiloxane and decamethylcyclopentasiloxane in the aging female Fischer 344 rat. Toxicol Lett 279:54–74

50. Wang DG, Norwood W, Alaee M, Byer JD, Brimble S (2013) Review of recent advances in research on the toxicity, detection, occurrence and fate of cyclic volatile methyl siloxanes in the environment. Chemosphere 93(5):711–725

51. Quinn AL, Dalu A, Meeker LS, Jean PA, Meeks RG, Crissman JW, Gallavan Jr RH, Plotzke KP (2007) Effects of octamethylcyclotetrasiloxane (D4) on the luteinizing hormone (LH) surge and levels of various reproductive hormones in female Sprague–Dawley rats. Reprod Toxicol 23(4):532–540

52. Quinn AL, Regan JM, Tobin JM, Marinik BJ, McMahon JM, McNett DA, Sushynski CM, Crofoot SD, Jean PA, Plotzke KP (2007) In vitro and in vivo evaluation of the estrogenic, androgenic, and progestagenic potential of two cyclic siloxanes. Toxicol Sci 96(1):145–153

53. He B, Rhodes-Brower S, Miller MR, Munson AE, Germolec DR, Walker VR, Korach KS, Meade BJ (2003) Octamethylcyclotetrasiloxane exhibits estrogenic activity in mice via ERα. Toxicol Appl Pharmacol 192(3):254–261

54. Siddiqui WH, Stump DG, Plotzke KP, Holson JF, Meeks RG (2007) A two-generation reproductive toxicity study of octamethylcyclotetrasiloxane (D4) in rats exposed by whole-body vapor inhalation. Reprod Toxicol 23(2):202–215

55. Meeks RG, Stump DG, Siddiqui WH, Holson JF, Plotzke KP, Reymolds VL (2007) An inhalation reproductive toxicity study of octamethylcyclotetrasiloxane (D4) in female rats using multiple and single day exposure regimens. Reprod Toxicol 23(2):192–201

56. McKim Jr JM, Wilga PC, Breslin WJ, Plotzke KP, Gallavan RH, Meeks RG (2001) Potential estrogenic and antiestrogenic activity of the cyclic siloxane octamethylcyclotetrasiloxane (D4) and the linear siloxane hexamethyldisiloxane (HMDS) in immature rats using the uterotrophic assay. Toxicol Sci 63:37–46

57. Jean PA, Plotzke KP (2017) Chronic toxicity and oncogenicity of octamethylcyclotetrasiloxane (D4) in the Fischer 344 rat. Toxicol Lett 279:75–97

58. Franzen A, Landingham CV, Greene T, Plotzke K, Gentry R (2016) A global human health risk assessment for decamethylcyclopentasiloxane (D5). Regul Toxicol Pharmacol 74:S25–S43

59. McKim Jr JM, Choudhuri S, Wilga PC, Madan A, Burns-Naas LA, Gallavan RH, Mast RW, Naas DJ, Parkinson A, Meeks RG (1999) Induction of hepatic xenobiotic metabolizing enzymes in female Fischer-344 rats following repeated inhalation exposure to decamethylcyclopentasiloxane. Toxicol Sci 50(1):10–19

60. Burns-Naas LA, Mast RW, Meeks RG, Mann PC, Thevenaz P (1998) Inhalation toxicology of decamethylcyclopentasiloxane (D5) following a 3-month nose-only exposure in Fischer 344 rats. Toxicol Sci 43(2):230–240

61. Jean PA, Plotkze KP, Scialli AR (2016) Chronic toxicity and oncogenicity of decamethylcyclopentasiloxane in the Fischer 344 Rat. Regul Toxicol Pharmacol 74S:S57–S66

62. Ecological Structural Activity Relationships (2004) Ver 0.99g. US EPA, Office of Pollution Prevention and Toxics, Syracuse. https://www.epa.gov/tsca-screening-tools/ecological-struc ture-activity-relationships-ecosar-predictive-model. Accessed 12 Feb 2018

63. Bondurant S, Ernster V, Herdman R (eds) (1999) Safety of silicone breast implants. National Academies Press, Washington. https://doi.org/10.17226/9602

64. OSPAR (2004) Hexamethyldisiloxane (HMDS) background document on hexamethyldisiloxane. Hazardous substances series No 201. OSPAR Commission, London

65. PubChem. https://pubchem.ncbi.nlm.nih.gov/compound. Accessed 14 Feb 2018

66. OECD (2011) Hexamethyldisiloxane (HMDS). SIDS initial assessment profile. CoCAM 1 (10–12 Oct 2011) US/ICCA. http://webnet.oecd.org/Hpv/UI/handler.axd?id=98264d1f-2476-42fb-ade8-0fc8485bae4c. Accessed 14 Feb 2018

67. OECD (2010) Octamethyltrisiloxane (L3). SIDS initial assessment profile. SIAM31 (20–22 Oct 2010) US/ICCA CoCAM. http://webnet.oecd.org/hpv/ui/handler.axd?id=83c0a20e-ecb8-4667-8f2d-7a06aaf70e91. Accessed 14 Feb 2018

68. Klaunig JE, Dekant W, Plotzke K, Scialli AR (2016) Biological relevance of decamethylcyclopentasiloxane (D5) induced rat uterine endometrial adenocarcinoma tumorigenesis: mode of action and relevance to humans. Regul Toxicol Pharmacol 74S:S44–S56

69. Dekant W, Scialli AR, Plotzke K, Klaunig JE (2017) Biological relevance of effects following chronic administration of octamethylcyclotetrasiloxane (D4) in Fischer 344 rats. Toxicol Lett 279S:42–53

70. UK Competent Authority (2015) UK proposes restriction on octamethylcyclotetrasiloxane (D4) and decamethylcyclopentasiloxane (D5) in personal care products that are washed off in normal use. http://echa.europa.eu/documents/10162/12e03ccd-7c84-4325-bde1-9daeb562a6be. Accessed 10 Mar 2018

71. Borm PJA, Tran L, Donaldson K (2011) The carcinogenic action of crystalline silica: a review of the evidence supporting secondary inflammation-driven genotoxicity as a principal mechanism. Crit Rev Toxicol 41(9):756–770

72. British Standards Institution (2007) Nanotechnologies – Part 2: Guide to safe handling and disposal of manufactured nanomaterials, PD 6699-2-2007, London

73. Allen RB, Kochs P, Chandra G (1997) Industrial organosilicon materials, their environmental entry and predicted fate. In: Chandra G (ed) Organosilicon materials. The handbook of environmental chemistry, vol 3. Springer, Berlin, pp 1–25

74. Kazuyuki O, Masaki T, Tadao M, Hiroshi K, Nobuo T, Akira K (2007) Behavior of siloxanes in a municipal sewage-treatment plant. J Jpn Sewage Works Assoc 44(531):125–138

75. Mueller JA, Di Toro DM, Maiello JA (1995) Fate of octamethylcyclotetrasiloxane (OMCTS) in the atmosphere and in sewage treatment plants as an estimation of aquatic exposure. Environ Toxicol Chem 14(10):1657–1666

76. Van Egmond R, Sparham C, Hastie C, Gore D, Chowdhury N (2013) Monitoring and modelling of siloxanes in a sewage treatment plant in the UK. Chemosphere 93:757–765

77. Dewil R, Appels L, Baeyens J (2006) Energy use of biogas hampered by the presence of siloxanes. Energy Convers Manag 47:1711–1722

78. Grümping R, Michalke K, Hirner AV, Hensel R (1999) Microbial degradation of octamethylcyclotetrasiloxane. Appl Environ Microbiol 65(5):2276–2278

79. Griessbach EFC, Lehmann RG (1999) Degradation of polydimethylsiloxane fluids in the environment – a review. Chemosphere 38(6):1461–1468
80. Ohannessian A, Desjardin V, Chatain V, Germain P (2008) Volatile organic silicon compounds: the most undesirable contaminants in biogases. Water Sci Technol 58(9):1775–1781
81. Lehmann RG, Miller JR, Kozerski GE (2000) Degradation of silicone polymer in a field soil under natural conditions. Chemosphere 41(5):743–749
82. Sabourin CL, Carpenter JC, Leib TK, Spivack JL (1996) Biodegradation of dimethylsilanediol in soils. Appl Environ Microbiol 62(12):4352–4360
83. Xu L, Shi Y, Cai Y (2013) Occurrence and fate of volatile siloxanes in a municipal wastewater treatment plant of Beijing, China. Water Res 47(2):715–724
84. Accettola F, Guebitz GM, Schoeftner R (2008) Siloxane removal from biogas by biofiltration: biodegradation studies. Clean Techn Environ Policy 10(2):211–218
85. Accettola F, Haberbauer M (2005) Control of siloxanes. In: Lens P, Westermann P, Haberbauer M, Moreno A (eds) Biofuels for fuel cells. Renewable energy from biomass fermentation. IWA Publishing, London, pp 445–454
86. Hobson JF, Atkinson R, Carter WPL (1997) Volatile methylsiloxanes. In: Chandra G (ed) Organosilicon materials. The handbook of environmental chemistry, vol 3. Springer, Berlin, pp 137–179
87. Atkinson R (1991) Kinetics of the gas-phase reactions of a series of organosilicon compounds with OH and NO_3 radicals and O_3 at 297 ± 2 K. Environ Sci Technol 25(5):863–866
88. Navea JG, Young MA, Xu S, Grassian VH, Stanier CO (2011) The atmospheric lifetimes and concentrations of cyclic methylsiloxanes octamethylcyclotetrasiloxane (D4) and decamethylcyclopentasiloxane (D5) and the influence of heterogeneous uptake. Atmos Environ 45(18):3181–3191
89. MacLeod M, Kierkegaard A, Genualdi S, Harner T, Scheringer M (2013) Junge relationships in measurement data for cyclic siloxanes in air. Chemosphere 93(5):830–834
90. Chandramouli B, Kemens RM (2001) The photochemical formation and gas-particle partitioning of oxidation products of decamethylcyclopentasiloxane and decamethyltetrasiloxane in the atmosphere. Atmos Environ 35:87–95
91. Navea JG, Xu S, Stanier CO, Young MA, Grassian VH (2009) Effect of ozone and relative humidity on the heterogeneous uptake of octamethylcyclotetrasiloxane and decamethylcyclopentasiloxane on model mineral dust aerosol components. J Phys Chem 113(25):7030–7038
92. Whelan MJ, Estrada E, van Egmond R (2004) A modelling assessment of the atmospheric fate of volatile methyl siloxanes and their reaction products. Chemosphere 57:1427–1437
93. Yucuis RA, Stanier CO, Keri C, Hornbuckle KC (2013) Cyclic siloxanes in air, including identification of high levels in Chicago and distinct diurnal variation. Chemosphere 92(8):905–910
94. Beyer A, Mackay D, Matthies M, Wania F, Webster E (2000) Assessing long-range transport potential of persistent organic pollutants. Environ Sci Technol 34(4):699–703
95. Scheringer M, MacLeod M, Wegmann F (2006) OECD P_{OV} and LRTP Screening Tool ver 2.0. ETH Zürich, Zürich. http://www.oecd.org/chemicalsafety/risk-assessment/45373514.pdf. Accessed 10 Mar 2018
96. Genualdi S, Harner T, Cheng Y, MacLeod M, Hansen KM, van Egmond R, Shoeib M, Lee SC (2011) Global distribution of linear and cyclic volatile methyl siloxanes in air. Environ Sci Technol 45(8):3349–3354
97. Xu S, Wania F (2013) Chemical fate, latitudinal distribution and long-range transport of cyclic volatile methylsiloxanes in the global environment: a modeling assessment. Chemosphere 93:835–843
98. Xu S (2014) Long range transport potential of volatile methylsiloxanes. In: Workshop of the latest development on the evaluation method of environmental fate and bioaccumulation, The Society of Silicon Chemistry Japan, symposium in Tokyo, pp 19–25

Main Uses and Environmental Emissions of Volatile Methylsiloxanes

Yuichi Horii and Kurunthachalam Kannan

Contents

Abstract The main uses and environmental emissions of cyclic methylsiloxanes (CMSs) and linear methylsiloxanes (LMSs), especially the three volatile CMSs (D4, D5, and D6), were reviewed. This chapter provides information on production, use, concentrations in various products, as well as emission of volatile methylsiloxanes

Y. Horii (✉)
Center for Environmental Science in Saitama, Kazo, Saitama, Japan
e-mail: horii.yuichi@pref.saitama.lg.jp

K. Kannan
Wadsworth Center, New York State Department of Health, Albany, NY, USA
e-mail: kurunthachalam.kannan@health.ny.gov

V. Homem and N. Ratola (eds.), *Volatile Methylsiloxanes in the Environment*,
Hdb Env Chem (2020) 89: 33–70, DOI 10.1007/698_2019_375,
© Springer Nature Switzerland AG 2019, Published online: 26 March 2019

(VMSs) into the environment. Many silicone-based products contain residues of CMSs as impurities, and hence the occurrence of VMSs in silicone-based materials (such as rubber products) has been described. CMSs are mainly used as intermediates in the production of silicone polymers, silicone fluids, elastomers, and resins, all of which have diverse industrial and consumer applications. CMSs are also used directly in personal care products (PCPs), as carriers. The concentrations and profiles of CMSs and LMSs in PCPs and household products from North America, Europe, and Asia varied widely across and within the product categories. The measured concentrations ranged from 0.01% in body wash to 70% (by weight) in deodorants. D5 was the predominant CMS found with high detection frequency in most PCPs. The correlations among VMSs in consumer products suggested incorporation of different blends of silicones to the raw material or as additives in those products. The industrial production of VMSs and direct use of PCPs result in significant emissions of VMSs into the environment. High production volume, high mobility, and environmental persistence of VMSs are causes of concern. VMSs, especially CMSs, were found globally in various environmental matrices including air, water, and sludge. This chapter also provides information on the concentrations and patterns of VMSs in the environment surrounding silicone factories, paper production facilities, and oil fields.

Keywords Environmental emissions, Personal care products, Silicone industry, Siloxanes, Wastewater treatment plants

Abbreviations

CMS	Cyclic methylsiloxane
D3	Hexamethylcyclotrisiloxane
D4	Octamethylcyclotetrasiloxane
D5	Decamethylcyclopentasiloxane
D6	Dodecamethylcyclohexasiloxane
GC/MS	Gas chromatography/mass spectrometry
L3	Octamethyltrisiloxane
L4	Decamethyltetrasiloxane
L5	Dodecamethylpentasiloxane
L6	Tetradecamethylhexasiloxane
LMS	Linear methylsiloxane
LOQ	Limit of quantification
MDL	Method detection limit
NA	Not available
ND	Not detected
PCPs	Personal care products
PDMS	Polydimethylsiloxane

SOA   Secondary organic aerosols
SVHC   Substances of very high concern
VMS   Volatile methylsiloxane
VOC   Volatile organic compound
WWTP   Wastewater treatment plant

1 Introduction

Volatile methylsiloxanes (VMSs) including cyclic methylsiloxanes (CMSs) and linear methylsiloxanes (LMSs) have been widely used in personal care and household products, in the production of silicone polymers, and in a range of industrial applications [1–3], owing to their low surface tension and high thermal and chemical stabilities. Siloxanes were generally regarded as "safe" in consumer products; however, studies suggested that exposure to CMSs (e.g., octamethylcyclotetrasiloxane, D4) at notable amounts could cause direct or indirect toxic effects, such as estrogen mimicry, connective tissue disorders, adverse immunologic responses, and fatal liver and lung damage [4–8].

Three CMSs identified as priority pollutants by the international agencies for regulation were D4 (where D refers primarily to the dimethylsiloxane unit and the integer refers to the number of Si-O bonds that make up the chain), decamethylcyclopentasiloxane (D5), and dodecamethylcyclohexasiloxane (D6). CMSs possess unique physicochemical properties, including high volatility (4.6–132 Pa at 25°C) [9], hydrophobicity (e.g., low water solubilities, 5.3–56 μg/L), and high octanol-water partition coefficients (log K_{ow}, 6.98–8.87) [10, 11]. In 2012, Environment Canada [12] proposed measures to prevent or minimize the releases of D4 into the aquatic environment during production and usage. The UK Environment Agency [13–15] assessed the risks of D4, D5, and D6 and classified D4 as very persistent and very bioaccumulative (vPvB) and as persistent, bioaccumulative and toxic (PBT) chemical. D5 was classified as vPvB by the Registration, Evaluation, Authorisation and Restriction of Chemicals (REACH) regulation [16]. The European Commission restricted the amount of D4 and D5 in wash-off-type personal care products (PCPs) to a concentration of $\geq 0.1\%$ by weight [17]. D4, D5, and D6 were also registered as substances of very high concern (SVHC) in June 2018 by the REACH [18]. Also, the US Environmental Protection Agency (EPA) entered into an Enforceable Consent Agreement for D4 in 2014 [19]. In Japan, D4 and D6 were placed on the list of chemicals that required routine monitoring, under the Evaluation of Chemical Substances and Regulation of Their Manufacture Act, in April 2018. The major use of VMSs, which includes CMSs and LMSs, is as an intermediate in the production of polymers. Such polymers contain residual amounts of VMSs, and their use contributes to the release of those chemicals into the environment.

This chapter focuses on the production, use, and concentrations in consumer and industrial products and occurrence in water, air, and sludge matrices as well as emissions of VMSs into the environment. Many silicone-based products contain CMSs as impurities, and therefore occurrence of VMSs in silicone-based materials such as rubber products is also reviewed.

2 Production

Silicones are high-performance materials which include silicone polymers, silanes, and siloxanes and are used in diverse applications. Siloxanes are a class of anthropogenic substances and do not occur naturally in the environment. Silicones are polymeric forms of organic silicon, consisting of a backbone of alternating silicon-oxygen [Si-O] units with organic side chains attached to each silicon atom [20]. A major raw material used in the production of silicones is silicon, which is the second most abundant element on the Earth's crust. A flowchart of processing steps used in silicone industries and downstream applications is shown in Fig. 1. The first step in the production process is hydrolysis to form CMSs and/or LMSs from chloromethylsilane. Polydimethylsiloxane (PDMS) itself is then formed by either ring-opening polymerization of CMSs or polycondensation of LMSs in the presence of an end blocker, such as hexamethyldisiloxane and heat under acidic or alkaline conditions [21].

Silicones have heat, cold, and weather resistances in addition to chemical stability and therefore are used in a large number of industrial processes, systems, components, and end products [2]. Silicones are highly stable and have lubricating properties. Silicones can take a variety of physical forms ranging from solids and semi-viscous pastes to liquids and oils. Due to their advantageous characteristics such as

Fig. 1 Flowchart of production stages in the silicone industry and downstream applications. Data from [13–15, 22]; the figure was prepared by the authors

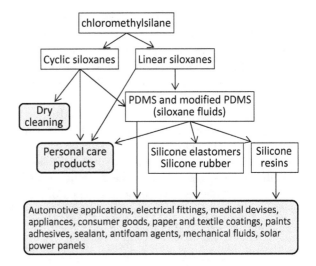

flexibility, resistance to moisture, stability, inertness, and permeability to gases, silicones find a wide array of industrial applications. Amec Foster Wheeler conducted a socioeconomic assessment of the contribution of silicone industry to the global economy [22]. It was acknowledged that silicones are used in a wide variety of applications from the key markets including transportation, construction materials, electronics, energy, healthcare, industrial processes, and personal care and consumer products (Table 1). In the transportation industry, for example, silicone is used in the manufacture of parts used in cars, aircrafts, and ships in the form of silicone rubber, resins, sealants, lubricants, plastic additives, coatings, and adhesives. Approximately 2.1 million tons of silicone-containing goods were produced and used in the key markets globally in 2013, and silicone production was an $11 billion industry (Fig. 2) [22, 23]. The major market of silicones was "industrial processes" (746,000 tons); this includes antifoaming products, moldings, and coatings, followed by "construction" (548,000 tons), and "personal care and consumer products" (357,000 tons). For personal care and consumer products, 55% of the sales volumes were related to hair and skin care. China was the largest manufacturer of siloxanes in the world in 2013, with silicone sales of over 442,000 tons, followed by the USA (412,000 tons) and Japan (145,000 tons) [22, 23]. Silicones are used in tens of thousands, if not millions, of products across the globe and can be found in fractional concentrations in final products, at ranges from 39% (transportation) to 68% (industrial processes). Based on the average concentrations in final products, the global production of silicones in 2013 was estimated at 1.2 million tons/year (Fig. 2). The data presented above did not include on-site industrial use of silicones. Based on the report of Global Silicones Council (GSC) in 2016 [22], over 3.2 million tons of silicone products including siloxanes, silane, and silicone polymers are

Table 1 The major uses of silicone in key markets

Category	Main use
Transportation	Car manufacturing, aircraft manufacturing, ship manufacturing (silicone rubber, resin, sealant, lubricant, plastic additives, coating, adhesives)
Construction materials	Sealants and adhesives, polyurethane foam for building, hydrophobic silicone additives, and exterior coating
Electronics	Semiconductors, printed circuit boards, light-emitting diodes, and wiring
Energy	Solar photovoltaic panels, wind turbines (adhesives), energy transmission and distribution (cable insulation and repair)
Healthcare	Medical devices (orthopedics and catheters), adhesives and coatings (transdermal patches, tapes), antifoaming agent in medications
Industrial processes	Oil and gas industry (antifoaming agents), pulp and paper industry (antifoaming agents), sealants, molds, coatings, hydraulic fluids, and lubricants
Personal care and consumer products	Hair care, skin care, makeup, polish, wax, household detergent, cooking utensils, sporting goods, baby products, furniture and bedding, sporting goods

Data from Ref. [22]

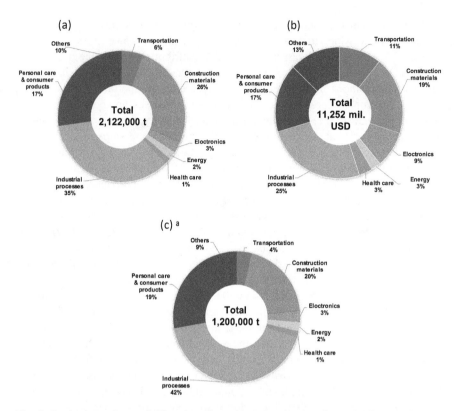

Fig. 2 Production volume of silicone products (**a**), its total value (**b**), and silicone volume in products (**c**) in downstream applications. [a]The volume was calculated based on average fraction of silicone in the products; Data from [22]; the figure was prepared by the authors

manufactured and used internally by the industry. Over one million tons of unformulated silicones were sold to customers. With regard to siloxanes including VMSs, the quantity manufactured for on-site use and quantity used in products marketed were 942,000 tons/year and 266,000 tons/year, respectively [22].

According to the Organisation for Economic Co-operation and Development (OECD) and the US EPA, D4, D5, and D6 are high production volume chemicals [24, 25]. According to the European Chemicals Agency (ECHA), the registered tonnage range of production of D4 and D5 in European Union (EU) countries was 100,000–1,000,000 and 10,000–100,000 tons/year, respectively [16]. Since 2002, the annual average production volume of CMSs was 800,000 tons in China and 470,000 tons in the USA [26]. The annual production and import of D5 and D6 in the USA increased tenfold in the last 25 years to >225,000 and >22,500 tons, respectively [27]. Japan is also a major consumer of silicones, with an annual silicone consumption rate of 117,000 tons in 2009 [28]. Canada does not produce CMSs, but the total quantity of annual import volume was in the range of 1,000–10,000 tons for D4 and D5 and 100–1,000 tons for D6 in 2006 [29–31]. The global silicone market is projected to reach a volume of 2.8 million tons by 2023, at an annual growth rate of more than 5% during 2018–2023.

3 Types of Siloxanes

3.1 Cyclic Methylsiloxanes (CMSs)

CMSs are distilled from the mixture of products found in hydrolysate, which is produced by the hydrolysis of chloromethylsilanes (Fig. 1). The predominant CMS produced commercially was D4, with smaller amounts of D3 and D5 [21]. The production ratio of D4 and D5 was generally 85% to 15%, respectively. CMSs were produced commercially for use as specialty chemicals in consumer and industrial products [2, 32]. CMSs are mainly used as intermediates for the production of silicone polymers. These silicone fluids, elastomers, gels, and resins in turn have a number of applications, as shown in Table 1. On-site intermediate and polymerization uses account for the vast majority of D4, with less than 5% used in PCPs. The use of D5 in PCPs was approximately a quarter of the total volume produced, and the remainder was used as an intermediate in industries and in polymerization [16, 33, 34].

Approximately 87% of the CMSs produced in the USA in 1993 were used as site-limited intermediates in the production of polymeric siloxanes [35], which were used in a wide range of applications. The remaining 13% (~20,000 tons) were used in PCPs, primarily as carriers in antiperspirants/deodorants, skin care, hair care, sun care, bath oil, and cosmetics [21]. D5 was also used as a replacement for perchloroethylene in dry cleaning processes [14]. The amount of D5 used in PCPs was estimated to be 25,000 tons/year, based on Cosmetics Europe [16, 23]. The supply volume of D5 was 750 tons/year in wash-off PCPs and 14,350 tons in leave-on PCPs. The direct use of D4 in PCPs has been declining since 2002, as it has been substituted by D5. However, D4 is also present in PCPs as an impurity from the use of D5 and other silicone polymers. The actual quantity of D4 used in PCPs is unknown, but the total quantity of D4 was estimated at 1.5% of D5, that is, up to 375 tons/year in EU [16].

3.2 Linear Methylsiloxanes (LMSs)

LMSs are manufactured by the stoichiometric co-hydrolysis of two chlorosilanes [35]. LMSs (e.g., octamethyltrisiloxane, L3) are primarily used as an ingredient in the production of PDMS polymers or mixtures. The PDMS formulations (polymer, oligomers, and mixtures) are generally referred to as dimethicone. L3 may also be added in its pure form to cosmetics, drugs, and natural health products as an antifoaming agent and/or skin conditioning agent [36].

L3 was used in Denmark from 2001 to 2010, with a total usage rate of 400 kg in 2010. Due to the low surface tension and excellent spreadability, volatile silicone fluids (linear, viscosity: <5 cSt) are used in antiperspirants, skin creams, skin lotions, suntan lotions, bath oils, and hair care products [21].

3.3 Polydimethylsiloxane (PDMS)

PDMS is formed by either the ring-opening polymerization of CMS or the polycondensation of LMS [35]. Approximately 138,000 tons of PDMSs were produced in or imported into the USA in 1993. Approximately 62% of this amount was used as site-limited intermediates in the production of elastomers, pressure-sensitive adhesives, and modified PDMS fluid [15]. The non-intermediate industrial uses of PDMSs are numerous, including antifoams, softness and wetting agents in textile manufacturing, components of polishes, and other surface-treatment formulations, lubricants, mold-release agents, paper coatings, and as dielectric fluids and heat-transfer liquids (i.e., polychlorinated biphenyl substitutes). PDMSs are also used in consumer applications, such as personal, household, and automotive care products [15, 35]. Low viscosity silicones (5–50 cSt) are used as an ingredient in PCPs such as skin creams, skin lotions, sunscreens, bath oils, and hair care products. These fluids are clear (transparent), tasteless, and odorless and provide nongreasy feel [21]. PDMSs are also used as food additives (e.g., antifoaming agent) during food processing.

3.4 Modified Siloxanes

Modified siloxanes with a general formula, $[R_2SiO]_n$, are substances in which R is usually a methyl group but can also be modified with hydrogen, vinyl, phenyl, or trifluoropropyl group (Fig. 3). The modified PDMSs are commonly manufactured by catalyzed ring-opening copolymerization of an appropriate functional monomer (ether cyclic or linear) with a cyclic oligomeric siloxane and end blocker [15, 35].

Methyl(hydrido)- and methyl(vinyl)siloxanes are used in the production of silicone elastomers. The methyl(hydrido)siloxanes are also used as waterproofing agents in textiles and wall boards. Methyl(phenyl)siloxanes are used as high-temperature oil baths, greases, diffusion pump fluids, and paint additives. Methyl (hydrido)- and methyl(phenyl)siloxanes are often used in cosmetics such as lipstick and foundations because of their good solubility in other organic agents and ultraviolet absorbing compounds and glossiness. Hydrogen and trifluoropropyl group provide greater solvent and fuel resistance to siloxane rubber (e.g., gasket materials) [15]. Silicone oil is stable over a wide temperature range (150°C for a long period of time), and the inclusion of diphenyl- or phenylmethylsiloxy group into the polymer increases thermal stability. The degradation of modified siloxanes can contribute to environmental releases of VMSs and volatile modified siloxanes.

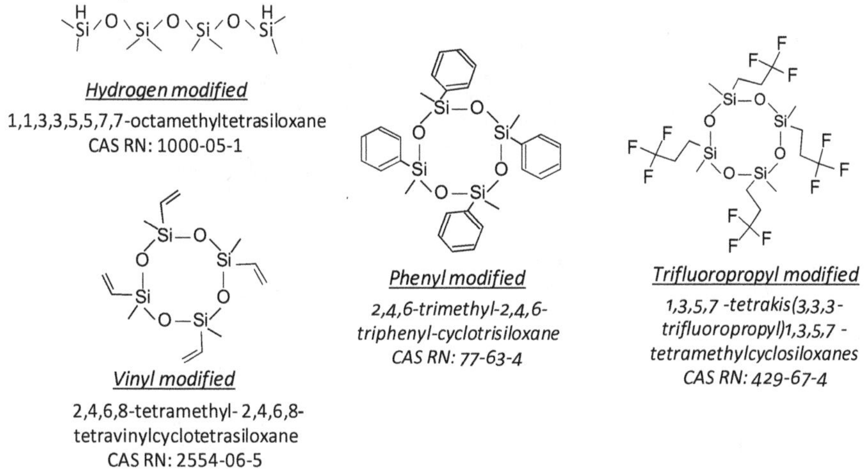

Hydrogen modified
1,1,3,3,5,5,7,7-octamethyltetrasiloxane
CAS RN: 1000-05-1

Vinyl modified
2,4,6,8-tetramethyl- 2,4,6,8-
tetravinylcyclotetrasiloxane
CAS RN: 2554-06-5

Phenyl modified
2,4,6-trimethyl-2,4,6-
triphenyl-cyclotrisiloxane
CAS RN: 77-63-4

Trifluoropropyl modified
1,3,5,7 -tetrakis(3,3,3-
trifluoropropyl)1,3,5,7 -
tetramethylcyclosiloxanes
CAS RN: 429-67-4

Fig. 3 Examples of modified siloxanes

4 Uses of Siloxanes

4.1 History

The organosilicon compound "tetraethylsilane" was first synthesized by Friedel and Crafts in 1863 [37]. A wide variety of CMSs and LMSs can now be found in many PCPs. The most commonly used silicones are CMSs (e.g., D5) and PDMSs of various viscosities. According to Skin Deep, an environmental working group, there are more than 75,000 PCPs in commerce, and over 16% of them contain CMSs, with D5 as the most widely used compound [38]. In the EU, the quantities of D4, D5, and D6 used annually in PCPs were 579, 17,300, and 1,989 tons, respectively, in 2004 [13–15]. It was estimated that over 50% of all new cosmetics launched in the last 10 years contained at least one type of silicone [22, 26].

Garaud [39] summarized the history of silicone uses in PCPs (Fig. 4). The first use of silicone in personal care applications was in the 1950s, when a mixture of PDMS was incorporated into a commercial formulation to provide skin protection [39, 40]. Since then, new applications of siloxanes were dynamically developed. Silicones made another breakthrough in the antiperspirant market during the 1970s: CMSs were used as volatile carriers for the antiperspirant active ingredient, enhancing the pleasant skin feel as well as nonstaining properties [41]. The application in hair care foams, which facilitated combing and styling, and in shampoos that improved the condition of the hair began in the 1980s [26], whereas high-molecular-weight PDMSs were formulated into two-in-one conditioning shampoos in the 1990s. Due to the low surface tension of VMSs, they impart smooth feeling of "silkiness" to cosmetics. Silicones can be found virtually in all types of PCPs.

Fig. 4 History of silicone use in personal care and chemical management of cyclic methyl siloxanes (CMSs). Data from [39], [a]Ref. [37]; [b]Ref. [41]; [c]Ref. [23]; [d]Ref. [17]; the figure was prepared by the authors

The direct use of D4 in PCPs has declined since 2002, as it was substituted by D5 [23]. Nevertheless, the US EPA reported that D4 represents a low risk to the aquatic environment [42]. Some CMSs were reported to meet regulatory criteria for large production volume, persistence in the environment, and bioaccumulation, by environmental programs in the UK [13–15], Canada [29–31], and the USA [19]. Therefore, the use of D5 in wash-off PCPs is expected to decline gradually. For instance, a restriction on D4 and D5 in wash-off cosmetic products will come into force in Europe in January 2020 [17]. Representatives of the cosmetics industry have indicated that, of the new products placed on the European market between March 2012 and March 2013, only 2% of 2,500 rinse-off shampoos and conditioners contained D5 [23]. Despite this, silicone polymers used in PCPs can contain CMSs as impurities. The use of silicone polymers in PCPs has been evaluated by Peter Fisk Associates [43] on behalf of the REACH program. Assuming that 20,000 tons of silicone polymers are annually used in the EU, with a maximum assumed concentration of 0.5% (w/w) for D4 and D5, the total potential emission rate was estimated at 100 tons/year [16].

4.2 Siloxane Concentrations in PCPs and Household Products

Concentrations of VMSs in PCPs and household products were reported for samples from the North American, European, and Asian markets [1, 38, 44, 45]. The categories of products studied include deodorants, skin care, hair care, nail care, sun care, cosmetics, and body wash (Table 2). The reported concentrations varied widely across and within the product categories measured that ranged from 0.01% in body wash to 70% in deodorant. It should be noted that the concentrations reported

Table 2 Reported concentrations (µg/g) of methylsiloxanes in personal care products

Product category	Country/region	D4	D5	D6	ΣLMS	Remarks	Ref
Hair care	USA/Japan	29 <0.35–82	5,890 <0.39–25,800	48 <0.33–162	0.78 <0.059–6.3 for L4–L14	Shampoo, conditioner, hairstyling	[1]
	Canada	NA <8–70	NA <8–1,690	NA <8–10	NA	Hair spray, hair mousse, hair gel	[45]
	Europe	5.4 <0.71–5.4	5,300 13–10,300	110 20–210	NA	Conditioner	[38]
	Portugal	91.5 3.87–267	18.4 1.0–39.9	18.0 0.69–42.0	0.54 0.12–1.34 for L2–L5	Shampoo	[46]
	China	13.8 <0.017–72.9	54.2 <0.005–1,110	8.1 <0.022–66.4	0.05 <0.005–328 for L4–L14	Shampoo, conditioner	[44]
Body wash	USA/Japan	<0.35	<0.39	<0.33	0.84 <0.059–7.6 for L4–L14	Body wash, facial cleanser, baby wash	[1]
	China	0.05 <0.017–0.49	0.13 <0.005–3.01	0.05 <0.022–1.46	1.22 <0.005–11.5 for L4–L14	Body wash, hand sanitizer, facial cleanser	[44]
Skin lotion	USA/Japan	7.3 <0.35–66	3,760 <0.39–47,300	606 <0.33–6,520	4,060 <0.059–73,000 for L4–L14	Body lotion, face cream, baby lotion, sunscreen	[1]
	Canada	NA <8–2,590	NA <8–23,400	NA <8–1,180	NA	Body lotion	[45]
	Europe	3.6 <0.71–15	15,100 2,300–28,300	900 <0.72–2,900	NA	Body lotion	[38]

(continued)

Table 2 (continued)

Product category	Country/region	D4	D5	D6	ΣLMS	Remarks	Ref
	Europe	310 <0.71–1,900	54,200 <0.67–214,000	5,600 56–70,800	NA	Face cream	[38]
	Portugal	23.0 ND–105	203 ND–754	118 0.11–471	0.28 ND–0.98 for L2–L5	Body lotion, milk, cream	[46]
	China	3.53 <0.017–24.2	37.6 <0.005–344	28.2 <0.022–154	610 <0.005–16,400 for L4–L14	Face cream, body lotion	[44]
Cosmetics	USA/Japan	49 <0.35–272	13,600 1.3–81,800	7,180 0.33–43,100	384 <0.059–2,290 for L4–L14	Lipstick, liquid foundation	[1]
	Europe	160 <0.71–390	107,000 21,000–213,000	55,200 2,400–151,000	NA	Liquid foundation	[38]
	China	6.83 <0.017–31.9	19.5 <0.005–65.4	67.1 <0.022–367	6,870 <0.005–52,600 for L4–L14	Liquid foundation, nail polisher	[44]
Deodorant/antiperspirant	Canada	NA <8–3,000	284,000 <8–683,000	3,600 <8–12,300	NA	Deodorant, antiperspirant	[45]
	Europe	86 19–200	110,000 35,700–285,000	2,200 510–5,300	NA	Spray type	[38]
	Portugal	2.87 ND–10.7	1.34 ND–3.58	0.74 ND–1.01	<LOQ ND–<LOQ for L2–L5	Roll-on deodorants, antiperspirants	[46]

NA not available, ND not detected, LOQ limit of quantification in the studies are "extractable fractions" of siloxanes found in the products by organic solvents (e.g., hexane).

Horii and Kannan [1] firstly determined concentrations of CMSs (D4–D7) and LMSs (L4–L14) in 76 personal care and household products collected from American and Japanese markets in 2008 (Table 2). The concentrations of CMSs in PCPs ranged from <0.35 to 9,380 µg/g for D4, from <0.39 to 81,800 µg/g for D5, from <0.33 to 43,100 µg/g for D6, and from <0.42 to 846 µg/g for D7. The highest concentrations of D5 and D6 were in cosmetics (liquid foundation) and the highest concentrations of D4 were in household sanitation products (furniture polish). All of the cosmetics contained D5 and D6, with the highest detection frequency of LMSs (83%). Concentrations of total LMSs (L4–L14) were <0.059 to 73,000 µg/g, with the highest levels found in skin lotions. Among LMSs, the mean concentration of L11 was the highest (341 µg/g), followed by L10 and L12.

In PCPs sold in Canada [45], D3, D4, D5, and D6 were found at concentrations of 0.8%, 4.8%, 14.3%, and 9.1%, respectively. D5 was the predominant CMS in Canadian PCPs, with the highest concentration of 680,000 µg/g found in an anti-perspirant sample. D5 was followed by D6 with the highest concentration of 98,000 µg/g in a baby diaper cream and D4 with the highest concentration of 11,000 µg/g in a body lotion. The most common siloxane-containing PCP category was antiperspirants with CMSs detected in 12 of the 13 samples analyzed (mean, 284,000 µg/g for D5 and 3,600 µg/g for D6), followed by body lotion and hair care products. The highest concentrations of D5 and D6 in lotions from Canada (35,300 µg/g and 6,260 µg/g) [45] were similar to those for the USA and Japan [1], whereas CMS concentrations in hair care products in Canada were generally lower than those reported in other studies.

The concentrations of D4, D5, and D6 were determined by Dudzina et al. [38] in PCPs collected from the European market (Table 2). D5 was the predominant siloxane found with a high detection frequency (47 out of 51 products). The mean concentration of D5 in all the products was 60,500 µg/g. CMS concentrations in deodorants/antiperspirants, cosmetics, and skin lotion were the highest. The respective median concentrations of D5, D6, and D4 were 142,000 µg/g, 2,300 µg/g, and 53 µg/g in deodorants/antiperspirants; 44,600 µg/g, 30,000 µg/g, and below the limit of quantification (LOQ) in skin care products; and 9,600, 180, and 5.5 µg/g in hair care products. D4 concentrations (mean, 180 µg/g) were two to three orders of magnitude lower than those of D5 and D6 (0.3% w/w for D4 and D5), which suggested D4 was an impurity from D5 or D6 mixtures. These findings agreed with the estimation of the ECHA [16] (see above) that assumed a maximum concentration of 0.5% (w/w) for D4 and D5 in siloxane polymers. Capela et al. [46] reported the concentrations of VMSs (D3–D6 and L2–L5) in best-selling brands of PCPs including moisturizers, deodorants, body and hair washes, toilet soaps, toothpastes, and shaving products, from Oporto region of Portugal (Table 2). VMSs were detected in 96% of the samples, at concentrations between 0.003 and 1,200 µg/g. Shampoos exhibited the highest concentration for CMSs and aftershaves for LMSs.

Lu et al. [44] investigated concentrations of CMSs (D4–D7) and LMSs (L4–L14) in PCPs collected from the Chinese market in 2011 (Table 2). Siloxanes were detected in 88% of the 158 products analyzed, and D5 and D6 were the most frequently detected siloxanes. CMSs were found in 72% of the products. The highest detection frequency of D4 and D5 (87% and 91%, respectively) was found in hair care products, whereas cosmetics showed 90% occurrence for D6. Interestingly, high detection frequencies of LMSs were found in skin lotion (89%) and cosmetics (80%) from China, whereas those values for the US and Japanese products were only 33% [1]. Thus, Chinese products showed a different profile of CMS and LMS. The concentrations of total siloxanes varied widely, by up to seven orders of magnitude, among the PCPs analyzed from China. The median concentrations of total siloxanes were high in cosmetics (417 µg/g) and skin lotions (120 µg/g). The concentrations of CMSs in PCPs in China were one to three orders of magnitude lower than those from the USA and Japan [1] and Europe [38], whereas the concentrations of LMSs tend to be higher in Chinese PCPs [44].

Several studies have shown a significant positive correlation between the pairs of methylsiloxanes (CMSs and LMSs) [1, 38, 46]. A large number of silicone blends have been developed for use in a variety of consumer products. These blends contain a mixture of several CMSs or LMSs or a combination of these two classes of siloxanes. Most of the silicone fluids contain low amounts of D4 at <1%. The correlations found among siloxanes in products can suggest the incorporation of different blends of silicones in raw materials or as additives.

4.3 Concentrations in Silicone Rubber Products

Owing to their thermal stability and chemical inertness, siloxanes are used as raw materials in the production of packaging materials, cookware, nipples (i.e., pacifiers), and childcare product (e.g., toys and teethers). This section summarizes concentrations and profiles of CMSs and LMSs in silicone rubber products of Chinese and American markets (Table 3). Xu et al. [47] investigated siloxanes in siliconized childcare products from China. They determined concentrations of cyclic and linear siloxanes in 190 products including hard toys, pacifiers, teethers, and soft rubber toys. The mean concentrations of target siloxanes (D4–D6 and L5–16) in all products analyzed ranged from 0.021 to 12.7 µg/g, with the detection frequencies ranging from 46 to 89%. The highest concentrations of CMSs were found in teethers, followed by pacifiers and soft rubber toys. For LMSs, long-chain compounds such as L15 and L16 were detected at high concentrations; for example, 7.13 µg/g for L15 and 4.69 µg/g for L16 in pacifiers and 13.3 µg/g for L15 and 14.5 µg/g for L16 in teethers.

Zhang et al. [48] also reported the concentrations of VMSs in silicone products including food-grade silicone fluids, silicone bakeware, and silicone nipples (no information available on the locations of markets). Three CMSs (D4, D5, and D6) were detected in all silicone nipples at concentrations ranging from 0.5 to

Table 3 Reported concentrations (µg/g) of methylsiloxanes in rubber and industrial products

Category	Source	D4	D5	D6	LMS	Ref
Hard toy	Chinese market	0.00049 <0.0006–0.009	0.002 <0.0004–0.013	0.002 <0.0007–0.025	0.024 for L5-L16 <LOQ–0.112	[47]
Pacifiers	Chinese market	1.46 0.012–45.1	3.23 0.031–66.3	5.66 0.062–74.1	68.9 for L5-L16 4.86–570.9	[47]
Teethers	Chinese market	0.312 0.005–1.46	3.26 0.016–10.4	4.96 0.018–16.4	91.7 for L5-L16 2.85–225	[47]
Soft rubber toys	Chinese market	0.143 0.03–0.969	0.515 <0.0004–1.93	1.24 0.024–8.55	8.03 for L5-L16 2.36–8.03	[47]
Food-grade silicone fluid	NA	145 30–306	139 28–346	62 10–150	2.4 for L3-L5 1.6–3.8	[48]
Bakeware	NA	23 1.4–65	1,550 9–3,451	3,180 645–4,692	5.5 for L3-L5 2.2–12	[48]
Nipple (pacifier)	NA	7.1 0.6–49	33 0.6–269	14 0.3–108	ND, for L3-L5	[48]
Home paint	Chinese facility	27 0.05–89	106 0.7–280	44 0.08–131	186 for L5-L16 186–10,620	[50]
Car shell paint/polish	Chinese facility	137 99–212	487 348–652	367 184–732	12,500 for L5-L16 6,700–16,086	[50]
Machine lubricant	Chinese facility	394 338–450	513 470–556	261 215–307	17.700 for L5-L16 17,061–18,390	[50]
Textile paint	Chinese facility	967 663–1,450	2,230 976–4,378	1,180 356–2,452	67,800 for L5-L16 59,300–72,600	[50]
Cookware	US market	1.5 <0.35–7.3	6.3 <0.39–15	96 <0.33–365	2.1 for L4-L14 <0.059–2,290	[1]
Sealant	US market	184 <0.35–551	214 <0.39–643	338 <0.33–1,010	8.8 for L4-L14 <0.059–26	[1]
Nipple (pacifier)	US market	0.74 0.62–0.87	81 5.7–159	342 12–741	14 for L4-L14 <0.059–33	[1]

NA not available, *LOQ* limit of quantification, *ND* not detected

269 µg/g, whereas no LMSs were detected in silicone nipples (Table 3). In bakeware products, concentrations of individual siloxanes ranged from not detected (ND, L3) to 7,030 µg/g (D6). Food-grade silicone fluids contained significant amounts of CMSs (30–306 µg/g for D4, 28–346 µg/g for D5, and 10–150 µg/g for D6) [48]. Horii and Kannan [1] examined concentrations of siloxanes (D4–D7 and L4–L14) in nipples, cookware, and sealants collected from the US market in 2008. All nipples (pacifiers) contained D4–D7 at concentrations of up to 0.87 µg/g for D4, 159 µg/g for D5, 714 µg/g for D6, and 846 µg/g for D7, and the highest concentrations of D7 were found in nipples.

The majority of the above studies did not analyze CMSs higher than D7, due to the lack of analytical standards. However, GC-MS analysis of monitoring of ions specific for methylsiloxanes (m/z 147, 221, 281, and 355) showed the presence of large amounts of high-molecular-weight CMSs in rubber materials [1]. In fact, the residues of additives in 23 kinds of rubber products for food contact use were investigated by Kawamura et al. [49], and they found that all of the samples contained 15–20 peaks of CMSs by GC-MS methods, corresponding to D6–D25 compounds. Based on the peak area of D6 standard, an estimate of the total concentrations of CMSs in cookware products was reported at 3,310–14,690 µg/g. D15–D18 were predominant in nipples, at concentrations ranging from 360 to 1,050 µg/g, whereas the concentrations of low-molecular-weight CMSs (D9) were relatively low (<50–360 µg/g).

4.4 Industrial Products

Xu et al. [50] investigated CMSs (D4–D6) and LMSs (L5–L16) in industrial products/additives such as home paints, car shell paint/polishes, and machine lubricants collected from various locations (Table 3). The total concentrations of CMSs (ΣCMS) in paint products ranged from 1.2 to 336 µg/g, with the detection frequencies of 75–79%, whereas the total concentrations of LMSs (ΣLMS) ranged from 186 to 10,620 µg/g (63–71% detection frequency). Concentrations of CMSs and LMSs in car shell paint/polishes and machine lubricant samples were one to two orders of magnitude higher than those of home paints. Generally, LMSs accounted for major proportions of total siloxanes in additives and products.

Silicone-based products are also used as alternatives to perfluorooctane sulfonic acid (PFOS) and its related chemicals in firefighting foams, coating additives, polymers for the impregnation of textile fabrics, leather, carpets, rugs, upholstery, and papers. For example, PFOS was used in "Scotch Guard," a water/stain repellent produced by the 3M company, but it has been stated that PFOS were replaced by siloxanes in those products in recent years. As mentioned above, some commercial silicone polymers used as alternatives to PFOS may contain VMSs. However, silicones can provide durable water repellency, but not oil or soil and stain repellency like fluorinated products [51].

5   Environmental Emission

Industrial manufacturing as well as direct use of PCPs containing VMSs can lead to emissions of these chemicals into the environment. The large production volume in combination with their high mobility/volatility and persistence are a matter of concern from the environmental emission point of view. VMSs, especially CMSs, were found globally in various environmental matrices including air, water, and sludge. This section provides information on the concentrations, patterns, and emission of VMSs in selected matrices including PCPs, WWTPs, landfills, silicone production, paper production, and oilfields. Although occurrence of VMSs in environmental matrices is discussed in other chapters, the concentration profiles of VMSs in the environment are briefly provided here for comparison with source materials.

5.1   Daily Usage Rates of Siloxanes from PCPs

The environmental emission of VMSs primarily occurs through the direct application of PCPs. Several studies have estimated usage rates (external exposure rate) of methylsiloxanes from the use of PCPs. The usage rates were estimated based on the concentrations measured in each product and the daily usage rates of PCPs, although several factors such as dermal permeation rates and retention times were applied for estimation of exposure dose. This section focuses on the usage rate and emission of VMSs from PCPs. Therefore, exposure rates of VMSs from several studies are calculated back with factor 1 (e.g., retention time factor) and are described as usage rate of VMSs hereafter (Table 4).

Horii and Kannan [1] reported that the daily exposure rate of total CMSs (D4–D7) and LMSs (L4–L14) was 307 mg/day for women in the USA. The highest exposure rate was 0.71 mg/day for D4 and 162 mg/day for D5, with hair conditioners contributing to the highest exposure. For D6, the daily exposure was high from liquid foundations (14 mg). LMS exposure was predominantly (50 mg/day) from the use of face creams, whereas D5 exposure was the highest from most PCPs. Among PCPs from Canada [45], the daily exposures of D4 and D5 were estimated at 96 and 306 mg/day from body lotion and 2.6 and 594 mg/day from antiperspirants, respectively. In total, 98.6 mg/day for D4 and 900 mg/day for D5 were estimated from the use of PCPs from the Canadian market.

Dudzina et al. [38] surveyed VMSs in European PCPs. Based on median concentrations of D4 and D5 in PCP categories, daily exposure of D4 and D5 in each category ranged from 0.0006 to 0.59 mg/capita/day and 1.2 to 673 mg/capita/day, respectively, and sunscreens were found to be high both for D4 and D5. The daily exposure rate of D4 and D5 for all categories studied, excluding sunscreens, was 0.08 mg/capita/day and 270 mg/capita/day, respectively. Capela et al. [46] reported dermal exposure of siloxanes (D3–D6 and L2–L5) through application of PCPs in

Table 4 Reported exposure rates of methylsiloxanes from personal care products

Category	Country	Exposure rate (mg/capita/day)				Ref
		D4	D5	D6	LMS	
Shampoo	USA	0.198	0.335	0.333	0.00087 for L4–L14	[1][a]
	China	0.090	0.179	0.10	NA	[44][b]
Hair conditioners	USA	0.712	162	1.31	0.022 for L4–L14	[1][a]
	Europe	0.001	10	NA	NA	[38][b]
	China	0.0523	0.192	0.164	0.002 for L4–L14	[44][b]
Body washes	USA	NA	ND	ND	0.022 for L4–L14	[1][a]
	China	0.0003 for total of D4–D7 and L4–L14				[44][b]
Body lotions	USA	0.0074	0.004	0.0011	0.0004 for L4–L14	[1][a]
	Canada	96	306	NA	NA	[45][c]
	Europe	0.003	68.4	NA	NA	[38][b]
	China	0.024	0.209	0.200	0.576 for L4–L14	[44][b]
Face creams	USA	0.075	43.4	6.1	49.9 for L4–L14	[1][a]
	Europe	0.001	21	NA	NA	[38][b]
	China	0.0015	0.00992	0.00278	0.384 for L4–L14	[44][b]
Cosmetics	USA	0.091	27.4	14.4	0.0033 for L4–L14	[1][a]
	Europe	0.020	16.4	NA	NA	[38][b]
	China	0.00312	0.0134	0.0367	1.19 for L4–L14	[44][b]
Antiperspirant/ deodorant	Canada	2.6	594	NA	NA	[45][c]
	Europe	0.050	153	NA	NA	[38][b]

Exposure rates of VMSs from several studies were calculated back with factor 1 and were described as usage rate of VMSs

NA not available

[a]Based on mean concentration in the category products

[b]Based on median concentrations in the category products

[c]Based on maximum concentrations in the category products

Portugal. The mean and maximum total daily dermal exposure for adults to siloxanes would be 25 and 89 µg/kg-bw/day (i.e., 1.5 and 5.3 mg/capita/day), respectively.

From the PCPs in the Chinese market [44], the daily exposure dose ranged from 0.0015 (face cream) to 0.09 mg/capita/day (shampoo), 0.0099 (face cream) to 0.21 mg/capita/day (body lotion), and 0.0028 (face cream) to 0.2 mg/capita/day (body lotion) for D4, D5, and D6, respectively. Interestingly, high levels of LMSs were found in face creams and cosmetics from China, whereas D5 was the predominant compound in those products from the USA and Europe.

VMS emission can be categorized based on the usage of PCPs: one category is leave-on products including skin care products, lipsticks, skin foundation, makeup products, makeup removers, deodorants, antiperspirants, sunscreen, and some hair care products; the other category is wash-off products that include shampoo and hair conditioners, makeup removers, and shower gel. Depending on the application, VMSs in products result in different rates of emissions into the aquatic environment [16] (Fig. 5). Given the high volatility of VMSs, leave-on products are expected to

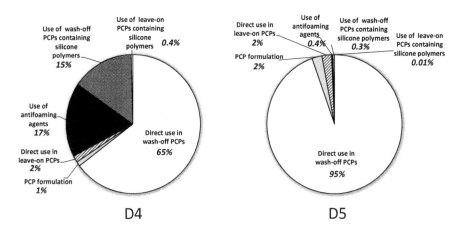

Fig. 5 Annual emission (%) of D4 and D5 in wastewater from European Union countries. Data from [16]; the figure was prepared by the authors

be released primarily into the atmosphere and a small proportion discharged into the sewer system. Gouin et al. [52] and Montemayor et al. [53] reported that ≤0.1% w/w of D5 contained in leave-on PCPs (such as antiperspirants and deodorants) is likely available for wash-off 8–24 h after application and most VMSs rapidly evaporate from skin. One study reported that 0.1% of the amount reached wastewater from leave-on products, in the reasonable worst case [16]. Therefore, VMS emissions to wastewater derived from use of leave-on PCPs can be low and contribute only 2% for D4 and D5 in the total emission (Fig. 5). The Canadian Board of Review for D5 reported that 40% of wash-off PCPs (such as hair conditioners) were released into wastewater after shower. Applying a worst-case scenario of 100% release yielded 11.3 tons/year for D4 and 750 tons/year for D5 from wash-off PCPs into the sewer system in EU. In other words, the direct use of VMSs in wash-off PCPs contributes to the emission of 65% for D4 and 95% for D5, and wash-off PCPs are the major sources of VMSs to wastewater. Although VMSs are used in PCP formulations, this only accounts for a small percentage of the total release into wastewater (Fig. 5). Moreover, the emission rates of wash-off PCPs for D4 and D5 are 4–240 times higher than those from their use as antifoaming agents.

5.2 Releases to the Atmospheric Environment

VMSs are released into the environment from different anthropogenic sources including industrial releases associated with silicone manufacturing and from the use of consumer products that contain VMSs. More than 90% of VMSs are released directly into air or quickly partitions from water to the atmosphere after discharge [54]. Airborne emissions from landfills and sewage treatment plants (aeration gas) are also important sources of VMSs to the atmosphere.

5.2.1   Indoor Environment

Use of VMSs in personal care and household products entails a major source to the indoor environment (chapter "Occurrence and Human Exposure of VMSs in Indoor Air" discussed VMSs in indoor air). The occurrence of VMSs in indoor environment including air and dust has been reported in several studies: home [55–60], office [57, 61–63], school [57, 64, 65], shops [57, 58], and cars [56]. The concentrations of D4, D5, and D6 in indoor air varied widely in the ranges of 0.0084–270 $\mu g/m^3$, 0.04–510 $\mu g/m^3$, and 0.015–180 $\mu g/m^3$, respectively (Fig. 6). The predominance of D5 in the indoor environment is in accordance with the composition in PCPs. LMS concentrations in indoor air were substantially lower than those of CMSs. D5 concentrations in commercial buildings in the USA ranged from 1.3 to 120 $\mu g/m^3$, higher than those of acetaldehyde, and one to two orders of magnitude higher than those of benzene, o-xylene, and chloroform. The profile and occurrence of VMSs in the indoor environment reflected the use of specific products and the existence of typical sources. Even within the same room type, large differences were found in the profiles of VMSs [55]. Because of the heavy use (high amount) of several indoor products including air fresheners, VMS profiles can differ depending on the consumer preferences and lifestyle choices. Emission of VMSs from electronics such as computer and printer can also affect the profiles of siloxanes in office room environments [62].

The concentration range of D5 in indoor air in a high school classroom in Albany, New York, was 0.1–1.0 $\mu g/m^3$ [65] and in two offices in Barcelona, Spain, was 2.4 ± 0.2 and 1.7 ± 0.2 $\mu g/m^3$ [61] (Fig. 6). The indoor air concentrations of CMSs (D4, D5, and D6) were 63 ± 33 $\mu g/m^3$ in a classroom at the University of California (Berkeley, USA), and D5 was the dominant compound with >90% of the total concentration [64]. The per capita emission rates were estimated at 0.048–30, 4.4–235, and 0.46–7.3 mg/day/capita for D4, D5, and D6, respectively, in the USA. These findings were consistent with those reported evaporation losses of D5 after the dermal application of PCPs [52, 53]. The diurnal variation in CMS emissions into air was consistent with the use of PCPs by students in the morning before going to classes.

5.2.2   Wastewater Treatment Plants (WWTPs)

Several studies showed that wastewater treatment plants (WWTPs) are important sources of VMSs into the aquatic and atmospheric environments. The concentration profiles of CMSs in select emission sources are shown in Fig. 6. Shoeib et al. [66] assessed emission of siloxanes (D3–D6 and L3–L5) into the atmosphere from eight WWTPs located in Ontario, Canada, using paired sorbent-impregnated polyurethane foam (SIP) passive air samplers. The types of WWTPs investigated include secondary activated sludge treatment plants in urban areas, secondary extended aeration in towns, and facultative lagoons in rural areas. The concentrations of D4, D5, and D6 ranged 0.14–0.348 $\mu g/m^3$, 0.115–1.29 $\mu g/m^3$, and 0.019–0.219 $\mu g/m^3$, respectively,

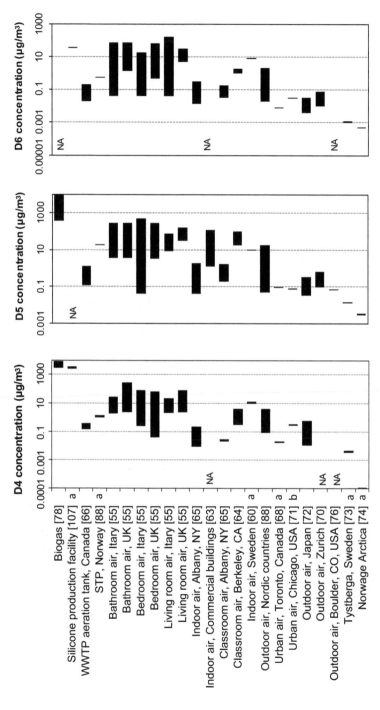

Fig. 6 Concentration ranges of D4, D5, and D6 in air samples reported worldwide. [a]Mean, [b]median

Table 5 Reported emission rates of methylsiloxanes to the atmosphere from various sources

Source/location	Emission	Remarks	Ref
WWTP, Ontario, Canada	540 kg/year for D4; 1,400 kg/year for D5; 2.6 kg/year for L3–L5	1,000,000 population served	[67]
WWTPs, Ontario, Canada	5–112 kg/year/tank for CMSs; 0.0026–0.396 kg/year/tank for LMSs	8 WWTPs; D3–D6 and L3–L5	[66]
Two land-fills, Ontario, Canada	20–70 kg/year for D4; 60–170 kg/year for D5; 1.1–1.2 kg/year for L3–L5	Unknown landfill sizes	[67]
Two office air, Chicago	271 mg/day/capita for D4; 603 mg/day/capita for D5	Per person emission rate; Engineering Arts and Sciences at the University of Iowa	[71]
Classroom air	0.048–30 mg/day/capita for D4; 4.4–235 mg/day/capita for D5; 0.46–7.3 mg/day/capita for D6	Per person emission rate; classroom at the University of California	[63]
US emissions	90 mg/day/capita for D4; 137 mg/day/capita for D5		[75]
Boulder, CO, USA	3–5.5 kg/day for D5	Emission in Boulder; 15.6 kg/day for benzene	[76]
Urban air (residential)	190 mg/day/capita for D5	Chicago, IL, USA	[69]
Urban air (residential)	310 mg/day/capita for D5	Zurich, Switzerland	[70]
Sales of personal care products	260 mg/day/capita for D5	Based on assumption of PCPs in the USA and Canada; 95% emission	[69]

in those WWTPs, and emissions to air were 5–112 kg/year/tank for the sum of D3–D6 (Table 5). In comparison to other compounds, the emission rates of siloxanes were two to three orders of magnitude higher than those of UV filters (0.016–2 kg/year/tank) and per- and polyfluoroalkyl substances (0.01–0.110 kg/year/tank). Cheng et al. [67] investigated atmospheric emissions of VMSs from WWTPs using passive air samplers and reported that the yearly emissions of VMSs from a large-scale WWTP serving about one million people into air were estimated at 2,100 kg for CMS (D3–D6), with predominance of D5 (1,400 kg), followed by D4 (540 kg). The normalized per capita emissions are on the order of up to 2 g/year/capita for CMSs and 0.25 g/year/capita for LMSs (L3–L6).

5.2.3 Outdoor Environment

CMSs were measured in air from urban and suburban sites in Chicago, USA; Toronto, Canada; Zurich, Switzerland; and Saitama, Japan [68–72] and in rural/

remote atmosphere (e.g., Arctic regions) [73, 74] (Fig. 6). The measured concentrations varied widely in the ranges of $0.00001–0.4$ $\mu g/m^3$ and $0.0029–19$ $\mu g/m^3$ and $0.00045–2.1$ $\mu g/m^3$ for D4, D5, and D6, respectively. D5 emission rates in urban areas were estimated to be 190 mg/day/capita in Chicago, Illinois, USA [69] and 310 mg/day/person in Zurich, Switzerland [70] (Table 5). The estimated emission rates based on the sales of PCPs in Canada were 260 mg/day/capita [53]. In the USA, a per capita emission estimate of 90 mg/day for D4 and 137 mg/day for D5 was reported [75]. These findings were comparable to per capita emissions of D5 in office and classroom settings, estimated at hundreds of mg/day/capita. Coggon et al. [76] reported diurnal variability and emission pattern of D5 from the application of PCPs in two North American cities. D5 and benzene (an indicator for vehicle exhaust) were measured. The total emission rate of D5 in Boulder, CO, USA was 3–5.5 kg/day, which was similar in magnitude to that of benzene (15.6 kg/day).

Air emissions of VMSs from manufacturing, processing, and/or formulating facilities occur through stacks and fugitive emissions; however, subsequent deposition of VMSs to surface waters and soils is expected to be negligible due to their low water solubility, high air-water partitioning, and a relatively low octanol-air partition coefficient (K_{oa}) [77]. It is also possible that CMSs and LMSs settle in bottom sediments. Information pertaining to the direct measurement of VMSs from manufacturing, processing, and/or formulating facilities is limited.

5.2.4   Landfill/Biogas

Since the 1980s, the emission of VMSs through gases released from landfills has been a topic of interest [78]. The formation of VMSs by degradation of high-molecular-weight silicone has been suggested. Similarly, biogas generated from sewage sludge treatment, as well as from landfills, not only contains methane [79] but also VMSs [80, 81]. Typical concentrations of total siloxanes in gases emanated from landfills were in the range of 3–24 mg/m^3 [79], which are three to four orders of magnitude higher than those in the ambient atmosphere. Landfill gases collected from a domestic waste disposal site contained VMS concentrations in the range of several mg/m^3 [78] (Fig. 6), with a dominance of D4 and D5. During the anaerobic digestion of sewage sludge, when the temperature reaches to 60°C, VMSs evaporate and end up in biogas, as confirmed by biogas measurements at various plants in the UK (2–12 mg/m^3 in digester gas) [82]. Cheng et al. [67] reported atmospheric emission of CMSs using passive air samplers in two landfills in Ontario, Canada. Mean on-site air concentration of ΣCMS (D3–D6) was 4,670 ng/m^3 which was 14 times higher than those from upwind sites. The annual emission of CMSs from landfills into the atmosphere was estimated at 80 and 250 kg/site (unknown landfill size), with D5 prevailing in the air from landfills, with a maximum emission rate of 170 kg/year (Table 5).

5.3 Releases into the Aquatic Environment

Although a major fraction of VMSs in personal care and consumer products is expected to be emitted into air, a small fraction can reach the aquatic environment through wastewater discharges. Down-the-drain discharge is considered a major route for VMSs into the aquatic environment. This section is focused on the concentrations, profiles, and emission rates of VMSs into the aquatic environment.

5.3.1 Emission Rates

The emission of VMSs into the aquatic environment was estimated based on the concentrations measured in WWTP effluents and the volume of wastewater discharged daily. Occurrence, fate, and removal of VMSs in WWTPs are discussed in the chapter "Occurrence, Fate and Removal of VMSs in Wastewater Treatment Plants"; therefore, the concentrations of VMSs in WWTPs are briefly described here. Several studies have reported the occurrence of CMSs in WWTP influent and effluent samples (Fig. 7). The concentrations of D4, D5, and D6 in influents from 11 WWTPs located in Ontario and Quebec, Canada, were in the ranges of 0.282–6.69 µg/L, 7.75–135 µg/L, and 1.53–26.9 µg/L, respectively, whereas those in effluents were <0.009–0.045 µg/L, <0.027–1.56 µg/L, and <0.022–0.093 µg/L, respectively [83]. The concentrations of D4, D5, and D6 in a municipal WWTP effluent in Beijing, China, were 0.25–0.55 µg/L, 0.50–1.00 µg/L, and 0.52–0.96 µg/L, respectively [84]. In WWTP effluents from Catalonia, Spain, the mean concentrations were 0.076 µg/L for D4 and 0.545 µg/L for D5 [85]. Horii et al. [86] determined concentrations of CMSs (D3–D6) and LMSs (L3–L5) in 25 effluents from sewage treatment plants (STPs) located in Tokyo Bay (Japan) watershed using a modified purge and trap method. The concentrations of ΣVMS in effluents varied widely, by up to two orders of magnitude among the WWTPs studied, ranging in concentrations from 99 to 2,500 ng/L. The mean concentration of D5 (540 ng/g) was the highest, followed by D6 (45 ng/L), D4 (27 ng/L), and D3 (21 ng/L). Bletsou et al. [87] surveyed the mass loading and fate of siloxanes in a large-scale WWTP in Greece, serving 3.7 million people. The total concentrations of CMSs (D3–D7) in influent and effluent were 5.14 µg/L and 2.11 µg/L, respectively; for LMSs (L3–L14), it was 15.1 µg/L and 1.47 µg/L, respectively. The overall reported concentrations of D4, D5, and D6 in WWTP effluents from North America, Europe, and Asia ranged from 0.001 to 1.29 µg/L, 0.0018 to 6.02 µg/L, and 0.0015 to 3.96 µg/L, respectively, except for industrial effluent [83–89]. The CMSs were reported to occur in WWTP effluents at significantly higher concentrations than those of LMSs (<MDL in most of cases), and D5 was the predominant VMS found in samples from most countries [83–89]. The removal efficiency of CMSs in WWTPs was generally >95%.

A mass balance analysis of CMSs in WWTPs in Canada (population served, 285,900, and treatment capacity, 165,000 m^3/day) showed that the average mass of D4, D5, and D6 entering and exiting the plant in influent and effluent, respectively,

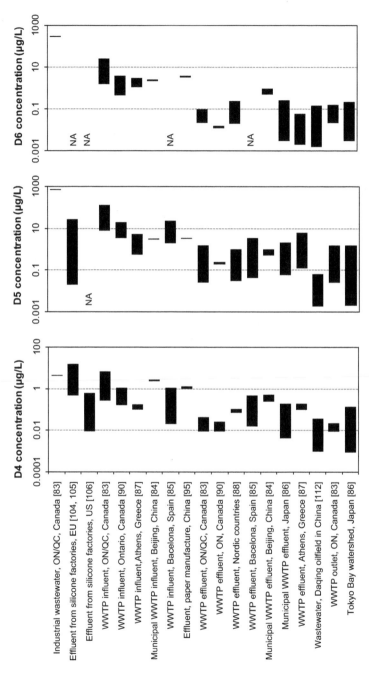

Fig. 7 Concentration ranges of D4, D5, and D6 in water samples reported worldwide

was 109, 2,050, 280, and 1.41, 27.0, 1.90 g/day [90]. The emission rates of CMSs in effluents varied widely depending on the receiving wastewater volume, 8.3 g/day for D5 and 2.4 g/day for D6 for WWTPs in the UK, with 179,800 population served: for WWTP in Greece (3.7 million population served), 96 g/day for D4, 1,330 g/day for D5, and 19 g/day for D6 (Table 6). The influx of VMSs in influents were found to be one to two orders of magnitude higher than the emission rates. Emission estimates normalized per capita were comparable among the WWTPs and ranged 0.0049–0.026 mg/day, 0.046–0.36 mg/day, and 0.0052–0.01 mg/day for D4, D5, and D6, respectively (Table 6). These studies suggested that the emission rates of VMSs from WWTPs into the aquatic environment were three to four orders of magnitude lower than the direct emission into the atmosphere from PCP application.

Horii et al. [86] estimated the emission of VMSs on the basis of sewage treatment volume and the geometric mean concentrations of VMSs. Flux-ΣVMS of large- and small-scale WWTPs varied widely from 0.0077 to 270 kg/year. The highest flux at 270 kg/year was found for a large-scale WWTP that served 1.7 million people with the highest sewage treatment volume (228 million m^3/year). The estimated annual emission into Tokyo Bay watershed was 87, 2,000, 120, and 2,300 kg for D4, D5, D6, and total 7 VMSs, respectively (Table 6). On a national scale, the total emission estimate was 7,600 kg for 7 VMSs. A widespread distribution of VMSs in Tokyo Bay watershed and a wide variation in the measured concentrations in surface waters were found, with the total concentrations of 7 VMSs (ΣVMS) in river water ranging from <MDL to 1,700 ng/L (median, 140 ng/L) (Fig. 7). High concentrations were found in samples collected near WWTP outlets, indicating that WWTP effluents and domestic wastewaters are the major sources of VMSs into rivers. The concentrations of D5 were shown to decrease with the increasing distance from a WWTP, possibly reflecting mixing/dilution of effluent in river/seawater and removal of D5 by volatilization from water surface and sedimentation onto the riverbed [91].

5.3.2 Pollution Prevention Plan in Canada

Pollution Prevention (P2) Plan in Canada is a process through which organizations can improve their environmental performance by strategically planning to reduce or eliminate pollution before creation. A P2 planning notice was selected as a risk management option for D4 [92]. The risk management objective in the notice was to meet, by the end of the implementation period, an 80% reduction in total D4 releases into the aquatic environment in comparison to the base year (i.e., 2013) levels. In 2013, seven facilities that manufactured or used ≥100 kg/year of D4 submitted a declaration that a P2 plan was prepared and implemented. By July 2017, five of the facilities declared that they had met the reduction target as a result of implementing P2 plan and had completed 1-year monitoring. The total D4 releases from all facilities at 227 kg in 2013 reduced to 101 kg in 2017, resulting in a reduction rate of D4 at 56%. The emission rate of D4 was comparable to those estimated for WWTPs in Tokyo Bay watershed (see above), although the P2 plan focuses mainly on industrial wastewater [86].

Table 6 Reported emission rates of methylsiloxanes into the aquatic environment from various sources

Source/ location	Emission	Remarks	Ref.
Effluent			
Tokyo Bay watershed	238 g/day for D4; 5,480 g/day for D5; 329 g/day for D6 0.0078 mg/day/capita for D4; 0.18 mg/day/capita for D5; 0.01 mg/day/capita for D6	78 STPs; 30,373,000 population served	[86]
WWTP, Ontario, Canada	1.41 g/day for D4; 27.0 g/day for D5; 1.90 g/day for D6 0.0049 mg/day/capita for D4; 0.094 mg/day/capita for D5; 0.0066 mg/day/capita for D6	285,900 population served; 165,000 m^3/day capacity; 70% home and 30% business	[90]
WWTP, UK	8.3 g/day for D5; 2.4 g/day for D6 0.046 mg/day/capita for D5; 0.013 mg/day/capita for D6	179,800 population served	[89]
WWTP, Greece	96 g/day for D4; 1,330 g/day for D5; 19 g/day for D6 0.026 mg/day/capita for D4; 0.36 g/day/capita for D5; 0.0052 mg/day/capita for D6	3,700,000 population served; 743,193 m^3/day average flow	[87]
Influent			
Input into WWTP, Ontario, Canada	109 g/day for D4; 2,050 g/day for D5; 280 g/day for D6 0.38 mg/day/capita for D4; 7.2 mg/day/capita for D5; 0.98 mg/day/capita for D6	285,900 population served; 165,000 m^3/day capacity	[90]
Input into WWTP, UK	480 g/day for D5; 238 g/day for D6 2.7 mg/day/capita for D5; 1.3 mg/day/capita for D6	179,800 population served	[89]
Input into WWTP, Greece	111 g/day for D4; 1,950 g/day for D5; 1,360 g/day for D6 0.03 mg/day/capita for D4; 0.53 mg/day/capita for D5; 0.37 mg/day/capita for D6	3,700,000 population served; 743,193 m^3/day average flow	[87]
Input into WWTP, Beijing, China	1,060 g/day for D4; 1,270 g/day for D5; 952 g/day for D6 1.3 mg/day/capita for D4; 1.6 mg/day/capita for D5; 1.2 mg/day/capita for D6	810,000 population served; 400,000 m^3/day capacity	[84]

5.4 Sludge

VMSs are hydrophobic with high log K_{oc} values (e.g., 6.12 for D5) and have a strong affinity for organic matter [93]. Therefore, VMSs are strongly adsorbed onto organic carbon or particulate matter (i.e., sludge) in wastewater treatment processes.

Sorption to sludge is a major removal mechanism of VMSs in WWTPs. Therefore, sewage sludge contains elevated concentrations of CMSs and LMSs. The concentration profiles of CMSs in WWTP sludge are summarized in Fig. 8 [84, 87, 88, 90, 94–97]. Similarly, concentrations reported for sediment and soil have also been compiled (Fig. 8) [83, 98–101]. Municipal and industrial wastewater sludge had a wide range of concentrations of up to over 100 µg/g dw with the predominance of D5, 0.03–4.37 µg/g dw for D4, 0.05–160 µg/g dw for D5, and 0.03–22.1 µg/g dw for D6, which were two to four orders of magnitude higher than those found in sediment from river, lake, and coastal area [84, 87, 88, 90, 94–97]. Lee et al. [94] conducted a nationwide survey of CMSs (D3–D7) and LMSs (L3–L17) in sewage sludge from WWTPs in Korea. The total concentrations of siloxanes in 40 sludge samples from domestic, industrial, and mixed WWTPs ranged from 0.05 to 142 µg/g dw, with mean concentration of 45.7 µg/g dw. The concentrations of CMSs and LMSs ranged from 0.05 to 48 µg/g dw and <LOQ to 128 µg/g dw, respectively. The concentrations of siloxanes in sludge from domestic WWTPs (66.1 µg/g dw) were significantly higher than those from industrial WWTPs (16.8 µg/g dw), which indicated that domestic releases from personal care and household products are an important source of siloxanes in WWTPs. The relative proportion of CMSs (58%) was greater than that of LMSs (42%), which is similar to that reported for sewage sludge from China [97]. Based on the amount of sludge produced in Korea (3.0 million tons/year in 2011), the estimated nationwide annual emission fluxes of siloxanes was 14,800 kg/year for CMSs and 18,500 kg/year for LMSs [94].

Wang et al. [90] measured CMSs in primary sludge and activated sludge from WWTPs located around Lake Ontario, Canada. The concentrations of D4, D5, and D6 in activated sludge were 1.69 µg/g dw, 67.8 µg/g dw, and 8.38 µg/g dw, respectively, and these concentrations were higher than those measured for primary sludge, indicating that CMSs were transferred from the aqueous phase to the particle phase in the aeration tank and secondary clarifier [90]. Furthermore, transformation of higher-molecular-weight VMSs in activated sludge process is expected.

In Canada, about 388,700 (dry weight) tons of biosolids are produced every year. Approximately 47% of these are incinerated, 43% are land-applied, and 4% are landfilled [102]. Wang et al. [83] measured CMSs in biosolid-amended soils collected from 11 farms in Canada, and the concentrations were <0.008–0.017 µg/g dw, <0.007–0.221 µg/g dw, and <0.009–0.711 µg/g dw for D4, D5, and D6, respectively. These concentrations were lower than those in sediment impacted by the wastewater effluent. D4 was below the MDL (0.008 µg/g dw) in most biosolid-amended soils, probably due to degradation and volatilization [103]. Sludge is also incinerated to some extent, and oligomeric and polymeric siloxanes are expected to degrade to inorganic silicon at incineration temperatures >800°C.

5.5   Siloxane Production Facilities

Despite the scrutiny of several regulatory agencies, the environment release of VMSs from silicone manufacturing industries is less known. The concentrations of

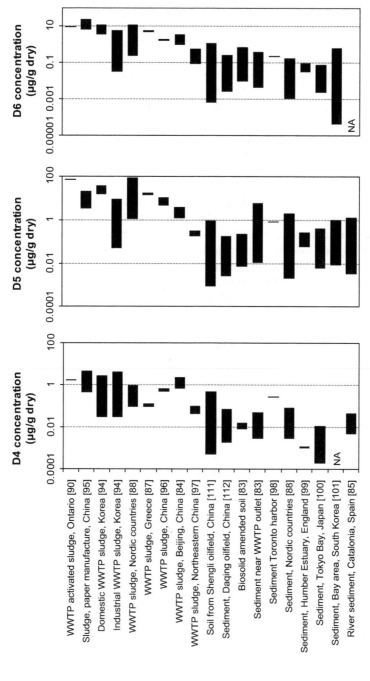

Fig. 8 Concentration ranges of D4, D5, and D6 in various solid samples reported worldwide

CMSs in WWTP effluents from silicone industries in Germany, France, and the UK varied widely from 0.5 to 16.4 μg/L for D4 and from <0.02 to 26.7 μg/L for D5 [104, 105] (Fig. 7). Nusz et al. [106] conducted an ecological risk assessment for D4 using exposure data collected from a nationwide environmental monitoring program of the USA. The median concentration of D4 measured in water downstream of on-site treatment of silicone facilities (0.08 μg/L) were high but of the same order of magnitude as those measured downstream from industrial and residential WWTPs (0.02 μg/L). The concentrations of VMSs in municipal WWTP effluents from North America, Europe, and Asia were mostly <1 μg/L and were two orders of magnitude lower than those reported for industrial wastewater effluents and comparable to surface water collected downstream of on-site treatment and industrial WWTPs.

Xu et al. [107] investigated the distribution of siloxanes around a siloxane production facility that produced CMS and LMS for more than 15 years, with an annual yield of about 10,000 tons. The mean concentrations of CMSs (D4–D6) and LMSs (L3–16) in indoor air collected from six workshops ranged from 34 μg/m^3 (D6) to 2,700 μg/m^3 (D4) and 1.2 μg/m^3 (L5) to 14.3 μg/m^3 (L6), respectively, which were one to three orders of magnitude higher than those found in residential indoor air (Fig. 6). Whereas the production volume of LMSs was about 70 times higher than that of CMSs in that facility, air concentrations of LMSs were one to five orders of magnitude lower than those of CMSs, likely due to low vapor pressures of LMSs [107]. That study also found that siloxane concentrations in air samples collected downwind decreased in an exponential fashion with the increase of the distance from the facility.

Xu et al. [50] also investigated the occurrence of siloxanes in indoor air and dust collected from industries including paint production, construction sites, automotive plants, and textile plants (see above for concentrations in the products). They found that the concentrations of siloxanes in indoor air and dust from the facilities were one to three orders of magnitude higher than those in residential homes in China. The concentrations of CMSs and LMSs in indoor air ranged 58.6–572 μg/m^3 and 0.05–11.2 μg/m^3, respectively, and in indoor dust ranged 2.74–50.4 μg/g and 184–6,700 μg/g, respectively. Although LMSs concentrations were higher than those of CMSs in many industrial additives/products, the concentrations of CMSs in indoor air were higher, while LMSs concentrations were higher in dust, reflecting the influence of differences in vapor pressure.

5.6 Paper Production

PDMS is used as a defoamer in paper production processes, such as pulp making, bleaching, and dewatering [95, 108]. The amount of PDMS applied as defoamer during pulp bleaching was 0.2–0.8 kg/tons, which can contain approximately 0.2–24 g of CMSs per ton of pulp [109]. China is leading the world in paper production capacity of more than 100 million tons/year. Free chlorine and sodium hypochlorite are still used for pulp bleaching in papermaking factories in China.

Xu et al. [95] investigated the occurrence of CMSs and chlorinated CMSs in a paper factory located in Dongying City, Shandong Province, China. PDMS was used in pulp washing, pulp bleaching, and pulp refining processes in that factory. The mean concentrations of total CMSs (D4–D6) collected from the processes ranged from 0.936 to 184 μg/L in aqueous samples (Fig. 7) and from 0.776 to 122 μg/g in solids (Fig. 8). Free chlorine was used in bleaching process, and monochlorinated CMSs, namely, monochloromethylheptamethylcyclotetrasiloxne [D3D(CH$_2$Cl)], monochloromethylnonamethylcyclopentasiloxane [D4D(CH$_2$Cl)], and monochloromethylundecamethylcyclohexasiloxane [D5D(CH$_2$Cl)], were detected in the bleaching process at the highest concentration of 287 μg/L in water and 270 μg/g in solid. The concentrations in the chlorination stage were found to be two to four orders of magnitude higher than those in hypochlorite stage (0.043 μg/L for water and 0.033 μg/g for solids). The mean mass flux from all papermaking processes (daily paper production volume; about 140 tons) was estimated at 40.6 g/day for D4, 174 g/day for D5, 93 g/day for D6, 8.76 g/day for D3D(CH$_2$Cl), 44.6 g/day for D4D(CH$_2$Cl), and 38.4 g/day for D5D(CH$_2$Cl). Approximately, 16.2% of D4, 19.0% of D5, and 27.7% of D6 in papermaking processes underwent monochlorination.

5.7 Oilfields

Emulsions are formed during oil exploitation due to the presence of natural surfactants, such as asphaltenes and resins. As water/oil phase separation is necessary before oil refining, surfactants are used to break water-in-oil emulsions, and formulations based on silicone copolymers are effective in promoting emulsion breaking [110]. The application of silicone de-emulsifiers and defoamers in oil production could result in the release of CMSs into the environment. Oil production also yields large amounts of sludge containing CMSs. Shi et al. [111] investigated the distribution, elimination, and rearrangement of CMSs in oil-contaminated soil of the Shengli Oil Field, which is the second largest in China. The concentrations of CMSs (sum of D4, D5, and D6) and LMSs (sum of L3–L16) in oil sludge were from 16.7 to 233 μg/g dw and 35.3–836 μg/g dw, respectively, which were one to three orders of magnitude greater than those reported for sludge from municipal WWTPs in China [97] (Fig. 8). Powell et al. [100] also analyzed soil samples collected from oilfields and found that the mean concentrations of CMSs in soils were ten times higher than those in soil samples collected from a reference area. The concentrations of CMSs in oilfield soils were positively correlated with total petroleum hydrocarbon (THP) concentrations, an important indicator of oil contamination. Powell et al. [100] also found a rising concentration of CMSs in soil from 2008 to 2013, due to the increasing usage of siloxanes as defoamers and de-emulsifiers, related to an increased oil production over time.

Zhi et al. [112] surveyed the occurrence and distribution of siloxanes (D4–D6 and L5–L16) in aqueous samples collected from the Daqing oilfield. For wastewater samples, the concentrations of total CMSs (D4–D6) and total LMSs (L5–L16) ranged from 137–1,911 ng/L to 49.4–190 ng/L, respectively (Fig. 7). CMSs were the predominant methylsiloxanes in wastewater samples from the oilfield. The authors also found that ΣCMS and ΣLMS concentrations in sediment samples collected from an old oilfield area were four to nine times higher than those in samples from a new oilfield area, with elevated concentrations of dimethylsilanediol (mean, 85.6 ± 98.3 ng/L) produced by the hydrolysis of methylsiloxanes.

6 Conclusions

This chapter summarizes main uses and environmental emissions of VMSs. The major use of VMSs, including CMSs and LMSs, is as an intermediate in the production of polymers. Silicone fluids, elastomers, rubbers, and resins are used in a wide range of downstream applications with a global production volume of 2.1 million tons in 2013, and it is projected to reach 2.8 million tons by 2023. Silicones can be found in fractional concentrations in the final products, and the silicone-based products contain residues of VMSs as impurities. Since D4, D5, and D6 have been registered as candidates of SVHC in June 2018 [18], the manufacturers are obliged to notify when these products contain $\geq 0.1\%$ SVHC by weight. Although the residual amounts of D4, D5, and D6 in the final products are unknown in many cases, such information would be used in a detailed estimation of emissions of D4, D5, and D6 into the environment through consumer product applications and products disposal.

VMSs are ubiquitous anthropogenic chemicals in the environment and are found in surface water, sediment, and outdoor air, including those from remote regions. Due to their strong association with PCP usage and relatively long atmospheric lifetime, VMSs can be useful tracers for monitoring PCP emissions into the atmospheric environments [76]. Recent studies suggested that PCP emissions could significantly contribute to VOC burdens in urban areas [76, 113]. Moreover, VMSs are considered important precursors of secondary organic aerosols (SOA) [113, 114], which are the major components of fine particulate matter in urban areas around the world.

Siloxanes are high-performance chemicals and have become indispensable in a wide range of industrial uses as well as in our daily lives. The health and environmental effects of these chemicals are a subject of another chapter. Whereas this chapter focused mainly on siloxanes, silicon, the inorganic element, has diverse applications in silicone industry. For example, hyper-purity silicon is produced using trichlorosilane as an intermediate, and these are mainly used for solar silicon and silicon wafers. For the new generation of solar power panels, silicone is used as a thin semiconductor. On the other hand, it is also true that a large amount of siloxanes are released into the environment through the industrial activities, the use of consumer products, and the disposal of silicone-based products. Waste management is a key factor for reducing siloxane emission into the environment. Since VMSs have unique physicochemical properties, with high volatility and high affinity to organic matter, prediction of their environmental fate by conventional approaches (e.g., bioconcentration factor, BCF) can be vague. As the chemistry of these chemicals is interesting and continuing to find newer applications, their usage is expected to grow. Appropriate management of environmental emission and development of appropriate risk management strategies are needed.

References

1. Horii Y, Kannan K (2008) Survey of organosilicone compounds, including cyclic and linear siloxanes, in personal-care and household products. Arch Environ Contam Toxicol 55:701–710
2. Wang DG, Norwood W, Alaee M, Byer JD, Brimble S (2013) Review of recent advances in research on the toxicity, detection, occurrence and fate of cyclic volatile methyl siloxanes in the environment. Chemosphere 93:711–725
3. Graiver D, Farminer KW, Narayan R (2003) A review of the fate and effects of silicones in the environment. J Polym Environ 11:129–136
4. Granchi D, Cavedagna D, Ciapetti G, Stea S, Schiavon P, Giuliani R, Pizzoferrato A (1995) Silicone breast implants: the role of immune system on capsular contracture formation. J Biomed Mater Res 29:197–202
5. Hayden JF, Barlow SA (1972) Structure-activity relationships of organosiloxanes and the female reproductive system. Toxicol Appl Pharmacol 21:68–79
6. He B, Rhodes-Brower S, Miller MR, Munson AE, Germolec DR, Walker VR, Korach KS, Meade BJ (2003) Octamethylcyclotetrasiloxane exhibits estrogenic activity in mice via ERalpha. Toxicol Appl Pharmacol 192:254–261
7. Lieberman MW, Lykissa ED, Barrios R, Ou CN, Kala G, Kala SV (1999) Cyclosiloxanes produce fatal liver and lung damage in mice. Environ Health Perspect 107:161–165
8. Quinn AL, Regan JM, Tobin JM, Marinik BJ, McMahon JM, McNett DA, Sushynski CM, Crofoot SD, Jean PA, Plotzke KP (2007) In vitro and in vivo evaluation of the estrogenic, androgenic, and progestagenic potential of two cyclic siloxanes. Toxicol Sci 96:145–153
9. Flaningam OL (1986) Vapor pressures of poly(dimethylsiloxane) oligomers. J Chem Eng Data 31:266–272
10. Varaprath S, Frye CL, Hamelink J (1996) Aqueous solubility of permethylsiloxanes (silicones). Environ Toxicol Chem 15:1263–1265
11. Xu S, Kropscott B (2012) Method for simultaneous determination of partition coefficients for cyclic volatile methylsiloxanes and dimethylsilanediol. Anal Chem 84:1948–1955
12. Environment Canada (2012) Notice requiring the preparation and implementation of pollution prevention plans in respect of cyclotetrasiloxane, octamethyl- (siloxane D4) in industrial effluents. Canada Gazette, Part I. http://www.gazette.gc.ca/rp-pr/p1/2012/2012-06-02/html/sup2-eng.html. Accessed 30 Jan 2019
13. Brooke DN, Crookes MJ, Gray D, Robertson S (2009) Environmental risk assessment report: octamethylcyclotetrasiloxane. Environment Agency of England and Wales, Bristol. https://assets.publishing.service.gov.uk/government/uploads/system/uploads/attachment_data/file/290565/scho0309bpqz-e-e.pdf. Accessed 30 Jan 2019
14. Brooke DN, Crookes MJ, Gray D, Robertson S (2009) Environmental risk assessment report: decamethylcyclopentasiloxane. Environment Agency of England and Wales, Bristol. https://assets.publishing.service.gov.uk/government/uploads/system/uploads/attachment_data/file/290561/scho0309bpqx-e-e.pdf. Accessed 30 Jan 2019
15. Brooke DN, Crookes MJ, Gray D, Robertson S (2009) Environmental risk assessment report: dodecamethylcyclohexasiloxane. Environment Agency of England and Wales, Bristol. https://assets.publishing.service.gov.uk/government/uploads/system/uploads/attachment_data/file/290562/scho0309bpqy-e-e.pdf. Accessed 30 Jan 2019

16. ECHA (2016) Background document to the opinion on the annex XV dossier proposing restrictions on octamethylcyclotetrasiloxane (D4) and decamethylcyclopentasiloxane (D5). ECHA/RAC/RES-O-0000001412-86-97/D, Helsinki. https://echa.europa.eu/documents/10162/13641/rest_d4d5_bd_en.pdf. Accessed 30 Jan 2019

17. European Commission (2018) Annex XVII to Regulation (EC) No 1907/2006 of the European Parliament and of the Council concerning the Registration, Evaluation, Authorisation and Restriction of Chemicals (REACH) as regards octamethylcyclotetrasiloxane (D4) and decamethylcyclopentasiloxane (D5), Brussels. https://eur-lex.europa.eu/legal-content/EN/TXT/PDF/?uri=CELEX:32018R1513&from=EN. Accessed 30 Jan 2019

18. ECHA (2018) Inclusion of substances of very high concern in the candidate list for eventual inclusion in Annex XIV, Doc: ED/61/2018, Helsinki. https://echa.europa.eu/documents/10162/eeed2c09-2263-25ad-49cd-a0926736c877. Accessed 30 Jan 2019

19. US-EPA (2014) Enforceable consent agreement for environmental testing for octamethyl-cyclotetrasiloxane (D4). EPA-HQ-OPPT-2012-0209. https://www.epa.gov/sites/production/files/2015-01/documents/signed_siloxanes_eca_4-2-14.pdf. Accessed 30 Jan 2019

20. Hobson JF, Atkinson R, Carter WPL (1997) Volatile methylsiloxanes. In: Chandra G (ed) Organosilicon materials. The handbook of environmental chemistry. Springer-Verlag, New York

21. O'Lenick AJ (2008) Silicones for personal care, 2nd edn. Allured Publishing Corporation, Carol Stream

22. GSC (2016) Socio-economic evaluation of the global silicones industry. AMEC Foster Wheeler Environment & Infrastructure UK Ltd, London. https://sehsc.americanchemistry.com/Socio-Economic-Evaluation-of-the-Global-Silicones-Industry-Final-Report.pdf. Accessed 30 Jan 2019

23. AMEC (2013) Market confidential final report. AMEC Foster Wheeler Environment & Infrastructure UK Ltd, London

24. OECD (2007) Manual for investigation of HPV chemicals

25. US-EPA (2007) High Production Volume (HPV) challenge program

26. Mojsiewicz-Pieńkowska K, Krenczkowska D (2018) Evolution of consciousness of exposure to siloxanes – review of publications. Chemosphere 191:204–217

27. US-EPA (2002) Non-confidential inventory update reporting production volume information. Toxic Substances Control Act (TSCA) Inventory

28. SIAJ (2009) Report of economic ripple and job creation effect of silicone in Japan (in Japanese). Silicone Industry Association of Japan, Tokyo. http://www.siaj.jp/ja/fact_sheet/pdf/keizaikoyo.pdf. Accessed 30 Jan 2019

29. Environment Canada (2008) Screening assessment for the challenge octamethylcyclo-tetrasiloxane (D4). http://www.ec.gc.ca/ese-ees/2481B508-1760-4878-9B8A-270EEE8B7DA4/batch2_556-67-2_en.pdf. Accessed 30 Jan 2019

30. Environment Canada (2008) Screening assessment for the challenge decamethylcyclopenta-siloxane (D5). https://www.cyclosiloxanes.org/uploads/Modules/Links/10.-environment-and-health-canada-screening-assessment-for-the-challenge-decamethylcyclopentasiloxane-(d5).pdf. Accessed 30 Jan 2019

31. Environment Canada (2008) Screening assessment for the challenge dodecamethylcyclohexa-siloxane (D6). https://www.ec.gc.ca/ese-ees/FC0D11E7-DB34-41AA-B1B3-E66EFD8813F1/batch2_540-97-6_en.pdf. Accessed 30 Jan 2019

32. Hunter MJ, Hyde JF et al (1946) Organo-silicon polymers; the cyclic dimethyl siloxanes. J Am Chem Soc 68:667–672

33. Reconsile Consortium (2014) Octamethylcyclotetrasiloxane (D4) chemical safety report

34. Reconsile Consortium (2014) Decamethylcyclopentasiloxane (D5) chemical safety report

35. Chandra G (1997) Organosilicon materials. The handbook of environmental chemistry, vol 3. Springer-Verlag, New York

36. Environment Canada (2015) Screening assessment for the challenge trisiloxane, octamethyl-(MDM). http://www.cela.ca/sites/cela.ca/files/780.draft%20RA%20MDM%20(Batch%2012).pdf. Accessed 30 Jan 2019

37. Friedel C, Crafts JM (1863). Liebigs Ann Chem 127:128

38. Dudzina T, von Goetz N, Bogdal C, Biesterbos JW, Hungerbuhler K (2014) Concentrations of cyclic volatile methylsiloxanes in European cosmetics and personal care products: prerequisite for human and environmental exposure assessment. Environ Int 62:86–94
39. Garaud JL (2007) Silicones in industrial applications. Dow Corning, Midland
40. DiSapio A, Fridd P (1988) Silicones: use of substantive properties on skin and hair. Int J Cosmet Sci 10:75–89
41. Abrutyn E, Bahr B (1993) Formulating enhancements for underarm applications. Cosmet Toiletries 108:51–54
42. Walker JD, Smock WH (1995) Chemicals recommended for testing by the TSCA interagency testing committee: a case study of octamethylcyclotetrasiloxane. Environ Toxicol Chem 14:1631–1634
43. Peter Fisk Associates (2012) Interim reports: technical support for D4 and D5 SEA
44. Lu Y, Yuan T, Wang W, Kannan K (2011) Concentrations and assessment of exposure to siloxanes and synthetic musks in personal care products from China. Environ Pollut 159:3522–3528
45. Wang R, Moody RP, Koniecki D, Zhu J (2009) Low molecular weight cyclic volatile methylsiloxanes in cosmetic products sold in Canada: implication for dermal exposure. Environ Int 35:900–904
46. Capela D, Alves A, Homem V, Santos L (2016) From the shop to the drain – volatile methylsiloxanes in cosmetics and personal care products. Environ Int 92–93:50–62
47. Xu L, Zhi L, Cai Y (2017) Methylsiloxanes in children silicone-containing products from China: profiles, leaching, and children exposure. Environ Int 101:165–172
48. Zhang K, Wong JW, Begley TH, Hayward DG, Limm W (2012) Determination of siloxanes in silicone products and potential migration to milk, formula and liquid simulants. Food Addit Contam Part A Chem Anal Control Expo Risk Assess 29:1311–1321
49. Kawamura Y, Nakajima A, Mutsuga M, Yamada T, Maitani T (2001) Residual chemicals in silicone rubber products for food contact use. Shokuhin Eiseigaku Zasshi 42:316–321. (in Japanese)
50. Xu L, Shi Y, Liu N, Cai Y (2015) Methyl siloxanes in environmental matrices and human plasma/fat from both general industries and residential areas in China. Sci Total Environ 505:454–463
51. UNEP (2016) Consolidated guidance on alternatives to perfluorooctane sulfonic acid and its related chemicals. https://www.informea.org/sites/default/files/imported-documents/UNEP-POPS-POPRC.12-INF-15-Rev.1.English.pdf. Accessed 30 Jan 2019
52. Gouin T, van Egmond R, Sparham C, Hastie C, Chowdhury N (2013) Simulated use and wash-off release of decamethylcyclopentasiloxane used in anti-perspirants. Chemosphere 93:726–734
53. Montemayor BP, Price BB, van Egmond RA (2013) Accounting for intended use application in characterizing the contributions of cyclopentasiloxane (D5) to aquatic loadings following personal care product use: antiperspirants, skin care products and hair care products. Chemosphere 93:735–740
54. Mackay D, Cowan-Ellsberry CE, Powell DE, Woodburn KB, Xu S, Kozerski GE, Kim J (2015) Decamethylcyclopentasiloxane (D5) environmental sources, fate, transport, and routes of exposure. Environ Toxicol Chem 34:2689–2702
55. Pieri F, Katsoyiannis A, Martellini T, Hughes D, Jones KC, Cincinelli A (2013) Occurrence of linear and cyclic volatile methyl siloxanes in indoor air samples (UK and Italy) and their isotopic characterization. Environ Int 59:363–371
56. Tran TM, Abualnaja KO, Asimakopoulos AG, Covaci A, Gevao B, Johnson-Restrepo B, Kumosani TA, Malarvannan G, Minh TB, Moon HB, Nakata H, Sinha RK, Kannan K (2015) A survey of cyclic and linear siloxanes in indoor dust and their implications for human exposures in twelve countries. Environ Int 78:39–44
57. Tran TM, Le HT, Vu ND, Minh Dang GH, Minh TB, Kannan K (2017) Cyclic and linear siloxanes in indoor air from several northern cities in Vietnam: levels, spatial distribution and human exposure. Chemosphere 184:1117–1124

58. Liu N, Xu L, Cai Y (2018) Methyl siloxanes in barbershops and residence indoor dust and the implication for human exposures. Sci Total Environ 618:1324–1330
59. Lu Y, Yuan T, Yun SH, Wang W, Wu Q, Kannan K (2010) Occurrence of cyclic and linear siloxanes in indoor dust from China, and implications for human exposures. Environ Sci Technol 44:6081–6087
60. Kaj L, Schlabach M, Andersson J, Cousins AP, Schmidbauer N, Brorström-Lundén E (2005) Results from the Swedish national screening programme 2004. Swedish Environmental Research Institute, Stockholm. https://www.ivl.se/download/18.343dc99d14e8bb0f58b542e/1443183072893/B2014.pdf. Accessed 30 Jan 2019
61. Companioni-Damas EY, Santos FJ, Galceran MT (2014) Linear and cyclic methylsiloxanes in air by concurrent solvent recondensation-large volume injection-gas chromatography-mass spectrometry. Talanta 118:245–252
62. McKone T, Maddalena R, Destaillats H, Hammond SK, Hodgson A, Russell M, Perrino C (2009) Indoor pollutant emissions from electronic office equipment. https://www.arb.ca.gov/research/seminars/mckone/mckone.pdf. Accessed 30 Jan 2019
63. Wu XM, Apte MG, Maddalena R, Bennett DH (2011) Volatile organic compounds in small- and medium-sized commercial buildings in California. Environ Sci Technol 45:9075–9083
64. Tang X, Misztal PK, Nazaroff WW, Goldstein AH (2015) Siloxanes are the most abundant volatile organic compound emitted from engineering students in a classroom. Environ Sci Technol Lett 2:303–307
65. Tran TM, Kannan K (2015) Occurrence of cyclic and linear siloxanes in indoor air from Albany, New York, USA, and its implications for inhalation exposure. Sci Total Environ 511:138–144
66. Shoeib M, Schuster J, Rauert C, Su K, Smyth SA, Harner T (2016) Emission of poly and perfluoroalkyl substances, UV-filters and siloxanes to air from wastewater treatment plants. Environ Pollut 218:595–604
67. Cheng Y, Shoeib M, Ahrens L, Harner T, Ma J (2011) Wastewater treatment plants and landfills emit volatile methyl siloxanes (VMSs) to the atmosphere: investigations using a new passive air sampler. Environ Pollut 159:2380–2386
68. Ahrens L, Harner T, Shoeib M (2014) Temporal variations of cyclic and linear volatile methylsiloxanes in the atmosphere using passive samplers and high-volume air samplers. Environ Sci Technol 48:9374–9381
69. Buser AM, Bogdal C, MacLeod M, Scheringer M (2014) Emissions of decamethyl-cyclopentasiloxane from Chicago. Chemosphere 107:473–475
70. Buser AM, Kierkegaard A, Bogdal C, MacLeod M, Scheringer M, Hungerbuhler K (2013) Concentrations in ambient air and emissions of cyclic volatile methylsiloxanes in Zurich, Switzerland. Environ Sci Technol 47:7045–7051
71. Yucuis RA, Stanier CO, Hornbuckle KC (2013) Cyclic siloxanes in air, including identification of high levels in Chicago and distinct diurnal variation. Chemosphere 92:905–910
72. Horii Y, Minomo K, Ohtsuka N, Motegi M, Takemine S, Yamashita N (2018) Development of a method for determination for atmospheric volatile methylsiloxanes and its application to environmental monitoring. Bunseki Kagaku 67:313–322. (in Japanese)
73. Kierkegaard A, McLachlan MS (2013) Determination of linear and cyclic volatile methylsiloxanes in air at a regional background site in Sweden. Atmos Environ 80:322–329
74. Krogseth IS, Kierkegaard A, McLachlan MS, Breivik K, Hansen KM, Schlabach M (2013) Occurrence and seasonality of cyclic volatile methyl siloxanes in Arctic air. Environ Sci Technol 47:502–509
75. Navea JG, Young MA, Xu S, Grassian VH, O'Stanier C (2011) The atmospheric lifetimes and concentrations of cyclic methylsiloxanes octamethylcyclotetrasiloxane (D4) and decamethyl-cyclopentasiloxane (D5) and the influence of heterogeneous uptake. Atmos Environ 45:3181–3191
76. Coggon MM, McDonald BC, Vlasenko A, Veres PR, Bernard F, Koss AR, Yuan B, Gilman JB, Peischl J, Aikin KC, DuRant J, Warneke C, Li SM, de Gouw JA (2018) Diurnal variability and emission pattern of decamethylcyclopentasiloxane (D5) from the application of personal care products in two north American cities. Environ Sci Technol 52:5610–5618

77. SEHSC (2012) Development of environmental monitoring proposal for certain cyclic siloxanes. Report submitted to US EPA, Silicones, Environmental, Health, and Safety Center
78. Schweigkofler M, niessner R (1999) Determination of siloxanes and VOC in landfill gas and sewage gas by canister sampling and GC-MS/AES analysis. Environ Sci Technol 33:3680–3685
79. Ajhar M, Travesset M, Yuce S, Melin T (2010) Siloxane removal from landfill and digester gas – a technology overview. Bioresour Technol 101:2913–2923
80. Eklund B, Anderson EP, Walker BL, Burrows DB (1998) Characterization of landfill gas composition at the fresh kills municipal solid-waste landfill. Environ Sci Technol 32:2233–2237
81. Ohannessian A, Desjardin V, Chatain V, Germain P (2008) Volatile organic silicon compounds: the most undesirable contaminants in biogases. Water Sci Technol 58:1775–1781
82. Dewil R, Appels L, Baeyens J (2006) Energy use of biogas hampered by the presence of siloxanes. Energy Convers Manag 47:1711–1722
83. Wang DG, Steer H, Tait T, Williams Z, Pacepavicius G, Young T, Ng T, Smyth SA, Kinsman L, Alaee M (2013) Concentrations of cyclic volatile methylsiloxanes in biosolid amended soil, influent, effluent, receiving water, and sediment of wastewater treatment plants in Canada. Chemosphere 93:766–773
84. Xu L, Shi Y, Cai Y (2013) Occurrence and fate of volatile siloxanes in a municipal wastewater treatment plant of Beijing, China. Water Res 47:715–724
85. Sanchis J, Martinez E, Ginebreda A, Farre M, Barcelo D (2013) Occurrence of linear and cyclic volatile methylsiloxanes in wastewater, surface water and sediments from Catalonia. Sci Total Environ 443:530–538
86. Horii Y, Minomo K, Ohtsuka N, Motegi M, Nojiri K, Kannan K (2017) Distribution characteristics of volatile methylsiloxanes in Tokyo Bay watershed in Japan: analysis of surface waters by purge and trap method. Sci Total Environ 586:56–65
87. Bletsou AA, Asimakopoulos AG, Stasinakis AS, Thomaidis NS, Kannan K (2013) Mass loading and fate of linear and cyclic siloxanes in a wastewater treatment plant in Greece. Environ Sci Technol 47:1824–1832
88. Kaj L, Schlabach M, Andersson J, Palm Cousins A, Schmidbauer N, Brorstrom Lunden E (2005) Siloxanes in the Nordic environment. Nordic Council of Ministers, Copenhagen. http://norden.diva-portal.org/smash/get/diva2:702777/FULLTEXT01.pdf. Accessed 30 Jan 2019
89. van Egmond R, Sparham C, Hastie C, Gore D, Chowdhury N (2013) Monitoring and modelling of siloxanes in a sewage treatment plant in the UK. Chemosphere 93:757–765
90. Wang DG, Aggarwal M, Tait T, Brimble S, Pacepavicius G, Kinsman L, Theocharides M, Smyth SA, Alaee M (2015) Fate of anthropogenic cyclic volatile methylsiloxanes in a wastewater treatment plant. Water Res 72:209–217
91. Sparham C, van Egmond R, O'Connor S, Hastie C, Whelan M, Kanda R, Franklin O (2008) Determination of decamethylcyclopentasiloxane in river water and final effluent by headspace gas chromatography/mass spectrometry. J Chromatogr A 1212:124–129
92. Environment and Climate Change Canada (2017) Final performance report: pollution prevention planning and siloxane D4 (2013–2017). https://www.canada.ca/content/dam/eccc/documents/pdf/p2/20180319-01-en.pdf. Accessed 30 Jan 2019
93. Panagopoulos D, Jahnke A, Kierkegaard A, MacLeod M (2015) Organic carbon/water and dissolved organic carbon/water partitioning of cyclic volatile methylsiloxanes: measurements and polyparameter linear free energy relationships. Environ Sci Technol 49:12161–12168
94. Lee S, Moon HB, Song GJ, Ra K, Lee WC, Kannan K (2014) A nationwide survey and emission estimates of cyclic and linear siloxanes through sludge from wastewater treatment plants in Korea. Sci Total Environ 497–498:106–112
95. Xu L, He X, Zhi L, Zhang C, Zeng T, Cai Y (2016) Chlorinated methylsiloxanes generated in the papermaking process and their fate in wastewater treatment processes. Environ Sci Technol 50:12732–12741
96. Li B, Li W-L, Sun S-J, Qi H, Ma W-L, Liu L-Y, Zhang Z-F, Zhu N-Z, Li Y-F (2016) The occurrence and fate of siloxanes in wastewater treatment plant in Harbin, China. Environ Sci Pollut Res 23:13200–13209

97. Zhang Z, Qi H, Ren N, Li Y, Gao D, Kannan K (2011) Survey of cyclic and linear siloxanes in sediment from the Songhua River and in sewage sludge from wastewater treatment plants, northeastern China. Arch Environ Contam Toxicol 60:204–211

98. Powell DE, Kozerski GE (2007) Cyclic methylsiloxane (cVMS) materials in surface sediments and cores for Lake Ontario. HES Study No. 10724–108, Dow Corning Corporation, Auburn

99. Kierkegaard A, van Egmond R, McLachlan MS (2011) Cyclic volatile methylsiloxane bioaccumulation in flounder and ragworm in the Humber estuary. Environ Sci Technol 45:5936–5942

100. Powell DE, Suganuma N, Kobayashi K, Nakamura T, Ninomiya K, Matsumura K, Omura N, Ushioka S (2017) Trophic dilution of cyclic volatile methylsiloxanes (cVMS) in the pelagic marine food web of Tokyo Bay, Japan. Sci Total Environ 578:366–382

101. Lee SY, Lee S, Choi M, Kannan K, Moon HB (2018) An optimized method for the analysis of cyclic and linear siloxanes and their distribution in surface and core sediments from industrialized bays in Korea. Environ Pollut 236:111–118

102. Apedaile E (2001) A perspective on biosolids management. Can J Infect Dis 12:202–204

103. Xu S, Chandra G (1999) Fate of cyclic methylsiloxanes in soils. 2. Rates of degradation and volatilization. Environ Sci Technol 33:4034–4039

104. Boehmer T, Gerhards R (2003) Octamethylcyclotetrasiloxane (D4), a compilation of environmental data. Centre Europeen des Silicones, Brussels

105. Boehmer T, Gerhards R (2003) Decamethycyclopentasiloxane (D5), a compilation of environmental data. Centre Europeen des Silicones, Brussels

106. Nusz JB, Fairbrother A, Daley J, Burton GA (2018) Use of multiple lines of evidence to provide a realistic toxic substances control act ecological risk evaluation based on monitoring data: D4 case study. Sci Total Environ 636:1382–1395

107. Xu L, Shi Y, Wang T, Dong Z, Su W, Cai Y (2012) Methyl siloxanes in environmental matrices around a siloxane production facility, and their distribution and elimination in plasma of exposed population. Environ Sci Technol 46:11718–11726

108. Habermehl J (2005) Silicone processing benefits pulp brownstock washing operations. China Pulp Paper Technology. https://consumer.dow.com/en-us/document-viewer.html?ramdomVar=6250614571543657171&docPath=/documents/en-us/tech-art/30/30-11/30-1147-01-silicone-processing-for-pulp-brownstock-washing.pdf. Accessed 30 Jan 2019

109. Chao SH (2012) Silicones in the pulp and paper industry. https://consumer.dow.com/en-us/document-viewer.html?ramdomVar=7674900645825177899&docPath=/documents/en-us/tech-art/26/26-24/26-2457-01-sdc-phase-3-chapter-4-.pdf. Accessed 30 Jan 2019

110. Daniel-David D, Pezron I, Dalmazzone C, Noïk C, Clausse D, Komunjer L (2005) Elastic properties of crude oil/water interface in presence of polymeric emulsion breakers. Colloids Surf A Physicochem Eng Asp 270–271:257–262

111. Shi Y, Xu S, Xu L, Cai Y (2015) Distribution, elimination, and rearrangement of cyclic volatile methylsiloxanes in oil-contaminated soil of the Shengli oilfield, China. Environ Sci Technol 49:11527–11535

112. Zhi L, Xu L, He X, Zhang C, Cai Y (2018) Occurrence and profiles of methylsiloxanes and their hydrolysis product in aqueous matrices from the Daqing oilfield in China. Sci Total Environ 631–632:879–886

113. McDonald BC, de Gouw JA, Gilman JB, Jathar SH, Akherati A, Cappa CD, Jimenez JL, Lee-Taylor J, Hayes PL, McKeen SA, Cui YY, Kim SW, Gentner DR, Isaacman-VanWertz G, Goldstein AH, Harley RA, Frost GJ, Roberts JM, Ryerson TB, Trainer M (2018) Volatile chemical products emerging as largest petrochemical source of urban organic emissions. Science 359:760–764

114. Wu Y, Johnston MV (2017) Aerosol formation from OH oxidation of the volatile cyclic methyl siloxane (cVMS) decamethylcyclopentasiloxane. Environ Sci Technol 51:4445–4451

Analytical Methods for Volatile Methylsiloxanes Quantification: Current Trends and Challenges

Vera Homem and Nuno Ratola

Contents

Abstract Silicon materials are widespread in our daily life and in numerous industrial applications, and this started raising concerns in the scientific community a couple of decades ago regarding the potential negative effects these chemicals could have in the environment and human health. Naturally, analytical methodologies were required to assess their presence around us. In particular, volatile methylsiloxanes (VMSs) have been the focus of research in this field, and their presence has been determined in many environmental matrices. However, this extended presence tends to provoke problems of external contamination during sampling and analysis, as, for instance, personal care products or chromatograph parts have VMSs in their formulations. Also, the volatility of these compounds advises against a large number of sample handling steps. This chapter reviews the analytical choices for the analysis of VMSs in water, air, sediments, soil and sewage sludge reported so far in literature, giving an overview of the sampling and sample processing precautions and the strategies employed for the extraction/clean-up (or lack thereof) before the typical analysis

V. Homem (✉) and N. Ratola (✉)
LEPABE – Laboratory for Process Engineering, Environment, Biotechnology and Energy, Faculty of Engineering, University of Porto, Porto, Portugal
e-mail: vhomem@fe.up.pt; nrneto@fe.up.pt

V. Homem and N. Ratola (eds.), *Volatile Methylsiloxanes in the Environment*,
Hdb Env Chem (2020) 89: 71–118, DOI 10.1007/698_2020_469,
© Springer Nature Switzerland AG 2020, Published online: 25 March 2020

by gas chromatography coupled with mass spectrometry detection (GC-MS), which in some cases presented different injection options.

Keywords Analytical methods, Environmental compartments, Sample handling, Volatile methylsiloxanes

1 Introduction

Volatile methylsiloxanes (VMSs) are low molecular weight organic compounds, with a Si-O backbone and methyl groups as side chains attached to the Si atoms [1]. These compounds have a significant vapour pressure (high volatility), low viscosity and tare poorly soluble in water (lipophilic compounds) [1]. According to their chemical structure, VMSs may be classified as linear (also known as dimethicone) and cyclic compounds (cyclomethicone).

VMS are mainly used in the industry as intermediates in the manufacturing of high molecular weight silicone polymers but are also incorporated in daily life products such as cosmetics and toiletries [2–4]. In these formulations, they are mainly used as carriers, conditioning agents and emollients [5]. Due to an unusual combination of properties, the production and use of VMSs have been increasing in the last years, which consequently implies a considerable release to the environment. Therefore, their ubiquity in a wide range of household products is likely to result in emissions to the sewage systems, reaching wastewater treatment plants (WWTPs). Since these facilities are not prepared to their total removal/degradation [6–11], WWTPs are, at the same time, undoubtedly important sources of emission to the environment. Consequently, the awareness of the scientific community and of some regulators has been growing [12], and VMSs have been detected in concentrations ranging from ng to few $\mu g\ L^{-1}$ in aqueous matrices [9, 13], ng g^{-1} in sediments [9, 13, 14], soils [15, 16] and biota [17–20] and usually from a few hundred to several thousand ng m^{-3} in air samples [16, 21–25].

Although some controversial issues have arisen on the potential risks associated with VMSs, recent studies suggest that due to their semi-volatile behaviour combined with their persistence in air, VMSs may have the ability to be transported over long distances [15, 24]. On the other hand, their lipophilic behaviour and low biodegradability [26] also point towards their potential to bioaccumulate and biomagnify [17]. Indeed, some of these contaminants are suspected of having hazardous effects to living organisms [27–29].

In order to clarify doubts regarding the hazards of these chemicals, it is essential to obtain data on their levels, fate and behaviour in the environment. With this purpose in mind, expedite, sensitive and robust analytical methodologies for the determination of VMSs in different matrices are mandatory.

This chapter intends to summarize the current knowledge on the analytical methods for the quantification of VMSs in environmental matrices (water, sewage sludge, soil, sediments and air), discussing the evolution from the first studies to the recent trends and the main challenges in their determination.

2 Analytical Methodologies

The complexity of most environmental matrices combined with the low concentrations in which the target compounds are usually found constitutes an important difficulty in the development of suitable analytical protocols for their analysis and quantification. In fact, very sensitive methodologies are required also for VMSs, and, for that reason, most authors have chosen to combine effective extraction strategies with the most powerful technique to identify and quantify volatile substances, gas chromatography coupled with mass spectrometry detection (GC-MS). Over the last few years, the number of publications on the subject of VMSs has been increasing [12] both regarding the strict development of analytical strategies and the application to a variety of studies such as monitoring in multiple matrices, temporal and spatial trends or toxicological assessments. It has to be mentioned that although there is not a standardized method for the analysis of VMSs, some of these studies share the same basic protocols, as sometimes these studies were carried out within the same research group. But in the case of VMSs, maybe even more important than using an adequate method is ensuring that the handling of samples and analytical materials does not compromise the final result on account of the ubiquitous presence of VMSs in the field, lab environment and instruments.

2.1 Background Contamination Issues

As previously mentioned, VMSs are used in a widespread range of daily life products. Therefore, it is essential to develop protocols that minimize the possibility of external cross-contaminations as well as quality assurance (QA) and control (QC) procedures to demonstrate the accuracy and precision of the analysis [30]. Many authors described contamination problems arising from the presence of VMSs in different sources, such as parts of the analysis equipment and lab material, but also external contamination related to the adopted sampling and extraction procedures [31]. For this reason, a number of measures have been applied to mitigate the negative effects associated.

As VMSs are incorporated in cosmetics and personal care products [2, 4, 32], their use (especially deodorants, hand/body moisturizer and make-up) should be avoided by all the personnel involved in sample collection and analysis [31, 33–35]. This practice is mentioned in most studies dealing with VMSs.

Contamination originating from different parts of GC systems is also a major concern, particularly when VMSs are analysed at trace levels. One of the most important potential sources is the inlet septa [2, 31]. Most of them are made of cross-linked polydimethylsiloxanes (PDMS), which may suffer degradation at high temperatures (>200–300°C), releasing D3 [31]. This is exactly the range of temperatures usually applied in the injection port of most GC systems. However, some authors chose to work with a low temperature (200°C) [36, 37] or a cooled injection system [38] to avoid this problem. Also, consecutive injections may release some septum fragments that accumulate inside the glass liner and consequently, introduce unwanted compounds into the system. The most common way reported in literature to avoid this situation is the use of a Merlin Microseal [13, 23, 39–42]. This is a septumless valve injection system made of an elastomer material based on fluorocarbon resistant to high temperatures, which will not release VMS compounds [43]. In the analysis of cyclic VMSs, it is also important to consider the accumulation in the glass liner of non-volatile silicone-based compounds present in the extracts. These deposits have the potential to generate VMSs in subsequent injections if the sample matrix is wet or contains any kind of catalyst. To reduce this risk, the frequent replacement of the glass liner is required [44]. To improve sample vaporization and/or to keep non-volatile material from entering the column, some authors opt to use GC liners containing glass wool. However, this procedure should be avoided for VMSs analysis, since silanized glass wool is a potential source of contamination. Companioni-Damas et al. [45] used it, but deactivated. Another important issue is the bleeding of the capillary GC columns. Most of them have PDMS stationary phases, which may result in the release of organosiloxanes during the chromatographic analysis. This behaviour can get worse as the column lifetime is approaching its end [46]. To overcome this drawback, low bleed or preferably ultra-inert columns should be employed and are reported in many studies [23, 36, 37, 42, 47]. Another alternative is the use of capillary columns with polar stationary phases [43]. Even if DB-5 MS or HP-5 MS columns are the most commonly used, columns based on polyethylene glycol such as DB-FFAP and DB-Wax [48–51] may be used. Also, some authors resorted to guard columns or retention gaps [39, 41, 45, 52]. The presence of water in the samples should also be avoided since D4 can be generated by the interaction of water molecules with the PDMS stationary phase [31, 33]. The use of drying agents such as anhydrous sodium or magnesium sulphate during the sample extraction and clean-up steps may prevent it.

The exposure of samples to the presence of these VMS compounds in lab material cannot be forgotten. For example, lubricants used in ground glass and vacuum systems, O-rings, pump fluids, caps and tubing containing silicone rubber, etc. can contribute significantly to a potential contamination. In particular, the septa used in vials which are in direct contact with the samples or extracts are a concern [33]. Most of these septa are made of a silicone rubber and coated with a layer of polytetrafluoroethylene (PTFE), commonly called Teflon. Chambers et al. [53] and Wang et al. [54] carried out studies that prove that the direct and prolonged contact of the sample with these septa can lead to the generation of cyclic VMSs. This situation can be avoided using butyl/PTFE septa, as suggested by some authors

[9, 45, 55, 56] or wrapping or completely replacing the vial septa with aluminium foil [15, 57, 58]. Lee et al. [59] used septa with fluorocarbon material instead. The tubing used within the GC equipment or in the nitrogen blowdown systems may be another source of VMS contamination, and some authors have decided to replace silicon tubing by PTFE or even steel alternatives [24, 59–62]. The gloves used by the lab personnel can also introduce some interferences in the analysis. Samples and extracts should thus be only manipulated using powder-free nitrile gloves [33, 34], which are the choice of the majority of authors, although Zhang et al. [63] used polyethylene gloves instead. The organic solvents commonly employed in VMS protocols may be prone to trap them from the containers or the surrounding air. Thus, solvents must be of high purity and pre-analysed to assess possible external contamination. More strict measures were adopted by Kierkegaard and McLachlan [25, 60], who treated hexane with sulphuric acid before use and by Tran et al. [58], who changed the solvent bottles daily. The non-calibrated glassware should be baked-out at a high temperature (>400°C), and the majority of studies refer this crucial step to remove possible contaminants, whereas Companioni-Damas et al. [45] treated glassware with chromium sulphuric acid and drying at 200°C. Some authors report the use of specific (mostly alkaline) detergents in washing steps [64, 65], and all general lab material is commonly pre-rinsed with an appropriate solvent. Finally, sample handling and extraction procedures should be performed inside clean rooms or clean air cabinets with appropriate air filtration to minimize background contamination [13, 25, 34, 39]. These facilities are quite expensive and are not available to all research groups, so an alternative would be to have a dedicated fume hood. As a consequence, and a complement of all these preventive steps, field and lab blank samples are mandatory throughout the whole analytical process, with each sampling procedure and batch of samples processed, respectively. These values are usually subtracted to the sample results obtained to determine the final levels of VMSs and are also used to determine the limits of detection (LODs) and quantification (LOQs).

2.2   Sampling

Sample collection is a crucial step in the development of analytical methodologies, and being traditionally the first stage of the protocol, it is essential to ensure that sampling is done properly and in a representative way, in order to generate reliable and meaningful results. And particularly in the analysis of VMSs, for the reasons mentioned above, it is imperative not to neglect a number of precautions to prevent external contamination. As a general rule, everyone involved in the collection, handling and analysis of the sampled material must avoid the use of personal care products and wear powder-free nitrile or polyethylene gloves.

Then, there is the choice of appropriate recipients to contain the samples for transport/shipping to the lab until analysis. According to Chainet et al. [31], siloxanes have a great affinity for glass, being able to adsorb to its surface. Therefore, samples should be collected and stored preferentially in PTFE or high-density polyethylene (HDPE) containers at low temperatures (also to avoid volatilization). However, according to the bibliographic survey (Tables 1, 2, 3, 4 and 5), this is not the usual procedure. For water samples (Table 1), the collection is usually performed in pre-cleaned amber glass or serum bottles (often baked at >400°C and rinsed with appropriate solvents) without headspace, to avoid the volatilization of the target compounds, and fitted with septa free of siloxanes (e.g. [9, 38, 48, 66]). Only in one study, this collection was carried out in HDPE containers [8]. Regarding sewage sludge collection (Table 2), distinct procedures were adopted. In most studies, amber glass bottles were used, as in water sampling, but some authors opted for polypropylene containers (bottles or bags) [8, 10, 11] or aluminium vessels [14]. Sewage sludge has different water contents, depending on the collection site, and can be either liquid or solid, which may be decisive in this choice. In the case of bags, samples were previously wrapped in aluminium foil [10]. Also solid are two other matrices that have been studied for their VMSs levels: sediments (Table 3) and soil (Table 4). Sediments, however, are underwater (sea, rivers or lakes) and need specific tools, usually metallic (stainless steel), for the collection. The most common were grabs like Van Veen, Birge-Eckman or bucket [13, 34, 56, 67, 68], but also corers were used [39, 51, 69, 70] or even scoops or spoons in shallow waters [52, 63, 67]. Similarly to water and sewage sludge, glass jars lined with Teflon or aluminium foil were the predominant storage container but also aluminium or stainless steel bottles [14, 71] or polypropylene and polyethylene bags [59, 72]. Samples were sometimes wrapped in aluminium foil as well [59, 69]. Soils were predominantly taken from the top layers, but there is usually no mention to the tools used, except in two cases, where a steel syringe [36] and a stainless-steel scoop [42] were reported. But as was the case of sediments, metallic tools are likely to be the obvious choice and the storage done in glass jars. Finally, air (Table 5) is a totally different matrix, and the sampling techniques are quite distinct from the previous ones. In the studies focusing on VMSs, active air sampling (AAS) was predominant for both outdoor and indoor environments, but passive air sampling (PAS) was also frequently employed. Both options rely almost exclusively on pre-cleaned or conditioned sorbent or filter materials that trap the target analytes when in contact with an air flow. While in AAS, a pump system is used, and the volume of air sampled is known with precision, for PAS an estimation of the air mass sampled has to be done via depuration compounds and specific calibration procedures [62, 73]. Naturally, the latter technique may entail a higher uncertainty in the results [74]. Regarding VMSs, the options of trapping materials for AAS reported in literature varied from cartridges with sorbents such as Tenax, ENV+, XAD, silica, charcoal or other carbon-derived materials to polyurethane foam (PUF) plugs for the gas-phase fraction, which contains the majority of the airborne VMSs. In some cases, the particulate fraction was also sampled (mainly in indoor environments), using in this case quartz fibre filters (QFF) [57, 75, 76] and glass-fibre filters (GFF) [21, 77]. In PAS the choices are mainly PUF disks impregnated with XAD sorbent (sorbent-impregnated PUFs – SIPs) [21, 42, 65, 78, 79] but also included the XAD alone [62] and charcoal sorbent [80]. In all cases, the

Table 1 Details of the analytical methods developed for the determination of VMSs in water samples

Analytes	Sampling details	Sample preparation	Extraction/clean-up method	Instrumental method	LOD (ng L^{-1})	%Rec	Reference
Wastewater							
D3-D6 L3-L5	Collection in a stainless steel bucket and stored in glass bottles without headspace at 4°C (extraction within 4 days)	–	600 mL sample Purge and trap: 120 min (purified ambient air through SPE cartridge Sep-Pak plus PS2), drying step with pure N$_2$ for 20 min, 3 mL DCM; extracts concentrated to 1 mL under N$_2$	GC-MS (EI, 30 m DB-5MS column)	0.6–3.0	^{13}C labelled: 83 ± 11–87 ± 12 Native: 70 ± 7–89 ± 7	[48]
D4-D6	Collection using wide-mouth glass jars with PTFE-lined lid. Transport to the lab in a cooler with blue ice	–	100 mL sample MASE, LDPE membrane; solvent, hexane	GC-MS (EI, 30 m DB-WAXetr or DB-5ms column)	17–20	81.6 ± 0.9 to 107 ± 13	[49]
D3-D6 L5-L14	Collection with brown glass bottles	Pre-filtered with glass-fibre filters (0.45 µm)	1,000 mL sample LLE: 100 mL DCM + 2 × 50 mL DCM	GC-MS (EI, 30 m DB-5MS column)	^{a}0.2–2	87.7 ± 13.0	[7]
D3-D6 L3-L5	Collection without headspace in serum bottles and quickly crimp sealed. Transport to the lab in a cooler with ice (extraction within 2 h)	–	100 mL sample MASE, LDPE membrane; extraction solvent, 0.5 mL Hex; addition of 2 mL ACN + 0.5 mL Hex; agitation, 30 min; centrifugation	GC-MS (EI, 30 m DB-35MS column)	–	72–97	[66]
D4-D6	Collection without headspace in serum bottles and quickly crimp sealed. Transport to the lab in a cooler with ice (extraction within 4 h)	–	100 mL sample MASE, LDPE membrane; extraction solvent, 0.5 mL Pen; agitation, incubator shaker, 60 min	GC-MS (EI, ?? m HP-5MS column)	9–27	^{13}C labelled: 102 ± 10 to 107 ± 12	[87]

(continued)

Table 1 (continued)

Analytes	Sampling details	Sample preparation	Extraction/clean-up method	Instrumental method	LOD (ng L^{-1})	%Rec	Reference
D3-D6 L2-L5	–	Pre-filtered with common lab filter paper	13 mL sample USA-DLLME, 13 µL CB (extraction solvent); agitation, 2 min; centrifugation	GC-MS (EI, 60 m low-bleed DB-624 column)	3–1,400	71 ± 23 to 99 ± 7	[47]
D3-D7 L3-L14	Collection in high-density PE bottles. Stored at −18°C until analysis	Filtered with prewashed glass-fibre filters	100 mL sample LLE, 50 mL Hex +25 mL Hex:DCM (1:1, v/v) + 25 mL Hex:EA (1:1); extract reduced to 3–5 mL; addition 0.5 mL IOC; evaporation under N$_2$; reconstitution in 0.5 mL Hex	GC-MS (EI, 30 m HP-5MS column)	[a]0.11–40	60.6 ± 3.7 to 134 ± 14	[8]
D3-D5 L3-L5	Samples stored in amber glass bottles without headspace at 4°C (extraction within 24 h)	–	500 mL sample LLE, 3 × 250 mL Hex; extract reduced to 5 mL (rotary evaporator); water removal step with Na$_2$SO$_4$ column; extract concentrated to 150 µL under N$_2$	GC-MS (EI, 30 m DB-5MS column)	3.2–13	40.3–114.6	[9]
D4-D6	Samples stored in sampling tubes previously cleaned with detergent	–	10 mL sample Simple dilution in 200 mL ultrapure water	HS-GC/MS (EI, 30 m DB-Wax column)	[a]200	67 ± 11 to 99 ± 4	[93]
Wastewater							
D4-D6	Collection without headspace in serum bottles and quickly crimp sealed. Stored at 4°C until analysis (extraction within 24 h)	–	100 mL sample MASE, LDPE membrane; extraction solvent, 0.5 mL Pen; agitation, incubator shaker, 60 min	GC-MS (EI, 30 m HP-5MS column)	9–27	100 ± 21 to 107 ± 29	[38]

Analytes	Sampling and storage		Sample preparation	Instrument	Range	Recovery (%)	Ref.
D3-D6 L3-L4	Collection into glass bottles. Stored at 4°C until analysis (extraction within 48 h)	–	40 mL sample HS-SPME, 65 μm PDMS/DVB fibre; addition of 0.1 g mL⁻¹ NaCl; extraction, 24°C, 45 min; thermal desorption, 200°C	GC-MS (EI, 30 m HP-5MS column)	2.6–7.8	78–96	[35]
D5	Collection into glass bottles and extraction within 24 h	–	Direct injection	HS-GC-MS (EI, 30 m DB-FFAP or DB-Wax column)	6.2	89.3–98.7	[55]
D4-D6 L2-L5	–	–	60–150 mL sample Purge and trap: 20 min for L2 and 2 h for all other VMSs (0.25 g Tenax TA and N$_2$ as purge gas)	TD-GC-MS (EI, 30 m CP-Sil 8CB column)	0.3–30	–	[91]
D3-D6 L2-L5	Collection in bottles with aluminium foil protecting the lids	–	60–150 mL sample Purge and trap: 20 min for L2 and 2 h for all other VMSs (0.25 g Tenax TA and N$_2$ as purge gas)	TD-GC-MS (EI, 30 m CP-Sil 8CB column)	0.04–27 (ng sample⁻¹)	–	[90]
D4-D6 L2-L5	Collection with pre-baked (400°C) bottles fitted with aluminium foil-lined screw cap	–	60–80 mL sample Purge and trap: 20 min for L2 and 2 h for all other VMSs (0.25 g Tenax TA and N$_2$ as purge gas)	TD-GC-MS (EI, 30 m CP-Sil 8CB column)	0.5–60	–	[89]
D4-D5	Collection in sampling jars and stored at 4°C	–	15 mL sample LLE: 10 mL Hep, 48 h agitation in rotary extractor	GC-MS (EI, 30 m DB-5MS column)	–	21–86	[86]
River water							
D3-D6 L3-L5	Collection in stainless steel bucket and stored in glass bottles with no headspace at	–	600 mL sample Purge and trap: 120 min (purified ambient air through a SPE cartridge Sep-Pak plus	GC-MS (EI, 30 m DB-5MS column)	0.6–3.0	^{13}C labelled: 83 ± 11 to 87 ± 12	[48]

(continued)

Table 1 (continued)

Analytes	Sampling details	Sample preparation	Extraction/clean-up method	Instrumental method	LOD (ng L^{-1})	%Rec	Reference
	4°C (extraction within 4 days)		PS2), drying step with pure N$_2$ for 20 min, 3 mL DCM; extracts concentrated to 1 mL under N$_2$			Native: 69 ± 2 to 89 ± 2	
D4-D6	–	–	100 mL sample MASE, LDPE membrane; extraction solvent, 0.5 mL Pen; agitation, incubator shaker, 60 min	LVI-GC-MS (EI, 30 m HP-5MS column)	9–27	100 ± 21 to 107 ± 29	[88]
D3-D5 L3-L5	Samples stored in amber glass bottles without headspace and fitted with septa free of siloxanes at 4°C (extraction within 24 h)	–	500 mL sample LLE: 3 × 250 mL Hex; extract reduced to 5 mL (rotary evaporator); water removal step with Na$_2$SO$_4$ column; extract concentrated until 150 µL, under N$_2$	GC-MS (EI, 30 m DB-5MS column)	–	–	[9]
D3-D6 L2-L5	Samples stored in glass bottles without headspace with black Viton septa	Samples filtered using 0.2 µm nylon syringe filters	20 mL sample Sample vortex mixed for 3 min and conditioned for 10 min; HS-SPME, 65 µm PDMS/DVB fibre, 25°C, 40 min, 750 rpm; thermal desorption, 240°C, 5 min	GC-MS (EI, 60 m DB-5MS column)	0.003–11	≈100	[40]
D5	Collection into glass bottles and extraction within 24 h	–	Direct injection	HS-GC-MS (EI, 30 m DB-FFAP column or DB-Wax column)	6.2	80.5–85.9	[55]

Lake water							
D4-D6	–	Collection in amber glass bottles with Teflon-lined caps stored at 4°C	100 mL sample LLE: 3 × 50 mL Hex, 30 min settling; extract reduced to 1 mL	GC-MS (EI, 30 m DB-5 column)	1.99-4.43	82 ± 21	[63]
Seawater							
D4-D7 L4-L14	–	Collection in glass bottles with Teflon-lined caps. Addition of 100 mL DCM to 1 L seawater and store at 4°C	1 L sample LLE: 3 × 100 mL DCM, 1 h settling; water removal with Na$_2$SO$_4$ column; extracts concentrated to 1 mL	GC-MS (EI, 30 m DB-5MS column)	0.79-10.76	83 ± 10 to 110 ± 11	[56]
D4-D6 L2-L5	–	–	60-150 mL sample Purge and trap: 20 min for L2 and 2 h for all other VMSs (0.25 g Tenax TA and N$_2$ as purge gas)	TD-GC-MS (EI, 30 m CP-Sil 8CB column)	0.3-30	–	[91]
D3-D6 L2-L5	–	Samples collected in bottles with aluminium foil protecting the lids	60-150 mL sample Purge and trap: 20 min for L2 and 2 h for all other VMSs (0.25 g Tenax TA and N$_2$ as purge gas)	TD-GC-MS (EI, 30 m CP-Sil 8CB column)	0.04-27 (ng sample^{-1})	–	[90]
D4-D6 L2-L5	–	Samples collected with bottles fitted with aluminium foil-lined screw cap	60-80 mL Purge and trap: 20 min for L2 and 2 h for all other VMSs (0.25 g Tenax TA and N$_2$ as purge gas)	TD-GC-MS (EI, 30 m CP-Sil 8CB column)	0.1-470	–	[89]

ACN acetonitrile, *CB* chlorobenzene, *DCM* dichloromethane, *EA* ethyl acetate, *EI* electron ionization, *GC-MS* gas chromatography-mass spectrometry, *Hep* heptane, *Hex* hexane, *HS-SPME* headspace solid-phase microextraction, *IOC* isooctane, *LDPE* low-density polyethylene, *LLE* liquid-liquid extraction, *LOD* limit of detection, *MASE* membrane-assisted extraction, *MeOH* methanol, *PDMS/DVB* polydimethylsiloxane/divinylbenzene, *PE* polyethylene, *Pen* pentane, *Rec* recovery, *TD* thermal desorption, *US* ultrasound, *USA-DLLME* ultrasound-assisted dispersive liquid-liquid microextraction
^{a}Limit of quantification (ng L^{-1})
"?" sample amount not provided

Table 2 Details of the analytical methods developed for the determination of VMSs in sludge samples

Analytes	Sampling details	Sample preparation	Extraction/clean-up method	Instrumental method	LOD (ng g^{-1} dw)	%Rec	Reference
D3-D6 L5-L14	Collection in stainless steel jars	Freeze-dried	1 g sample SLE: 3 × 10 mL Hex:EA (1:1 v/v), oscillator plate	GC-MS (EI, 30 m DB-5MS column)	[a]200–2,000	80.3 ± 10.2	[7]
D3-D6 L3-L5	Collection in amber glass jars. Transport to the lab in a cooler with ice (extraction within 2 h)	—	—	GC-MS (EI, 30 m DB-35MS column)	–	65–91	[66]
D4-D6	Collection in amber glass jars. Transport to the lab in a cooler with ice (extraction within 4 h)	—	2 g sample (wet weight) SLE, 10 mL ACN + 15 mL Hex, 30 min; centrifugation; water removal step with Na$_2$SO$_4$; addition 10 mL Hex to the initial sample; agitation, 10 min; water removal with Na$_2$SO$_4$; combination of extracts	GC-MS (EI, ?? m RTx-5Sil column)	560–970 (ww)	[13]C labelled: 90 ± 3 to 137 ± 9	[87]
D3-D6 L3-L17	Collection in pre-cleaned PP bottles and stored at −20°C until analysis	Freeze-dried	0.5 g sample SLE: 10 mL Hex + 10 mL Hex: DCM (1:1, v/v) + 10 mL Hex:EA (1:1, v/v), 3 x 60 min; centrifugation; extracts concentrated to 1 mL under N$_2$	GC-MS (EI, 30 m HP-5MS)	[a]0.03–36.2	[13]C labelled: 81–143	[11]
D4-D6 L3-L16	Samples packed in aluminium foil and sealed in PP bags	Freeze-dried and homogenized (sieved) and stored at −20°C	0.1 g sample USE: 3 × 10 mL EA:Hex (1:1, v/v), 15 min; centrifugation; extracts concentrated to 2 mL (N$_2$ stream); water removal step with Na$_2$SO$_4$ column; extracts concentrated to 1 mL under N$_2$	GC-MS (EI, 30 m HP-5MS)	0.5–1.7	69–104	[10]

D3-D6 L2-L5	Stored at 4°C	–	50 mL sample SLE, 50 mL Hex + 50 mL acet; agitation, 4 h; centrifugation	GC-MS (EI, 60 m HP-5MS)	–	–	[94]
D3-D7 L3-L14	Collection in high-density PE bags and stored at −18°C until analysis	Sample dried and homogenized in a mortar with 25–30 g of anhydrous Na$_2$SO$_4$	5 g sample (wet weight) SLE, 25 mL Hex +25 mL Hex: DCM (1:1, v/v) + 25 mL Hex:EA (1:1, v/v); agitation, 3 × 1 h; centrifugation; extracts concentrated to 3–5 mL (rotary evaporator); addition, 0.5 mL IOC; evaporation under N$_2$; reconstitution with 0.5 mL Hex	GC-MS (EI, 30 m HP-5MS column)	[a]0.006–9.9 (ww)	53.9 ± 7.4 to 102 ± 18	[8]
D3-D6 L3-L5	Collection in glass containers without headspace and stored at 4°C until analysis	Samples mixed with Na$_2$SO$_4$ and kept at 4°C for 3 h	0.5 g sample (wet weight) SLE, 2 × 3 mL Hex +0.2 g activated Cu, 10 min; centrifugation; cooled at 4°C for 30 min; SPE, 100 mg silica cartridge, 1.5 mL Hex	CSR-LVI-GC-MS (EI, 60 m DB-5MS column)	0.004–0.14 (ww)	80 ± 12 to 103 ± 10	[41]
D3-D6 L3-L4	Collection into glass. Stored at 4°C until analysis (extraction within 48 h)	Freeze-dried	0.2 g sample USE: 3 × 3 mL acet, 30 min; centrifugation; extracts reduced to 200 μL (N$_2$ stream); reconstitution with 40 mL ultrapure H$_2$O HS-SPME, 65 μm PDMS/DVB fibre; addition of 0.1 g mL^{-1} NaCl; extraction, 24°C, 45 min; thermal desorption, 200°C	GC-MS (EI, 30 m HP-5MS column)	<1	75–93	[35]
D4-D6 L3-L5	–	Freeze-dried, homogenized and ground in an agate ball mixer mill (d < 630 μm)	1 g sample SLE, 20 mL EtOH/NaCH$_3$COO buffer + 400 μL DEA-DDC, agitation, 2.5 h; 20 mL Hex, agitation, 1 h; centrifugation; 5 mL Hex added to the initial samples,	GC-MS (EI, 60 m DB-5MS column)	5–60	71–91	[96]

(continued)

Table 2 (continued)

Analytes	Sampling details	Sample preparation	Extraction/clean-up method	Instrumental method	LOD $(ng\ g^{-1}\ _{dw})$	%Rec	Reference
			agitation, 1 h; centrifugation; organic phase combined and concentrated to 5 mL SPE: 2 g aluminium oxide (activated with 10% water), Hex: EA (90:10, v:v); extracts concentrated to 900 μL under N_2; Hex added until 1 mL				
D4-D7 L4-L16	Collection and storage in aluminium containers	Freeze-dried and homogenized and stored at −20°C	1 g sample SLE, 3 × 25 mL Hex:EA (1:1, v/v), 30 min; centrifugation; extracts reduced to 2–3 mL (rotary evaporator); addition 5 mL IOC; extracts concentrated to 1 mL (N_2 stream); SPE, 0.5 g silica gel, 12 mL DCM:Hex (1:4, v/v); extracts concentrated under N_2	GC-MS (EI, 30 m Rxi-5MS column)	[a]0.28–2	78.7 ± 11.3 (mean)	[14]
D4-D6 L2-L5	–	–	2 g sample (wet weight) Purge and trap: dilution with 20 mL deionized water and 1 mL slurry diluted to 10 mL added to a purge and trap apparatus (0.25 g Tenax TA and N_2 as purge gas); samples purged for 20 min (analysis of L2) or for 2 h (all other VMSs)	TD-GC-MS (EI, 30 m CP-Sil 8CB column)	0.7–180	–	[91]

D4-D5	Samples stored at 4°C and extracted within 24 h	Samples were homogenized	50 mL sample SLE: 10 mL Hex, 10 min vortex; centrifugation	GC-FID (? m VF-1MS column)	–	73.7–100.8	[95]
D3-D6 L2-L5	Collection in bottles with aluminium foil protecting the lids	–	2 g sample (wet weight) Purge and trap: dilution with 20 mL deionized water and 1 mL of this slurry diluted to 10 mL added to a purge and trap apparatus (0.25 g Tenax TA and N_2 as purge gas); samples purged for 20 min (analysis of L2) or for 2 h (all other VMSs)	TD-GC-MS (EI, 30 m CP-Sil 8CB column)	40–3,900 (ng sample^{-1})	–	[90]
D4-D6 L2-L5	Collection bottles fitted with aluminium foil-lined screw cap	–	2 g sample (wet weight) Purge and trap: dilution with 20 mL deionized water and 1 mL of this slurry diluted to 10 mL added to a purge and trap apparatus (0.25 g Tenax TA and N_2 as purge gas); samples purged for 20 min (analysis of L2) or for 2 h (all other VMSs)	TD-GC-MS (EI, 30 m CP-Sil 8CB column)	0.1–470	–	[89]
D4-D5	Collection in sampling jars and stored at 4°C	–	15 mL sample SLE: Dilution with diluted to 25% (v/v) with deionized water, 10 mL Hep, 48 h agitation in rotary extractor	GC-MS (EI, 30 m DB-5MS column)	–	53–95	[86]

Acet acetone, *ACN* acetonitrile, *d* diameter, *DEA-DDC* diethyl-ammonium-diethyl-dithiocarbamate, *DCM* dichloromethane, *EA* ethyl acetate, *EI* electron ionization, *EtOH* ethanol, *GC-MS* gas chromatography-mass spectrometry, *Hep* heptane, *Hex* hexane, *HS-SPME* headspace solid-phase microextraction, *IOC* isooctane, *LOD* limit of detection, *PE* polyethylene, *PDMS/DVB* polydimethylsiloxane/divinylbenzene, *PP* polypropylene, *Rec* Recovery, *SLE* solid-liquid extraction, *SPE* solid-phase extraction, *TD* thermal desorption, *USE* ultrasound extraction, *ww* wet weight

[a]Limit of quantification $(ng\ L^{-1})$

"?" sample amount not provided

Table 3 Details of the analytical methods developed for the determination of VMSs in sediment samples

Analytes	Sampling details	Sample preparation	Extraction/clean-up method	Instrumental method	LOD (ng g^{-1} dw)	%Rec	Reference
Marine sediments							
D4-D6 L2-L5	Collection with glass bottles, later placed in plastic bags	Diluted in MilliQ water and homogenized by shaking	2 g sample (wet weight) Purge and trap: 2 h (N$_2$ stream through 0.25 g Tenax TA adsorbent tubes)	TD-GC-MS (EI, 30 m CP-Sil 8CB column)	0.02–1.9	–	[90]
D4-D6 L2-L5	Collection with glass bottles, later placed in plastic bags	Diluted in MilliQ water and homogenized by shaking	2 g sample (wet weight) Purge and trap extraction for 2 h (N$_2$ stream through 0.25 g Tenax TA adsorbent tubes)	TD-GC-MS (EI, 25 m Agilent Ultra2 column)	–	–	[91]
D4-D6	Sampling with box corer or multicorer, fractions 0–1 and 0–3 cm. Samples wrapped in aluminium foil and frozen	–	? g sample USE: wet sediment extracted with hexane	GC-MS (EI)	8–59	–	[69]
D3-D6 L2-L5	Collected from the 0–2 cm top layer fraction	–	? g sample SLE: wet sediment extracted with hexane by vortex; centrifugation	GC-MS (EI)	0.19–17	–	[97]
D4-D6	Collection with Van Veen grab and transferred to glass jars and frozen immediately at −20°C	Homogenized in an Ultra Turrax homogenizer	0.3 to 0.5 g sample SLE: 1 mL Hex vortex extraction for 30 min; centrifugation	GC-MS (EI, 30 m DB-WAX ETR column)	0.5–0.9 (ww)	^{13}C labelled: 83 ± 7 to 91 ± 11	[34]
D4-D6	Collection with a Van Veen grab and transferred to glass jars. Frozen immediately at −20°C	Homogenized in an Ultra Turrax homogenizer. Wetted	*NILU* 0.5 g sample SLE: 1 mL Hex vortex extraction, 30 min; centrifugation	*NILU* GC-MS (EI, 30 m DB-WAX	*NILU* 0.5–0.8 (ww) *Dow*	*NILU* ^{13}C labelled: 83 ± 7 to	[50]

	Sampling	Pretreatment	Extraction	Instrument	Concentration	Recovery	Ref
		with ACN for extraction	*Dow Corning* 6 g sample SLE: 2 × Hex in a 2:1 solvent to sample weight ratio by vortex, 30 min; centrifugation; extracts combined; volume reduced with N$_2$ to 1–2 mL in a 25°C water bath. *Evonik* 5 g sample SLE: 2 × 10 mL Pent for 30 min in orbital shaker; centrifugation; extracts combined; SPE: 0.1 g MgSO$_4$ + 0.8 g activated Florisil; 4 mL aliquot of extract eluted with 5 mL petroleum ether and MTBE (99:1), volume reduced to 0.5 mL using N$_2$	ETR column) *Dow Corning* GC-MS (EI, 30 m DB-WAXetr column) *Evonik* GC-MS (EI, 50 m 1909IF-115 (FFAP) column)	1.2–1.3 (ww) *Evonik* 2.4–2.5 (ww)	91 ± 11 *Dow* ^{13}C labelled: 96 ± 8 to 96 ± 9 *Evonik* ^{13}C labelled: 70–90	[56]
Marine sediments							
D4-D7 L4-L17	Collected with a bucket grab, then packed in glass bottles and stored at −20°C	Freeze-dried and homogenized. Stored at −20°C	2 g sample SLE: Shaken for 30 min with 5 mL of Hex:EA (1:1 v/v); centrifugation; extract rotary evaporated and solvent exchanged (final volume 1 mL) to IOC	GC-MS (EI, 30 m DB-5MS column)	0.17–4.45 (ww)	80 ± 11 to 109 ± 15	[56]
D4-D6	Collection with a Van Veen grab; top 2 cm removed with stainless steel spoon and transferred to glass jars. Frozen at −20°C	Homogenized in an Ultra Turrax homogenizer	0.5 g sample SLE: 1 mL Hex vortex extraction, 30 min; centrifugation	GC-MS (EI, 30 m DB-WAX ETR column)	1.4–2.7 (ww)	^{13}C labelled: 74 ± 11 to 81 ± 12	[67]

(continued)

Table 3 (continued)

Analytes	Sampling details	Sample preparation	Extraction/clean-up method	Instrumental method	LOD (ng g⁻¹dw)	%Rec	Reference
D4-D6	Collected with grab sampler. Sediment removed with acrylic core tube and upper 1 cm placed in stainless steel container and stored on ice in the dark	–	6 g sample (wet weight) SLE: Shaken for 30 min with 6 mL ACN and 10 mL Hex; centrifugation: to remove the Hex fraction, inserted into a glass vial with Na_2SO_4; another 10 mL Hex added, process repeated; extracts combined and reduced to 1 mL under N_2	GC-MS	0.52–0.90 (ww)	101 ± 3 to 103 ± 4	[71]
D4-D6	Collected with a double corer or a Van Veen grab (0–2 cm layer) and stored in glass containers. Samples sent to labs and stored at −18°C	–	–	GC-MS	*Dow* 0.30–0.42 (ww) *Evonik* 3.7 (ww)	–	[70]
D4-D6 L8-L16	Collected using a bucket grab sampler (0–5 cm layer). Stored at −20°C	Sediments centrifuged for 10 min to discard supernatant water	1.2 g sample (wet weight) SLE, vortexed for 5 min with 10 mL EA/Hex (1:1); USE, 15 min; process repeated 3×, extract combined and concentrated to 5 mL under N_2. Drying: 1 g Na_2SO_4 column, then concentrated to 1 mL under N_2	GC-MS (EI, 30 m HP-5MS column)	[a]1.4–3.1 (ww)	89–96	[68]
Marine and river sediments							
D4-D7 L4-L17	Samples packed in aluminium foil and sealed in PP bags	Freeze-dried and homogenized. Stored at −20°C	5 g sample SLE: 60 min in an orbital shaker with (1) 10 mL Hex (2) 10 mL Hex:DCM (1:1 v/v) (3) 10 mL Hex:EA (1:1 v/v);	GC-MS (EI, 30 m HP-5MS)	0.02–0.29 (ww)	63 ± 10 to 144 ± 3	[59]

	Collection	Preparation	Extraction	Instrument	Conc.	Recovery	Ref.
			centrifugation; all extracts combined and concentrated to 1 mL under N_2				
Estuarine sediments							
D5	Collection with a Van Veen grab, sieved and transferred to glass jars	Centrifuged and dried by mixing with diatomaceous earth	1 g sample SLE: 10 mL of ACN:Hex (1:1 v/v) for 60 min; centrifugation	GC-MS (EI, 30 m ZB-5HT column)	1	77 ± 5	[13]
D4-D6	Collected with stainless steel spoon from top layer (1–2 cm); transferred to 500 mL glass jars and stored at 10°C in the dark	—	10 g sample SLE: 10 mL Acet +2 mL Pent, rotated for 30 min and centrifuged; process repeated and extract combined; purge and trap: Pent extract purged with purified N_2 at room temperature and gas stream passed through a sample trap with 20 mg ENV+. After 20 h, extract removed and eluted with 0.5 mL (ENV+ cartridges) or 0.8 mL (hand-packed columns) of Hex	LVI-GC-MS (EI, 30 m DB-5MS column, with a 5 m retention gap)	3.7–34	–	[52]
River sediments							
D4-D7 L4-L16	Collected and stored in aluminium containers	Freeze-dried and homogenized. Stored at −20°C in the dark	5 g sample SLE: Shaken with 25 mL Hex: EA (1:1 v/v) × 3 for 30 min each; centrifugation; extract concentration, solvent exchange to IOC and purification with silica gel column eluted with 12 mL DCM:Hex (1:4 v/v); extracts concentrated before analysis	GC-MS (EI, 30 m Rxi-5MS column)	a0.28–2.0	79 ± 11	[14]

(continued)

Table 3 (continued)

Analytes	Sampling details	Sample preparation	Extraction/clean-up method	Instrumental method	LOD (ng g^{-1} dw)	%Rec	Reference
D4, D5	Collection with a Van Veen grab, sieved and transferred to glass jars	Centrifuged and dried by mixing with diatomaceous earth	2.5 g sample ASE: extracted with EA, one cycle; extracts dried with Na$_2$SO$_4$, transferred into a 50 mL volumetric flask and made to volume with EA	GC-MS (EI, 30 m ZB-5HT column)	7–37	78 ± 14 to 89 ± 7	[13]
D3-D5 L3-L5	Collection in glass jars. Stored at 4°C until analysis (within 24 h)	Centrifuged, ground in an agate mortar and homogenized	5 g sample USE: 3 mL of Hex/EA (1:1 v/v) for 15 min; centrifugation; solvent reduction	GC-MS (EI, 30 m DB-5MS column)	0.3–0.9	81 ± 1 to 105 ± 3	[9]
D4	Collected with stainless steel tools, samples transferred to PE bags, homogenized and placed in glass jars with a Teflon-lined lid until analysis	–	Samples were extracted with either Hex or tetrahydrofuran	GC-MS	0.8 (ww)	83	[72]
River and lake sediments							
D4-D6	Collection with a stainless steel scoop into glass jars. Stored at 4°C until analysis (within 24 h)	–	1 g sample SLE: 60 min in an orbital shaker with 10 mL ACN/Pent (1:1 v/v); centrifugation	LVI-GC-MS (EI, 30 m HP-5MS column)	3–11	69 ± 10 to 74 ± 8	[38]
Lake sediments							
D4-D6	Collected with a stainless steel mini-box core and stored in a dark cooler	Samples homogenized and stored at 4°C	6 g samples (wet weight) SLE, 6 mL ACN and 12 mL Hex added and vortexed for 30 min; centrifugation, to remove the Hex layer; extract dried with 0.25 MgSO$_4$; process repeated, extracts combined and volume reduced to 1–2 mL under N$_2$	GC-MS (EI, 30 m DB-WAXetr column)	4.5–10.4 (ww)	98 ± 16 to 105 ± 10	[51]

D4-D6	Collected with a gravity corer, transferred to glass jars and stored at −20°C	Samples centrifuged (10 min)	1.5 g sample (wet weight) SLE: 4 mL Hex:ACN (1:1 v/v), vortexed and put in a rolling mixer for 24 h; centrifugation	GC-MS (EI, 30 m Rxi-5MS column, with 5 m guard column)	0.3–1.0	85 ± 12 to 100 ± 11	[39]
D4-D6	Collected from the top layer (1–2 cm) with stainless steel spoon. Samples packed in amber glass bottles. Stored at 4°C	Centrifuged, ground in an agate mortar and homogenized in an orbital shaker	5 g sample SLE: Shaken with 25 mL Hex: EA (1:1 v/v) × 3 for 30 min each; centrifugation; extracts combined and concentration by rotary evaporation to 1–2 mL, solvent exchange to IOC and volume reduction to 1 mL under N_2	GC-MS (EI, 30 m DB-5 column)	0.89–1.05	82.5 ± 21	[63]

Acet acetone, *ACN* acetonitrile, *ASE* accelerated solvent extraction, *DCM* dichloromethane, *EA* ethyl acetate, *EI* electron ionization, *GC-MS* gas chromatography-mass spectrometry, *Hex* hexane, *IOC* isooctane, *LOD* limit of detection, *LVI* large-volume injection, *MTBE* methyl tert-butyl ether, *Pent* pentane, *PP* polypropylene, *Rec* Recovery, *SLE* solid-liquid extraction, *TD* thermal desorber, *USE* ultrasound extraction, *ww* wet weight

[a]Limit of quantification ($ng\ g^{-1}$)

"?" sample amount not provided

Table 4 Details of the analytical methods developed for the determination of VMSs in soil samples

Analytes	Sampling details	Sample preparation	Extraction/clean-up method	Instrumental method	LOD (ng g^{-1} dw)	%Rec	Reference
D4-D6 L2-L5	Sampling using glass bottles, later placed in plastic bags	Diluted in MilliQ water and homogenized	2 g sample (wet weight) Purge and trap extraction for 2 h (N$_2$ stream through 0.25 g Tenax TA adsorbent tubes)	TD-GC-MS (EI, 30 m CP-Sil 8CB column)	–	–	[90]
D4-D6 L5-L14	Collected from 0 to 10 cm into glass containers	Air-dried, sieved (2 mm) and stored at −18°C	5 g sample SAESC: 20 mL glass columns with 2 g anhydrous Na$_2$SO$_4$ and the soil; sonication with 2 × 5 mL Hex (15 min). Extract collected in a vacuum manifold and concentrated to 1 mL under N$_2$	GC-MS (EI, 30 m ZB-5MS column)	0.4–1.1	85–108	[98]
D3-D6 L3-L6	Collected from 0 to 5 cm into glass containers with no headspace	Samples mixed with 2 g of anhydrous sodium sulphate and kept at 4°C for 3 h	0.5 g sample (wet weight) SLE, 2 × 3 mL Hex + 0.2 g activated Cu, 10 min; centrifugation; cooled at 4°C for 30 min; SPE, 100 mg silica cartridge, elution with 1.5 mL Hex	CSR-LVI-GC-MS (EI, 60 m DB-5MS column with a 5 m guard column)	0.005–0.15	84–103	[41]
D4-D6 L3-L16	Collected with a steel syringe and placed into glass container. Stored at −18°C	Samples sieved (500 μm mesh) inside a sealed plastic bag at 4°C	0.2 g sample SLE, vortex at 2500 rpm with 10 mL EA/Hex (1:1), 5 min; centrifugation; process repeated 2×; combined extract concentrated to 2 mL under N$_2$; clean-up, 1 g	GC-MS (EI, 30 m low-bleed HP-5MS column)	[a]0.5–1	81–94	[36]

Analytes	Sampling/storage	Sample prep	Extraction	Instrument	Conc.	Recovery (%)	Ref
			Na$_2$SO$_4$, eluted with 5 mL EA/Hex (1:1); extract concentrated to 1 mL under N$_2$				
D4–D6	Collected from 0 to 10 cm layer into glass containers	Samples pulverized, thoroughly mixed and composited	1 g sample SLE: 60 min in an orbital shaker with 10 mL ACN/Pent (1:1 v/v); centrifugation	LVI-GC-MS (EI, 30 m HP-5MS column)	7–9	70 ± 12 to 78 ± 11	[38]
D3–D6 L3–L17	Stored in sealed glass tubes without headspace. Stored at −18°C	Ground if hard and sieved (500 μm mesh)	0.2 g sample SLE: vortex (5 min) with 3 × 10 mL of Hex:EA (1:1 v/v) and centrifugation (10 min); extracts combined, volume reduced under N$_2$ to 5 mL and centrifuged; supernatant purified through 1 g anhydrous Na$_2$SO$_4$ cartridge; eluate concentrated to 1 mL under N$_2$	GC-MS	a0.5–2.5	76 ± 3 to 94 ± 5	[61]
D3–D6 L3–L6	Top soil (0–5 cm) collected	Ground in agate mortar	3 g sample USE: 25 min with 3.0 mL of Hex	GC-MS (EI, 30 m DB-5MS)	0.002–0.11	70 ± 11 to 95 ± 5	[15]
D3–D6 L2–L5	Collected from 0 to 10 cm layer with stainless steel scoop into amber glass containers	Sieved (2 mm) and stored at −20°C	2.5 g sample USE, 15 min with 10 mL Hex:DCM (1:1 v/v); filtration, 0.2 μm PTFE filter;	GC-MS (EI, 30 m DB-5MS ultra-inert column)	0.003–0.020	69 ± 17	[42]

(continued)

Table 4 (continued)

Analytes	Sampling details	Sample preparation	Extraction/clean-up method	Instrumental method	LOD (ng g^{-1} dw)	%Rec	Reference
			QuEChERS clean-up with (1) 6 g MgSO$_4$ + 1.5 g CH$_3$COONa and (2) 900 mg MgSO$_4$ + 300 mg PSA + 150 mg C$_{18}$; extract reduced to near dryness under N$_2$ and redissolved in 150 μL Hex				
D4-D6	Surface soil collected, immediately frozen and transported to the lab	Stored at 4°C	5 g sample SLE: Shaken with 3 × 25 mL Hex:EA (1:1 v/v), 30 min; centrifugation; extracts combined and concentration by rotary evaporation to 1–2 mL, solvent exchange to IOC and volume reduction to 1 mL under N$_2$	GC-MS (EI, 30 m DB-5 column)	0.89–1.05	82.5 ± 21	[63]

CSR concurrent solvent recondensation, *DCM* dichloromethane, *EA* ethyl acetate, *EI* electron ionization, *GC-AED* gas chromatography-atomic emission detection, *GC-MS* gas chromatography-mass spectrometry, *Hex* hexane, *IOC* isooctane, *LOD* limit of detection, *LVI* large-volume injection, *Pent* pentane, *PTFE* polytetrafluoroethylene, *Rec* recovery, *SAESC* sonication-assisted extraction in small columns, *SLE* solid-liquid extraction
[a]Limit of quantification (ng g−1)

Table 5 Details of the analytical methods developed for the determination of VMSs in outdoor and indoor air and dust samples

Analytes	Sampling details	Sample preparation	Extraction/clean-up method	Instrumental method	LOD (ng m^{-3})	%Rec	Reference
Outdoor air							
D3-D5	AAS with SPE Tenax GC cartridges, with GFF to remove particles; air flow of 15–40 L min^{-1}; tubes transported in centrifuge tubes with glass wool	–	Tubes desorbed directly in the GC-MS at 270°C for 6.5 min; cryogenic trap at −100°C	TD-GC-MS (60 m and 120 m SCOT columns)	Qualitative identification	–	[77]
D3-D6 L3-L5	PAS with SIP disks with XAD-4, deployed 2 m above ground/water. Disks stored in amber glass aluminium-lined jars at −20°C until analysis	–	Soxhlet, petroleum ether/Acet (1:1), 18 h; extracts concentrated with rotary evaporator and N$_2$ to 1 mL; clean-up, 1 g Na$_2$SO$_4$ column, eluted with 3 × 1 mL petroleum ether/Acet (1:1); IOC added and extract reduced to 0.5 mL with N$_2$	GC-MS (EI, 60 m DB-5 column)	<0.01–18.18	25–90	[65]
D3-D5	AAS with SPE stainless steel tubes with silica, carbon sieve and charcoal, collected 1.2 m above ground at 0.5 L min^{-1}	–	–	TD-GC-MS (EI, 30 m HP-5MS column)	–	–	[101]
D4-D6 L2-L5	AAS with SPE Tenax TA tubes, air flow of 0.1 L min^{-1}	–	Tubes desorbed directly in the GC-MS at 275°C for 20 min	TD-GC-MS (EI, 30 m DB1701 column)	–	–	[90]
D4, D5	AAS with direct sampling with a Teflon-coated pump, gas flow controller, electronic control box and glass transfer tube	–	–	APCI-MS/MS (positive mode)	4,000–6,000	93–98	[82]

(continued)

Table 5 (continued)

Analytes	Sampling details	Sample preparation	Extraction/clean-up method	Instrumental method	LOD (ng m^{-3})	%Rec	Reference
D5	AAS with SPE ENV+ cartridges, air flow 0.9–3.0 L min^{-1}. Cartridges capped, wrapped in aluminium foil and stored at −18°C	–	SPE: Cartridges eluted with 0.6 mL Hex	LVI-GC-MS (EI, 30 m DB-5MS column)	0.12	^{13}C labelled: 99 ± 11	[60]
D5	AAS with SPE ENV+ cartridges, air flow 0.9–4.5 L min^{-1}. Cartridges capped, placed in glass tubes and stored at −18°C	–	SPE: Cartridges eluted with 0.6 mL Hex	LVI-GC-MS (EI, 30 m DB-5MS column)	–	–	[102]
D3-D6 L3-L5	PAS with SIP disks with XAD-4	–	–	GC-MS (EI, 60 m DB-5 column)	0.011–7.1	20–100	[100]
Outdoor air							
D3-D6 L3-L5	PAS with SIP disks with XAD-4	–	ASE, three cycles of petroleum ether/Acet (85:15); SPE, ENVI-Carb columns, 4 mL DCM/Hex (1:4); extracts concentrated by rotary evaporation and N$_2$ to 0.5 mL, IOC as keeper	GC-MS (EI, 60 m DB-5 column)	0.0008–0.9	48 ± 18 to 51 ± 15	[79]
D4-D6	AAS with SPE ENV+ cartridges, air flow 11–12.5 L min^{-1}. Cartridges capped, wrapped in aluminium foil, packed in Al-laminated PE bag and stored at −18°C	–	SPE: Cartridges eluted with 1.5–1.8 mL Hex	GC-MS (EI, 30 m DB-5MS column)	–	–	[103]

Analyte	Sampling		Extraction	Analysis	Concentration	Recovery	Ref.
D3-D6 L3-L6	AAS with SPE ENV+ cartridges, air flow 12 L min⁻¹. Cartridges capped, wrapped in aluminium foil and stored at −18°C	—	SPE: Cartridges eluted with 1.3 mL Hex +1.3 mL DCM	GC-MS (EI, 30 m DB-5MS column)	a0.004–0.270	~45 to ~93; 13C labelled: ~78 to ~83 (inferred from graph)	[25]
D3-D6	AAS with SPE ENV+ cartridges set 2 m above ground, air flow 18.0 ± 0.4 L min⁻¹. Cartridges capped, wrapped in aluminium foil and stored at −18°C in PE bottles	—	SPE: Cartridges eluted with 3 mL Hex	GC-MS (EI, 30 m DB-WAX ETR column)	0.01–0.14	66 ± 4 to 90 ± 5; 13C labelled: 99 ± 1 to 108 ± 1	[24]
D3-D6 L3-L5	PAS with XAD-2 resin, 10 g placed in mesh cylinders under stainless steel housings 1.5 m above ground; AAS with SPE, ENV+ cartridges, air flow 4.7–5 L min⁻¹. Samples wrapped in aluminium foil and stored at −20°C	—	PAS with XAD-2 resin, SLE in separation funnel with 22 mL Hex, shaken; extract taken, 2 × 10 mL Hex added and shaken; extracts combined AAS with SPE, Cartridges eluted with 3 mL Hex	GC-MS (EI, 30 m HP-5MS column)	PAS, <0.171–12.200; AAS, <0.024–2.540	13C-labelled PAS, 84 ± 8 to 179 ± 8; AAS, 85 ± 12 to 107 ± 21	[62]
D3-D6 L3-L5	PAS with SIP disks, PUF disks impregnated with XAD-2; placed in stainless steel chambers 2 m above ground; AAS with GFF and PUF/XAD-2, sequential alignment with air flow of ~230 L min⁻¹ for 24 h. Samples stored in the dark at −20°C until analysis (max 4 weeks)	—	PAS, SIPs extracted by ASE with 2 cycles of petroleum ether/Acet (83:17); AAS, PUF/XAD-2 plugs Soxhlet extracted with petroleum ether/Acet (85:15) for 6 h; GFFs extracted 3 × with DCM by USE; all extracts concentrated by rotary evaporation and N_2 to 0.5 mL, IOC as keeper	GC-MS (EI, 60 m DB-5MS column)	PAS, 0.01–3.30; AAS, 0.004–1.0	13C labelled PAS, 49 ± 9 to 68 ± 11; AAS, 57 ± 20 to 70 ± 21	[21]

(continued)

Table 5 (continued)

Analytes	Sampling details	Sample preparation	Extraction/clean-up method	Instrumental method	LOD (ng m^{-3})	%Rec	Reference
Outdoor air							
D3-D6 L2-L5	PAS with SIP disks, deployed at 1–2 m above ground in a 'flying saucer' protective chamber; samples stored in clean stainless steel tins, frozen and kept from light	–	Soxhlet: DCM:Hex (1:1) overnight; extracts passed through a Na$_2$SO$_4$ glass column concentrated by rotary evaporation and N$_2$ to near dryness and redissolved in 150 µL of Hex	GC-MS (EI, 30 m DB-5ms ultra-inert column)	4.7–10.2 (ng SIP^{-1})	87 ± 8	[42]
D3-D6 L3-L5	PAS with SIP disks, deployed at 2 m above water/ground; samples stored in clean glass jars at −20°C until analysis	–	ASE: two cycles of petroleum ether/Acet (83:17); extracts concentrated by rotary evaporation and N$_2$ to 0.5 mL, IOC as keeper	GC-MS (EI, 60 m DB-5 column)	0.020–0.720	^{13}C labelled: 67 ± 10 to 78 ± 6	[78]
D3-D6 L2-L5	AAS with Carbotrap/Carbopack/Carboxen SPE cartridges, air flow 0.07 L min^{-1} for 24 h. Samples packed in aluminium foil and sealed in PP bags	–	Primary desorption of the cartridges at 300°C with helium at 0.05 L min^{-1}; cold trap at −30°C; secondary desorption heating from −30 to 300°C, held for 10 min	TD-GC-MS (EI, 60 m DB-624 column)	0.6–0.9	–	[64]
D5	Sampling conducted by drawing air through a Teflon inlet; air pumped at 5 L min^{-1} by external pump	–	Collection of air (1 min accumulation period) directly by the PTR-TOF-MS equipment	PTR-TOF-MS	–	–	[81]
Outdoor and indoor air							
D4, D5	PAS with charcoal sorbent; samples sealed with a plastic top and mailed to the lab	–	–	GC-MS (EI, 12 m cross-linked dimethyl silicone column)	50	–	[80]

Analyte		Sampling method	Sample preparation	Instrument	Concentration	Recovery	Ref.
D5	–	AAS with SPE Tenax TA tubes (added of a Carbosieve S-III sorbent portion at the outlet end), air flow of 0.005 L min^{-1}	Tubes desorbed directly in the GC-MS at 235°C for 6.5 min; cryogenic trap at −100°C	TD-GC-MS	–	–	[104]
D4-D6 L3-L16	–	AAS with SPE ENV+ cartridges, air flow 0.5 L min^{-1} for 24 h (outdoor) and 8 h (indoor). Cartridges sealed and stored at −18°C until analysis	SPE: Cartridges eluted with 10 mL Hex; extracts concentrated with N$_2$ to 1 mL	GC-MS (EI, 30 m low-bleed HP-5MS column)	Outdoor, [a]0.14–0.36; indoor, [a]0.4–1.0	95–99	[36]
D4-D6	–	AAS with SPE cartridges, air flow of 3–6 L min^{-1}. Cartridges wrapped in aluminium foil, stored in amber glass jars and frozen	SPE: Cartridges eluted with 1.5 mL Hex	GC-MS (EI, ?? m RTX-5MS column)	0.430–3.840	99–114	[22]
Outdoor and indoor air							
D3-D6 L2-L5	–	AAS with SPE Isolute ENV+ cartridges, air flow 1.5 L min^{-1}. Cartridges sealed with PTFE endcaps and stored frozen at −18°C until analysis (in 24 h)	SPE: Cartridges eluted with 3 mL Hex	CSR-LVI-GC-MS (EI, 60 m DB-5MS column with a 5 m guard column)	0.010–0.180	96 ± 4 to 104 ± 3	[45]
Outdoor and indoor air and dust							
D4-D6 L4-L16	–	*Air:* AAS with SPE ENV+ cartridges, air flow 0.5 L min^{-1} for 24 h. *Dust:* not specified	*Air* SPE: Cartridges eluted with Hex; extracts concentrated to GC vials with N$_2$ *Dust* SLE: Hex/EA; extracts concentrated to GC vials with N$_2$	GC-MS (EI, 30 m low-bleed HP-5MS column)	*Air* [a]0.17–1.8; *Dust* [a]0.5–2.4 ng g^{-1}	95–99	[99]

(continued)

Table 5 (continued)

Analytes	Sampling details	Sample preparation	Extraction/clean-up method	Instrumental method	LOD (ng m^{-3})	%Rec	Reference
Indoor air							
D3-D6 L2-L5	AAS with SPE Tenax GR and graphitized carbon black tubes; deployed at 1.5 m above the floor, air flow 0.12 L min^{-1}. Tubes stored at 4°C until analysis	–	Tubes desorbed directly in the GC-MS at 320°C for 60 min	TD-GC-MS (EI, 60 m inert DB 624 column)	7–40	82 ± 6 to 92 ± 8	[23]
D3-D6 L2-L5	AAS with SPE Tenax GR and graphitized carbon black tubes; deployed at 1.5 m above the floor, air flow 0.1 L min^{-1}	–	Tubes desorbed directly in the GC-MS at 320°C for 60 min	TD-GC-MS (EI, 60 m inert DB 624 column)	7–40	–	[105]
D4-D6	Sampling conducted by drawing air through a Teflon solenoid valve, 2 m above the floor; an in-line PTFE membrane filter before inlet to remove particulate matter	–	Collection of air directly into the PTR-TOF-MS equipment	PTR-TOF-MS (H$_3$O$^+$ as the primary reagent ion)	–	–	[83]
D3-D7 L3-L11	AAS with PUF plugs and QFFs at 5 L min^{-1} for 12–24 h. Samples kept at −18°C until analysis (max 3 weeks)	–	PUFs: SLE in an orbital shaker with DCM/Hex (3:1) for 30 min twice (100 + 80 mL solvent); extracts concentrated with rotary evaporator to 5 mL and to 1 mL with N$_2$. QFFs: SLE in an orbital shaker with DCM/Hex (3:1) for 3 x 5 min (20 mL solvent each time); extracts concentrated as for PUFs	GC-MS (EI, 30 m HP-5MS column)	Vapour phase, 0.06–0.83; particulate phase, 0.8–32 ng g^{-1}	Vapour phase: 66–123 Particulate phase, 78–125; ^{13}C labelled, 83–122	[57]

Indoor air

Analytes	Sampling		Procedure	Instrument	Value	Recovery	Ref.
D3-D6	AAS with SPE Tenax GR and graphitized carbon black tubes; deployed at 1.5 m above the floor, air flow 0.12 L min⁻¹. Tubes sealed and stored at 4°C	—	Tubes desorbed directly in the GC-MS at 320°C for 15 min	TD-GC-MS (EI, 60 m Inert DB 624 column)	2–40	–	[106]
D3-D7 L3-L11	AAS with ORBO-1000 PUF plugs and QFFs at 4 L min⁻¹ for 12 to 24 h. Samples wrapped in aluminium foil, kept at −18°C until analysis (max. 2 weeks)	—	PUFs: SLE in an orbital shaker with DCM/Hex (3:2) for 30 min twice (100 + 80 mL solvent); extracts concentrated with rotary evaporator to 5 mL and to 1 mL with N₂. QFFs: SLE in an orbital shaker with DCM/Hex (3:2) for 3 × 5 min (20 mL solvent each time); extracts concentrated as for PUFs	GC-MS (EI, 30 m DB-5MS column)	[a]Vapour phase: 0.12–0.7 [a]Particulate phase: 1.5–9.0 ng g⁻¹	Vapour phase, 83–106; particulate phase, 90–103	[75]

Indoor air and dust

Analytes	Sampling	Sample note	Procedure	Instrument	Value	Recovery	Ref.
D4-D6	*Air:* AAS with PUF plugs and QFFs at 5 L min⁻¹ for 20–24 h. *Dust:* Collected with brooms from floor and furniture, packed with aluminium foil	*Dust* samples sieved (150 μm), stored in a dark glass jar and sealed at 4°C until analysis	*Air* PUFs, SLE, shake with 2 x DCM/Hex (3:2) for 30 min (100 + 80 mL solvent); extracts concentrated with rotary evaporator to 5 mL and to 1 mL with N₂. QFFs: SLE in an orbital with DCM/Hex (3:2) for 3 x 5 min (5 mL solvent); extracts concentrated as for PUFs *Dust*	GC-MS (EI, 30 m DB-5MS column)	<u>*Air*</u> [a]Vapour phase: 0.7 [a]Particulate phase: 1.5 ng g⁻¹ *Dust* [a]5.0 ng g⁻¹	–	[76]

(continued)

Table 5 (continued)

Analytes	Sampling details	Sample preparation	Extraction/clean-up method	Instrumental method	LOD (ng m^{-3})	%Rec	Reference
			0.3 g sample SLE, shaken with 5 mL DCM/Hex (3:2), 5 min; centrifugation; process repeated 2×; extracts concentrated to 1 mL under N$_2$; filtration, 0.22 µm PTFE				
Dust							
D4-D7 L4-L14	Collected with vacuums and brushes; dust placed in aluminium foil, sealed in a plastic bag and stored at −20°C	Non-dust particles removed, and samples sieved (500 µm mesh)	0.3–0.5 g sample SLE, shaken with 5 mL Hex, 15 min; centrifugation; process repeated 3× with EA/Hex (1:1); each extract concentrated to 1–2 mL; clean-up, 0.2 g Na$_2$SO$_4$ + 0.5 silica gel, elution with 6 mL Hex + 5 mL DCM/Hex (1:1); extracts combined and concentrated to 0.5 mL under N$_2$	GC-MS (EI, 30 m DB-5MS column)	^{a}0.18–1.47 ng g^{-1}	68 ± 2 to 108 ± 16	[84]
Dust							
D4-D7 L4-L14	Collected with vacuums and brushes; dust placed in aluminium foil, sealed in a plastic bag and stored at −20°C	Non-dust particles removed, and samples sieved (500 µm mesh)	0.3–0.5 g sample SLE, shaken with 5 mL Hex, 15 min; centrifugation; process repeated 3× with EA/Hex (1:1); each extract concentrated to 1–2 mL under rotary evaporation; clean-up, 0.2 g Na$_2$SO$_4$ + 0.5 silica gel, elution with 6 mL Hex + 5 mL DCM/Hex (1:1); extracts combined and concentrated to 0.5 mL under N$_2$	GC-MS (EI, 30 m DB-5MS column)	^{a}0.18–1.47 ng g^{-1}	68 ± 2 to 108 ± 16	[84]

Analytes	Collection/storage	Sieving	Extraction	Instrument	LOD/LOQ	Recovery (%)	Ref.
D4–D6 L3–L16	Collected with vacuums and brushes and transferred to glass containers and sealed. Stored at −18°C	Samples sieved (500 μm mesh) inside a sealed plastic bag at 4°C	0.2 g sample SLE, vortex with 10 mL EA/Hex (1:1), 5 min; centrifugation; process repeated 2×; combined extract concentrated to 2 mL under N_2; clean-up, 1 g Na_2SO_4, elution with 5 mL EA/Hex (1:1); extract concentrated to 1 mL under N_2	GC-MS (EI, 30 m low-bleed HP-5MS column)	[a]0.5–1 ng g^{-1}	81–94	[36]
D3–D7 L4–L14	Collected with vacuums and brushes; dust stored in PE bags or glass jars at 4°C until analysis	Samples sieved (150 μm)	0.3–0.5 g sample SLE: shaken with 5 mL DCM/Hex (3:1), 5 min; centrifugation; process repeated with 3 mL DCM/Hex (3:1) and 3 mL Hex; extracts concentrated to 1 mL under N_2; filtration, 0.2 μm cellulose filter	GC-MS (EI, 30 m HP-5MS column)	[a]2.0–6.0 ng g^{-1}	67–121; ^{13}C labelled: 75–118	[58]
D4–D6 L4–L16	Collected with wipes and brushes from furniture; dust sieved (60 μm), placed in aluminium foil, sealed in a PE bag and stored at −20°C	Samples sieved (500 μm mesh)	0.2 g sample SLE: vortex with 10 mL EA/Hex (1:1), 5 min followed by USE, 15 min; centrifugation; process repeated 3×; combined extract dried in a 1 g Na_2SO_4 cartridge, elution with 5 mL EA/Hex (1:1); extract concentrated to 1 mL under N_2	GC-MS (EI, 30 m low-bleed HP-5MS column)	[a]0.5–1.8 ng g^{-1}	82 ± 7 to 94 ± 5	[37]

AAS active air sampling, Acet acetone, APCI atmospheric pressure chemical ionization, ASE accelerated solvent extraction, CSR concurrent solvent recondensation, DCM dichloromethane, EA ethyl acetate, EI electron ionization, GC-MS gas chromatography-mass spectrometry, GFF glass-fibre filters, Hex hexane, IOC isooctane, LOD limit of detection, LVI large-volume injection, MeOH methanol, PAS passive air sampling, PE polyethylene, PLE pressurized liquid extraction, PP polypropylene, PTR proton transfer reactions, PUF polyurethane foam, QFF quartz fibre filters, Rec recovery, SIP sorbent-impregnated polyurethane foam disks, SLE solid-liquid extraction, SPE solid-phase extraction, TD thermal desorption, TOF time-of-flight, USE ultrasound extraction

[a]Limit of quantification (pg m^{-3} for air and ng g^{-1} for dust)

"?" sample amount not provided

cartridges and filters were wrapped in aluminium foil and placed in bags or glass jars. In three studies, the air was pumped directly into the gas chromatograph, thus avoiding the troublesome sample handling steps prone to add external contamination to the analysis [81–83]. In indoor environments, sometimes dust is also collected (Table 5), commonly collected with wipes, brushes, brooms or vacuums, wrapped in aluminium foil and stored in plastic bags [37, 84] or glass jars [36, 58].

Regarding the transport to the lab, for all the matrices described above, there was the intention of minimizing contaminations or loss by volatilization, so it was common to refrigerate the samples during transportation and keep them frozen in the lab until further handling for analysis, which should be performed as soon as possible after collection, for the same reasons.

2.3   Sample Preparation and Extraction

Environmental samples are often complex, containing high amounts of possible interferences and most likely low levels of the target pollutants. Therefore, the implementation of robust analytical methodologies is mandatory. For this purpose, sample preparation, involving extraction, clean-up and pre-concentration steps, is required, and it is crucial to achieve reliable results. However, this is usually a laborious, high cost and time-consuming effort in the analytical procedures [85]. In the last decades, there was a clear improvement in the extraction and clean-up techniques, making them faster, cheaper, safer, more environmentally friendly and sensitive. This evolution was also perceived for VMS analysis. And in the next subsections, the main protocols applied for the determination of VMSs in different environmental matrices will be presented and discussed. It is important to emphasize that most of the studies found in the literature focused on the study of cyclic VMSs, namely, D4, D5 and D6, while only a few studies were carried out on the determination of linear VMSs of higher molecular weight (L6-L14).

2.3.1   Water Samples

Studies for the determination of VMSs in aqueous matrices are still scarce. An overview of the analytical methodologies published so far is shown in Table 1. In terms of extraction techniques, liquid-liquid extraction (LLE) is the most common procedure (30%) [7–9, 56, 63, 86], followed by membrane-assisted solvent extraction (MASE; 25%) [38, 49, 66, 87, 88] and purge and trap approach (P&T; 20%) [48, 89–91]. Two microextraction methods were also used: solid-phase microextraction (SPME; 10%) [35, 40] and dispersive liquid-liquid microextraction (DLLME; 5%) [47].

LLE is a classical approach based on the transference of the target analytes from one solvent to another, immiscible in the former. Due to the lipophilic nature of VMSs, this extraction technique has typically been carried out using non- or low-polar solvents immiscible in water, such as hexane [9, 63], heptane

[86], dichloromethane [7, 56] and mixtures of these solvents [8]. These solvents are usually added in large amounts to a high sample volume (>500 mL). This procedure is also time-consuming, requiring a contact time between the two phases up to 48 h [86]. However, this methodology seems to be appropriate for the extraction of the VMSs, leading to low limits of detection (LODs, 0.1–13 ng L^{-1}) and high recovery rates (21–134%; mean $\approx$ 80%) (Table 1). The lowest recoveries were found for the most volatile compounds (D3 and L3) and may be explained by volatilization losses. In fact, this conventional extraction procedure requires considerable sample handling, which may contribute to external contamination and loss of the target analytes. As an alternative, MASE is also employed for this type of compounds and matrices [38, 49, 66, 87, 88]. This technique is based on small-scale LLE, i.e. the target analytes are extracted from an aqueous sample to an organic acceptor phase, permeating through a low-density polyethylene membrane. This approach is especially useful when the samples contain a high number of suspended particles, like wastewater and some river waters. Although this extraction approach led to better recoveries (72–107%) than LLE, the LODs were generally slightly higher (9–27 ng L^{-1}). In fact, this methodology has the advantage of using smaller volumes of sample and organic solvents. This situation often leads to a lower pre-concentration factor, which may negatively affect the sensitivity. Similarly, non-polar solvents such as hexane [49, 66] and pentane [38, 87, 88] were used. The third most used technique is purge and trap [48, 89–91]. In this case, the sample is purged with an inert gas, causing the volatilization of the compounds with lower vapour pressure. These compounds are retained in an adsorbent trap, such as Tenax TA and styrene-divinylbenzene copolymer packed in a cartridge. To analyse VMSs, they should be chemically or thermally eluted from the sorbent. In the first case, a low-polar solvent like dichloromethane [48] should be used, and, in the second, a thermal desorber unit may be connected to the GC-MS system [89–91]. The application of this methodology for the determination of VMSs yielded high recovery rates (70–90%) and LODs in the same order of magnitude of the previous methods (0.3–60 ng L^{-1}). This methodology has the advantage of being applied to larger sample volumes to achieve greater sensitivity and do not use solvents, thus reducing possible matrix effects. It is particularly useful to determine poorly water-soluble compounds in different types of aqueous matrices.

One of the most recent trends in the analytical field is the application of microextraction techniques, using reduced volumes of extractive phase. Companioni-Damas et al. [40] and Xu et al. [35] proposed the use of SPME for the analysis of VMSs, using headspace. In fact, this is a solventless methodology, in which the analytes are absorbed onto a fused-silica fibre coated with an appropriate sorbent layer, and their desorption usually occurs by heating the exposed fibre in the injection port of the GC system. Both studies proposed the headspace configuration since VMSs are volatile compounds and the use of semipolar PDMS/divinylbenzene (DVB) fibres. Although the LODs were similar to the other techniques, the use of a fibre coated with PDMS may release cyclic VMSs with the high temperatures reached, leading to false positives, due to the fibre bleeding. To avoid this situation, other coatings should be investigated, like polyacrylate, new molecularly imprinted polymers and polymeric ionic liquids [92]. Finally, Cortada et al. [47] proposed the employment of an ultrasound-assisted dispersive liquid-liquid microextraction for the determination of cyclic and linear VMSs. This methodology has the main advantages of the conventional LLE, but it is considered environmentally friendly due to the low amount of solvents used. In this study, a few microlitres of chlorobenzene were added as extraction solvent, and the mixture was sonicated in an ultrasonic bath for 2 min. The recoveries achieved ranged from 71 to 99% and the limits of detection between 3 and 1,400 ng L^{-1}, which are slightly higher than those obtained in the other methodologies. The variability of the results is around 20%, which is the typical value for this kind of matrix (wastewater).

As can be seen in Table 1, these methodologies can be used to extract the VMSs in different types of aqueous matrices with no additional clean-up steps.

2.3.2 Sewage Sludge

To assess the levels of linear and cyclic VMSs in sewage sludge samples (primary, activated and digested), some analytical methodologies have been suggested (Table 2). Solid-liquid extraction (SLE) is the most common extraction technique used for this kind of analysis (67%) [7, 8, 11, 14, 41, 86, 87, 94–96], followed by purge and trap (20%) [89–91] and ultrasound extraction (USE, 13%) [10, 35]. Solid-phase extraction (SPE) was frequently employed as a clean-up procedure after a SLE approach [14, 41, 96], and in a single case, SPME was used for the same purpose [35].

SLE is based on the diffusion of the target analytes from the solid matrix to an appropriate solvent, in which they are highly soluble. This is a simple approach, but like in LLE, it uses high volumes of solvent (sometimes multiple extraction steps are required to recover the analytes) and may require high extraction times (performed under constant stirring). In addition, the contact area between solid and liquid is a crucial factor. Larger contact areas usually produce better extraction rates and, consequently, higher percentages of recovery. For this reason, some authors opted to powder the sludge prior to extraction [8, 10, 14, 95]. It is also usual to freeze-dry the sample and only then increase its surface area by grinding or crushing. Finally, the sludge is homogenized by sieving, which helps in getting reproducible results. After the SLE, the supernatant containing the target analytes is recovered by centrifugation. Due to the lipophilic nature of VMSs, this extraction technique has been performed using non- or low-polar solvents, such as hexane [41, 95], acetone [35] and mixtures of hexane with other solvents, namely, ethyl acetate and dichloromethane [7, 8, 10, 11, 14]. The use of SLE alone obtained recoveries between 53 and 143% and LODs within a wide range (0.03–2,000 ng g^{-1} dw). Although the combination of SPE clean-up with the SLE is less common for VMSs in sludge, three studies apply this solution [14, 41, 96]. SPE is a clean-up technique used to isolate the target analytes from a liquid extract through their retention in a solid adsorbent and subsequent recovery by elution with an appropriate solvent. This technique allows not only the removal of interferents (clean-up step) but also the pre-concentration of the analytes. Similarly to SLE, the most critical point is the selection of the solvent for the elution step, which may require a large volume. Nevertheless, SPE is usually efficient, has high extraction yields and can be automated. In the extraction/clean-up of VMSs from sludge using SPE, aluminium oxide [96] and silica [14, 41] were used as adsorbents and hexane or mixtures of hexane with dichloromethane or ethyl acetate as elution solvents. The best results were achieved with silica, which may be explained by the low polarity of the target compounds. In fact, aluminium oxide is a polar adsorbent that preferably retains weak and moderately polar compounds, while silica is less polar. Companioni-Damas et al. [41] performed assays using a silica cartridge and eluting the target compounds with hexane, reaching LODs between 0.004 and 0.140 ng g^{-1} dw and recovery of 80–103%. And in general, SLE followed by SPE showed better results than SLE alone, allowing the removal of possible interferences.

USE was employed in two studies to extract the VMSs from sewage sludge [10, 35]. In this method, an appropriate solvent is added to the sludge sample and submitted to ultrasound waves. Large amounts of energy are released, creating cavitation bubbles and promoting the penetration of the solvent into the sample matrix, increasing the mass transfer of the analytes to the solvent phase. This extraction is faster than other conventional methods (usually less than 15 min), employs lower amounts of solvent and usually leads to high recoveries. However, this approach is more suitable for non-volatile or semi-volatile organic compounds (SVOCs), given that the energy used favours the heating of the medium, which can produce losses of the compounds with higher vapour pressure through volatilization. Liu et al. [10] used low-polar solvents (hexane/ethyl acetate) and achieved low detection limits (0.5–1.7 ng g^{-1} dw) and high recoveries (69–104%), whereas Xu et al. [35] combined USE with an additional step of SPME, trying to avoid coextraction of interferents. In this case, acetone was used as extraction solvent, and sonication was done for 30 min. Then, the solvent phase was collected after centrifugation, and the volume was reduced and diluted in ultrapure water. The resulting solution was subjected to SPME (PDMS/DVB fibre), using the headspace mode. Again, this configuration was chosen due to the complexity of the matrix (avoid coextraction of interferences) and the high volatility of the analytes. This combined approach led to results like those found by USE (limits of detection <1 ng g^{-1} dw; recovery: 75–93%). Although the results obtained are promising, the use of PDMS-based fibre coatings may be a source of VMS contamination.

As for wastewater, the purge and trap approach was also applied [89–91], and the conditions were very similar (purge gas, nitrogen; adsorbent trap, Tenax TA; thermal desorber unit connected to the GC-MS system) but resulting in higher LODs (0.1–470 ng g^{-1} dw). As aforementioned, purge and trap method may be applied to larger sample volumes and achieve higher sensitivities while reducing the analytical steps.

It is also important to notice that in most of the protocols described, the authors decided to include an additional step, in which the extracts are treated with a drying agent (mainly sodium sulphate) to remove traces of moisture. This is a very important action, since the injection of extracts containing traces of water can lead to the degradation of the stationary phase of the GC column, producing bleeding [31].

2.3.3 Sediments and Soils

Sediments and soils are solid matrices, and the extraction procedures rely essentially on SLE. In fact, for sediments (Table 3), 68% of the studies used this approach [13, 14, 39, 50, 51, 56, 59, 63, 67–69, 71, 88, 97], followed by purge and trap (9%) [90, 91] and USE (9%) [9, 69]. A combination of purge and trap and SLE [52], another of SLE and USE [68] and accelerated solvent extraction (ASE) [13] were employed in one study each. In the case of soils, there are fewer works regarding VMSs (Table 4), and 56% of them used SLE [36, 41, 61, 63, 88], followed by USE (22%) [15, 42] and purge and trap [90] and sonication-assisted extraction in small columns (SAESC) [98] with one study each.

Sediments are collected underwater, and researchers divide themselves into an analysis with the wet sample or after a drying process (freeze-drying, centrifugation or mixing with diatomaceous earth are the reported options). Processes of homogenization are also frequently used, and in one case, the sediments were wetted with acetonitrile before SLE [50]. The extraction technique not mentioned before is ASE, which is a form of pressurized liquid extraction (PLE), where the extraction and clean-up steps are performed under pressure with an organic solvent through a cell containing the sample and the sorbents. In fact, Sparham et al. [13] used this approach for river sediments (with ethyl acetate) and SLE (with acetonitrile:hexane 1:1) for estuarine sediments. This was the only study that resorted to different methodologies for different sediment origins. ASE had slightly better recoveries but higher LODs (7–37 versus 1 ng g^{-1}). SLE relied mainly in vortex shaking, and the solvents used were hexane, ethyl acetate, acetonitrile, dichloromethane or combinations of them. Pentane was also used in two cases [50, 52]. Regarding the LODs, SLE approaches reached values between 0.02–0.29 ng g^{-1} [59] and 3.7–34 ng g^{-1} [52], although this last study had a combination of SLE and purge and trap followed by large-volume injection GC-MS. The other purge and trap approach with LOD information had values similar to the lower limits [90]. Sanchís et al. [9] obtained comparable values with USE using hexane:ethyl acetate (1:1) (LODs: 0.3–0.9 ng g^{-1}). In terms of recoveries, all approaches reached values between 69 ± 10 and 105 ± 10%. Lee et al. [59] showed the widest range (63 ± 10 to 144 ± 3%) with their SLE method with three consecutive extractions with hexane, hexane:dichloromethane (1:1) and hexane:ethyl acetate (1:1), but they were also the study that analysed the higher number of VMSs (D4-D7 and L4-L17), which may have accounted for that result.

The preparation of soil samples for VMS analysis is similar to the sediments, but in most cases, the samples were collected already dry, so they mostly underwent a process of homogenizing and sieving, as granulometry may affect the potential for the uptake of SVOCs. In one case, soils were diluted in water as preparation for purge and trap extraction [90]. SLE and USE were employed using the same solvents as for sediments, with a predominance for hexane alone or in combination with ethyl acetate. In one of the studies using USE, Ramos et al. [42] included a QuEChERS clean-up (based on a first stage of drying and salting out and a second one of dispersive SPE with appropriate sorbents) after an extraction with hexane: dichloromethane (1:1). This procedure yielded low LODs (0.003–0.020 ng g^{-1}) and good recoveries (69 ± 17%). Sánchez-Brunete et al. [98] chose a SAESC method, which is a tandem combination of USE (with hexane) and SPE in small columns, filled in this case with sodium sulphate and the sample. The LODs were slightly higher (0.04–1.1 ng g^{-1}), but so were the recoveries (85–108%). SLE protocols had LODs typically one order of magnitude higher than USE ones and with comparable recoveries, ranging from 76 ± 3% to 103%. USE is likely more effective in removing the target analytes from complicated matrices from soils than SLE, but usually the authors who decided to have sequential extractions with fresh solvent obtained better recoveries in the process.

2.3.4 Outdoor Air, Indoor Air and Dust

Air sampling is necessarily different from the previous matrices, as in this case gaseous and often also particulate phases need to be collected from the atmosphere. Most VMSs are present in the gas phase, but as the molecules become larger, also particulate fractions may appear. Another important question is that due to their volatility, collecting air is prone to external contamination when VMSs are the target analytes. For that reason, many studies tried to reduce the sample handling and preparation steps. In fact, two of them collect the air and inject it directly to a proton transfer reactions (PTR)-TOF-MS chromatograph [81, 83]. However, the majority of protocols use SPE sorbent cartridges (indicated for active sampling) and either elute them with mostly hexane (35%) or thermally desorb it directly into the GC-MS (24%) (see Table 5). When other materials like foams or filters were used (both for active or passive sampling), SLE (14%), Soxhlet extraction (10%) and ASE (10%) are the preferred choices. In these cases, the preferred solvents were mixtures of hexane and dichloromethane for Soxhlet and of petroleum ether and acetone, mostly used in ASE [21, 78, 79]. When active and passive sampling were used in the same study, different approaches for each case were employed, as in Krogseth et al. [62] (SLE and SPE, respectively). Ahrens et al. [21] even employed different methods for PUF plugs (ASE) and GFFs (USE) used in active sampling while choosing Soxhlet extraction for the SIP passive samplers. Dust collected in indoor air studies (Table 5), being a solid matrix, was analysed exclusively with SLE approaches (with Liu et al. [37] adding an extraction step by USE), using solvents like hexane alone or in combination with ethyl acetate (predominantly) or dichloromethane [36, 37, 58, 76, 84, 99].

In terms of LODs, in general the protocols with direct injection or desorption do not reach as low detection concentrations as those with elution or extraction procedures (differences can reach up to three orders of magnitude from the ng to pg m^{-3} level). Rauert et al. [79] reached the lowest LOD with 0.8 pg m^{-3}. This may be due to the fact that sample clean-up and concentration steps can improve the resolution of the chromatograms. Also, active sampling methods have usually lower LODs than passive sampling, when both are used in the same study. Recoveries, however, have overall wider ranges than the previous matrices, with percentages that can reach as low as 20% [100] and as high as 125% [57]. Low recoveries are more frequent in the most volatile VMSs, due to potential external contamination or losses in the process. Still, SPE approaches seem to produce the better and most consistent results. In dust analysis, the LOQs vary from 0.18 to 6.0 ng g^{-1} and the recoveries from 67 to 121%, which are similar to the solid matrices mentioned above.

2.4   Instrumental Determination

Being semi-volatile compounds, the instrumental technique most suitable for the determination of VMSs is gas chromatography (GC). Different detectors may be coupled to this methodology, but the most used is the mass spectrometer (MS) (Tables 1, 2, 3, 4 and 5). This hyphenated technique combines separation by gas-liquid chromatography with mass spectrometry detection, which allows the identification of the different target analytes by the mass-charge ratios of the fragments originated during the ionization of the compounds [107]. The main advantages of using GC-MS for VMSs analysis are the ability to separate complex mixtures, to clearly identify and quantify the target analytes even at trace levels, simplicity, high reproducibility and resolution. In addition to the mass detector, the VMSs may be also detected by the flame ionization detector (FID) [95]. This instrumental methodology requires low cost and maintenance of the equipment and has a high sensitivity to almost combustible compounds but presents a low selectivity and does not provide enough information to identify unequivocally the analytes, for which reason it is sometimes complemented with a GC-MS analysis for compound confirmation. The versatility of GC-MS can also be seen as it can be adapted to different extraction procedures or injection set-ups different from the most common injection of a sample organic extract. In the case of VMSs, some of these techniques aim for a reduction in sample handling to mitigate external contamination and have been applied to different environmental samples. For instance, thermal desorption was used in all the environmental matrices reviewed in this chapter (water, sludge, sediments, soil and air) but in particular for water and air samples. Another technique used is large-volume injection (LVI), which allows the injection of amounts of sample higher than the usual

1–2 μL, thus reducing the need of a strong sample pre-concentration, which could raise the losses by volatilization. LVI was also employed in all matrices for VMSs analysis [38, 52, 60, 88, 102], sometimes linked to concurrent solvent recondensation (CRS) [41, 45]. Then, there are other alternatives that were only used occasionally, like headspace injection (HS-GC-MS) for water samples [55, 93], where the samples are heated and the headspace air above them is injected, or proton transfer reactions time-of-flight mass spectrometry (PTR-TOF-MS) [81, 83] and atmospheric pressure chemical ionization with tandem mass spectrometry (APCI-MS/MS) [82] for air. Contrary to all the other options, where LODs ranged from pg to low ng levels (regardless of the chromatographic technique employed), chemical ionization yielded much higher values for D4 and D5 (4,000–6,000 ng m^{-3}). This approach is easy to use and allows direct sampling of gaseous samples and can be a valid choice for more concentrated VMSs gaseous samples, with very good recoveries of 93–99%. All these options have their strong and weak points, but the main challenge is to have a low detection without external contamination introduced by the chromatograph parts, as mentioned previously. And as can be seen in Tables 1, 2, 3, 4 and 5, low bleed or inert capillary columns (or with guard columns) play an important part in that process.

3 Conclusions and Perspectives

The challenge of VMS analysis in environmental matrices is how to lower even further the LODs while avoiding the external contaminations that will ultimately compromise them. As reported in this chapter, all the analytical protocols employed have this issue in high consideration. The reduction of sampling handling steps is crucial, also to avoid losses by volatilization of these SVOCs. Relying almost exclusively in GC-MS quantification, techniques like thermal desorption or LVI were used in all sample types studied and marginally also HS-GC-MS or PTR-TOF and chemical ionization. When extraction procedures were included, the choices were in line with the matrix to analyse. For solid matrices (including the filters and sorbents used in active and passive air sampling), SLE was the predominant choice, while for water samples, purge and trap and LLE predominated, and in air samples, SPE was the most used technique. Other extraction solutions resorted to sonication or pressurized liquid extraction equipment, and QuEChERS clean-up was also attempted. A wide range of LODs/LOQs (from the ng to the pg level) and recoveries (from 20 to 144%) were achieved depending on the protocols, matrices and VMS congeners analysed. Future perspectives include the presence of larger VMS molecules more often in the analytical protocols or the possibility of using and developing faster and portable equipment for an accurate and expedite detection on the field. Chemical ionization devices may be a solution, since until now the development of sensor-like approaches was not reported, probably due to the fact that most of them are based on silicon materials.

Acknowledgements This work was also financially supported by (1) Project UID/EQU/00511/2019 'Laboratory for Process Engineering, Environment, Biotechnology and Energy – LEPABE' funded by national funds through FCT/MCTES (PIDDAC); (2) Projects POCI-01-0145-FEDER-029425 'AGRONAUT - Agronomic impact of sludge amendment using a comprehensive exposure viewpoint' and POCI-01-0145-FEDER-032084 'LANSILOT - LAunching New SILOxane Treatments: assessing effluent, sludge and air quality and improving biogas production in WWTPs', funded by FEDER funds through COMPETE2020 – Programa Operacional Competitividade e Internacionalização (POCI) and by national funds (PIDDAC) through FCT/MCTES; (3) Project 'LEPABE-2-ECO-INNOVATION – NORTE-01-0145-FEDER-000005', funded by Norte Portugal Regional Operational Programme (NORTE 2020), under PORTUGAL 2020 Partnership Agreement, through the European Regional Development Fund (ERDF). V. Homem acknowledges the Assistant Researcher contract (Individual Scientific Employment Stimulus 2017 - CEECIND/00676/2017) funded by Fundação para a Ciência e Tecnologia (FCT).

References

1. Rucker C, Kummerer K (2015) Environmental chemistry of organosiloxanes. Chem Rev 115:466–524
2. Horii Y, Kannan K (2008) Survey of organosilicone compounds, including cyclic and linear siloxanes, in personal-care and household products. Arch Environ Contam Toxicol 55:701–710
3. Lu Y, Yuan T, Wang W, Kannan K (2011) Concentrations and assessment of exposure to siloxanes and synthetic musks in personal care products from China. Environ Pollut 159:3522–3528
4. Dudzina T, von Goetz N, Bogdal C, Biesterbos JW, Hungerbuhler K (2014) Concentrations of cyclic volatile methylsiloxanes in European cosmetics and personal care products: prerequisite for human and environmental exposure assessment. Environ Int 62:86–94
5. Lassen C, Hansen CL, Mikkelsen SH, Maag J (2005) Siloxanes – consumption, toxicity and alternatives. Danish Environmental Protection Agency, Odense
6. Capela D, Ratola N, Alves A, Homem V (2017) Volatile methylsiloxanes through wastewater treatment plants – a review of levels and implications. Environ Int 102:9–29
7. Li B, Li WL, Sun SJ, Qi H, Ma WL, Liu LY, Zhang ZF, Zhu NZ, Li YF (2016) The occurrence and fate of siloxanes in wastewater treatment plant in Harbin, China. Environ Sci Pollut Res Int 23:13200–13209
8. Bletsou AA, Asimakopoulos AG, Stasinakis AS, Thomaidis NS, Kannan K (2013) Mass loading and fate of linear and cyclic siloxanes in a wastewater treatment plant in Greece. Environ Sci Technol 47:1824–1832
9. Sanchis J, Martinez E, Ginebreda A, Farre M, Barcelo D (2013) Occurrence of linear and cyclic volatile methylsiloxanes in wastewater, surface water and sediments from Catalonia. Sci Total Environ 443:530–538
10. Liu N, Shi Y, Li W, Xu L, Cai Y (2014) Concentrations and distribution of synthetic musks and siloxanes in sewage sludge of wastewater treatment plants in China. Sci Total Environ 476-477:65–72
11. Lee S, Moon HB, Song GJ, Ra K, Lee WC, Kannan K (2014) A nationwide survey and emission estimates of cyclic and linear siloxanes through sludge from wastewater treatment plants in Korea. Sci Total Environ 497-498:106–112
12. Mojsiewicz-Pienkowska K, Krenczkowska D (2018) Evolution of consciousness of exposure to siloxanes-review of publications. Chemosphere 191:204–217

13. Sparham C, van Egmond R, Hastie C, O'Connor S, Gore D, Chowdhury N (2011) Determination of decamethylcyclopentasiloxane in river and estuarine sediments in the UK. J Chromatogr A 1218:817–823
14. Zhang Z, Qi H, Ren N, Li Y, Gao D, Kannan K (2011) Survey of cyclic and linear siloxanes in sediment from the Songhua River and in sewage sludge from wastewater treatment plants, Northeastern China. Arch Environ Contam Toxicol 60:204–211
15. Sanchis J, Cabrerizo A, Galban-Malagon C, Barcelo D, Farre M, Dachs J (2015) Unexpected occurrence of volatile dimethylsiloxanes in Antarctic soils, vegetation, phytoplankton, and krill. Environ Sci Technol 49:4415–4424
16. Ratola N, Ramos S, Homem V, Silva JA, Jimenez-Guerrero P, Amigo JM, Santos L, Alves A (2016) Using air, soil and vegetation to assess the environmental behaviour of siloxanes. Environ Sci Pollut Res Int 23:3273–3284
17. Borga K, Fjeld E, Kierkegaard A, McLachlan MS (2012) Food web accumulation of cyclic siloxanes in Lake Mjosa, Norway. Environ Sci Technol 46:6347–6354
18. Wang DG, de Solla SR, Lebeuf M, Bisbicos T, Barrett GC, Alaee M (2017) Determination of linear and cyclic volatile methylsiloxanes in blood of turtles, cormorants, and seals from Canada. Sci Total Environ 574:1254–1260
19. Kierkegaard A, Bignert A, McLachlan MS (2013) Cyclic volatile methylsiloxanes in fish from the Baltic Sea. Chemosphere 93:774–778
20. Sanchis J, Llorca M, Pico Y, Farre M, Barcelo D (2016) Volatile dimethylsiloxanes in market seafood and freshwater fish from the Xuquer River, Spain. Sci Total Environ 545-546:236–243
21. Ahrens L, Harner T, Shoeib M (2014) Temporal variations of cyclic and linear volatile methylsiloxanes in the atmosphere using passive samplers and high-volume air samplers. Environ Sci Technol 48:9374–9381
22. Yucuis RA, Stanier CO, Hornbuckle KC (2013) Cyclic siloxanes in air, including identification of high levels in Chicago and distinct diurnal variation. Chemosphere 92:905–910
23. Pieri F, Katsoyiannis A, Martellini T, Hughes D, Jones KC, Cincinelli A (2013) Occurrence of linear and cyclic volatile methyl siloxanes in indoor air samples (UK and Italy) and their isotopic characterization. Environ Int 59:363–371
24. Krogseth IS, Kierkegaard A, McLachlan MS, Breivik K, Hansen KM, Schlabach M (2013) Occurrence and seasonality of cyclic volatile methyl siloxanes in Arctic air. Environ Sci Technol 47:502–509
25. Kierkegaard A, McLachlan MS (2013) Determination of linear and cyclic volatile methylsiloxanes in air at a regional background site in Sweden. Atmos Environ 80:322–329
26. Arespacochaga N, Valderrama C, Raich-Montiu J, Crest M, Mehta S, Cortina JL (2015) Understanding the effects of the origin, occurrence, monitoring, control, fate and removal of siloxanes on the energetic valorization of sewage biogas – a review. Renew Sust Energ Rev 52:366–381
27. Velicogna J, Ritchie E, Princz J, Lessard ME, Scroggins R (2012) Ecotoxicity of siloxane D5 in soil. Chemosphere 87:77–83
28. McKim JM, Wilga PC, Kolesar GB, Choudhuri S, Madan A, Dochterman LW, Breen JG, Parkinson A, Mast RW, Meeks RG (1998) Evaluation of octamethylcyclotetrasiloxane(D4) as an inducer of rat hepatic microsomal cyto-chrome P450, UDP-glucuronosyl transferase, and epoxide hydrolase: a 28-day inhalation study. Toxicol Sci 41:29–41
29. Brooke DN, Crookes MJ, Gray D, Robertson S (2009) Environmental risk assessment report: Octamethylcyclotetrasiloxane. Environment Agency of England and Wales, Bristol
30. U.S.E.P. Agency (2012) Quality assurance, quality control, and quality assessment measures, in: water: monitoring & assessment. United States Environmental Protection Agency, Washington, D.C

31. Chainet F, Lienemann C-P, Courtiade M, Ponthus J, Xavier Donard OF (2011) Silicon speciation by hyphenated techniques for environmental, biological and industrial issues: a review. J Anal At Spectrom 26:30–51
32. Capela D, Alves A, Homem V, Santos L (2016) From the shop to the drain – volatile methylsiloxanes in cosmetics and personal care products. Environ Int 92-93:50–62
33. Varaprath S, Stutts DH, Kozerski GE (2006) A primer on the analytical aspects of silicones at trace levels-challenges and artifacts – a review. Silicon Chem 3:79–102
34. Warner NA, Evenset A, Christensen G, Gabrielsen GW, Borga K, Leknes H (2010) Volatile siloxanes in the European Arctic: assessment of sources and spatial distribution. Environ Sci Technol 44:7705–7710
35. Xu L, Shi Y, Cai Y (2013) Occurrence and fate of volatile siloxanes in a municipal wastewater treatment plant of Beijing, China. Water Res 47:715–724
36. Xu L, Shi Y, Wang T, Dong Z, Su W, Cai Y (2012) Methyl siloxanes in environmental matrices around a siloxane production facility, and their distribution and elimination in plasma of exposed population. Environ Sci Technol 46:11718–11726
37. Liu N, Xu L, Cai Y (2018) Methyl siloxanes in barbershops and residence indoor dust and the implication for human exposures. Sci Total Environ 618:1324–1330
38. Wang DG, Steer H, Tait T, Williams Z, Pacepavicius G, Young T, Ng T, Smyth SA, Kinsman L, Alaee M (2013) Concentrations of cyclic volatile methylsiloxanes in biosolid amended soil, influent, effluent, receiving water, and sediment of wastewater treatment plants in Canada. Chemosphere 93:766–773
39. Krogseth IS, Whelan MJ, Christensen GN, Breivik K, Evenset A, Warner NA (2017) Understanding of cyclic volatile methyl siloxane fate in a high latitude Lake is constrained by uncertainty in organic carbon-water partitioning. Environ Sci Technol 51:401–409
40. Companioni-Damas EY, Santos FJ, Galceran MT (2012) Analysis of linear and cyclic methylsiloxanes in water by headspace-solid phase microextraction and gas chromatography-mass spectrometry. Talanta 89:63–69
41. Companioni-Damas EY, Santos FJ, Galceran MT (2012) Analysis of linear and cyclic methylsiloxanes in sewage sludges and urban soils by concurrent solvent recondensation-large volume injection-gas chromatography-mass spectrometry. J Chromatogr A 1268:150–156
42. Ramos S, Silva JA, Homem V, Cincinelli A, Santos L, Alves A, Ratola N (2016) Solvent-saving approaches for the extraction of siloxanes from pine needles, soils and passive air samplers. Anal Methods 8:5378–5387
43. Companioni-Damas EY (2017) Problemas en el análisis de metilsiloxanos volátiles (VMS): origen y soluciones. Química Nova 40:192–199
44. Varaprath S, Seaton M, McNett D, Cao L, Plotzke KP (2000) Quantitative determination of Octamethylcyclotetrasiloxane (D4) in extracts of biological matrices by gas chromatography-mass spectrometry. Int J Environ Anal Chem 77:203–219
45. Companioni-Damas EY, Santos FJ, Galceran MT (2014) Linear and cyclic methylsiloxanes in air by concurrent solvent recondensation-large volume injection-gas chromatography-mass spectrometry. Talanta 118:245–252
46. Zeeuw JD (2005) How to minimize septum problems in GC. Am Lab 37:18–19
47. Cortada C, dos Reis LC, Vidal L, Llorca J, Canals A (2014) Determination of cyclic and linear siloxanes in wastewater samples by ultrasound-assisted dispersive liquid-liquid microextraction followed by gas chromatography-mass spectrometry. Talanta 120:191–197
48. Horii Y, Minomo K, Ohtsuka N, Motegi M, Nojiri K, Kannan K (2017) Distribution characteristics of volatile methylsiloxanes in Tokyo Bay watershed in Japan: analysis of surface waters by purge and trap method. Sci Total Environ 586:56–65

49. Knoerr SM, Durham JA, McNett DA (2017) Development of collection, storage and analysis procedures for the quantification of cyclic volatile methylsiloxanes in wastewater treatment plant effluent and influent. Chemosphere 182:114–121

50. Warner NA, Kozerski G, Durham J, Koerner M, Gerhards R, Campbell R, McNett DA (2013) Positive vs. false detection: a comparison of analytical methods and performance for analysis of cyclic volatile methylsiloxanes (cVMS) in environmental samples from remote regions. Chemosphere 93:749–756

51. Powell DE, Woodburn KB (2009) Trophic Dilution of Cyclic Volatile Methylsiloxanes (cVMS) materials in a temperate freshwater lake. Health and Environmental Sciences/Dow Corning Corporation, Auburn, pp 1–61

52. Kierkegaard A, van Egmond R, McLachlan MS (2011) Cyclic volatile methylsiloxane bioaccumulation in flounder and ragworm in the Humber estuary. Environ Sci Technol 45:5936–5942

53. Chambers DM, McElprang DO, Mauldin JP, Hughes TM, Blount BC (2005) Identification and elimination of polysiloxane curing agent interference encountered in the quantification of low-picogram per milliliter methyl tert-butyl ether in blood by solid-phase microextraction headspace analysis. Anal Chem 77:2912–2919

54. Wang YX (2006) How pierced PTFE/silicone septa affect GC-MS experiments. Am Lab 38:10–12

55. Sparham C, Van Egmond R, O'Connor S, Hastie C, Whelan M, Kanda R, Franklin O (2008) Determination of decamethylcyclopentasiloxane in river water and final effluent by headspace gas chromatography/mass spectrometry. J Chromatogr A 1212:124–129

56. Hong WJ, Jia H, Liu C, Zhang Z, Sun Y, Li YF (2014) Distribution, source, fate and bioaccumulation of methyl siloxanes in marine environment. Environ Pollut 191:175–181

57. Tran TM, Kannan K (2015) Occurrence of cyclic and linear siloxanes in indoor air from Albany, New York, USA, and its implications for inhalation exposure. Sci Total Environ 511:138–144

58. Tran TM, Abualnaja KO, Asimakopoulos AG, Covaci A, Gevao B, Johnson-Restrepo B, Kumosani TA, Malarvannan G, Minh TB, Moon HB, Nakata H, Sinha RK, Kannan K (2015) A survey of cyclic and linear siloxanes in indoor dust and their implications for human exposures in twelve countries. Environ Int 78:39–44

59. Lee SY, Lee S, Choi M, Kannan K, Moon HB (2018) An optimized method for the analysis of cyclic and linear siloxanes and their distribution in surface and core sediments from industrialized bays in Korea. Environ Pollut 236:111–118

60. Kierkegaard A, McLachlan MS (2010) Determination of decamethylcyclopentasiloxane in air using commercial solid phase extraction cartridges. J Chromatogr A 1217:3557–3560

61. Shi Y, Xu S, Xu L, Cai Y (2015) Distribution, elimination, and rearrangement of cyclic volatile Methylsiloxanes in oil-contaminated soil of the Shengli oilfield, China. Environ Sci Technol 49:11527–11535

62. Krogseth IS, Zhang X, Lei YD, Wania F, Breivik K (2013) Calibration and application of a passive air sampler (XAD-PAS) for volatile methyl siloxanes. Environ Sci Technol 47:4463–4470

63. Zhang Y, Shen M, Tian Y, Zeng G (2018) Cyclic volatile methylsiloxanes in sediment, soil, and surface water from Dongting Lake, China. J Soils Sediments 18:2063–2071

64. Gallego E, Perales JF, Roca FJ, Guardino X, Gadea E (2017) Volatile methyl siloxanes (VMS) concentrations in outdoor air of several Catalan urban areas. Atmos Environ 155:108–118

65. Cheng Y, Shoeib M, Ahrens L, Harner T, Ma J (2011) Wastewater treatment plants and landfills emit volatile methyl siloxanes (VMSs) to the atmosphere: investigations using a new passive air sampler. Environ Pollut 159:2380–2386

66. Wang DG, Du J, Pei W, Liu Y, Guo M (2015) Modeling and monitoring cyclic and linear volatile methylsiloxanes in a wastewater treatment plant using constant water level sequencing batch reactors. Sci Total Environ 512–513:472–479

67. Warner NA, Nost TH, Andrade H, Christensen G (2014) Allometric relationships to liver tissue concentrations of cyclic volatile methyl siloxanes in Atlantic cod. Environ Pollut 190:109–114

68. Zhi L, Xu L, He X, Zhang C, Cai Y (2019) Distribution of methylsiloxanes in benthic mollusks from the Chinese Bohai Sea. J Environ Sci (China) 76:199–207

69. Bakke T, Boitsov S, Brevik EM, Gabrielsen GW, Green N, Helgason LB, Klungsøyr J, Leknes H, Miljeteig C, Måge A, Rolfsnes BE, Savonova T, Schlabach M, Skaage BB, Valdersnes S (2008) Mapping selected organic contaminants in the Barents Sea 2007. SPFO-report 1021/2008, TA-2400/2008. Norwegian Pollution Control Authority, Oslo

70. Powell DE, Schoyen M, Oxnevad S, Gerhards R, Bohmer T, Koerner M, Durham J, Huff DW (2018) Bioaccumulation and trophic transfer of cyclic volatile methylsiloxanes (cVMS) in the aquatic marine food webs of the Oslofjord, Norway. Sci Total Environ 622–623:127–139

71. Powell DE, Suganuma N, Kobayashi K, Nakamura T, Ninomiya K, Matsumura K, Omura N, Ushioka S (2017) Trophic dilution of cyclic volatile methylsiloxanes (cVMS) in the pelagic marine food web of Tokyo Bay, Japan. Sci Total Environ 578:366–382

72. Nusz JB, Fairbrother A, Daley J, Burton GA (2018) Use of multiple lines of evidence to provide a realistic toxic substances control act ecological risk evaluation based on monitoring data: D4 case study. Sci Total Environ 636:1382–1395

73. Moeckel CH, Harner T, Nizzetto L, Strandberg B, Lindroth A, Jones KC (2009) Use of depuration compounds in passive air samplers: results from active sampling supported field deployment, potential uses, and recommendations. Environ Sci Technol 43:3227–3232

74. Holt E, Bohlin-Nizzetto P, Boruvkova J, Harner T, Kalina J, Melymuk L, Klanova J (2017) Using long-term air monitoring of semi-volatile organic compounds to evaluate the uncertainty in polyurethane-disk passive sampler-derived air concentrations. Environ Pollut 220:1100–1111

75. Tran TM, Le HT, Vu ND, Minh Dang GH, Minh TB, Kannan K (2017) Cyclic and linear siloxanes in indoor air from several northern cities in Vietnam: levels, spatial distribution and human exposure. Chemosphere 184:1117–1124

76. Tran TM, Tu MB, Vu ND (2018) Cyclic siloxanes in indoor environments from hair salons in Hanoi, Vietnam: emission sources, spatial distribution, and implications for human exposure. Chemosphere 212:330–336

77. Pellizzari ED, Bunch JE, Berkley RE, McRae J (1976) Determination of trace hazardous organic vapor pollutants in ambient atmospheres by gas chromatography/mass spectrometry/computer. Anal Chem 48:803–807

78. Shoeib M, Schuster J, Rauert C, Su K, Smyth SA, Harner T (2016) Emission of poly and perfluoroalkyl substances, UV-filters and siloxanes to air from wastewater treatment plants. Environ Pollut 218:595–604

79. Rauert C, Shoieb M, Schuster JK, Eng A, Harner T (2018) Atmospheric concentrations and trends of poly- and Perfluoroalkyl Substances (PFAS) and Volatile Methyl Siloxanes (VMS) over 7 years of sampling in the Global Atmospheric Passive Sampling (GAPS) network. Environ Pollut 238:94–102

80. Shields HC, Fleischer DM, Weschler CJ (1996) Comparisons among VOCs measured in three types of U.S. commercial buildings with different occupant densities. Indoor Air 6:2–17

81. Coggon MM, McDonald B, Vlasenko A, Veres P, Bernard F, Koss AR, Yuan B, Gilman JB, Peischl J, Aikin KC, DuRant J, Warneke C, Li S-M, de Gouw JA (2018) Diurnal variability and emission pattern of decamethylcyclopentasiloxane (D5) from the application of personal care products in two North American cities. Environ Sci Technol 52:5610

82. Badjagbo K, Furtos A, Alaee M, Moore S, Sauvé S (2009) Direct analysis of volatile methylsiloxanes in gaseous matrixes using atmospheric pressure chemical ionization-tandem mass spectrometry. Anal Chem 81:7288–7293

83. Tang X, Misztal PK, Nazaroff WW, Goldstein AH (2015) Siloxanes are the Most abundant volatile organic compound emitted from engineering students in a classroom. Environ Sci Technol Lett 2:303–307

84. Lu Y, Yuan T, Yun SH, Wang W, Wu Q, Kannan K (2010) Occurrence of cyclic and linear siloxanes in indoor dust from China and implications for human exposures. Environ Sci Technol 44:6081–6087

85. Ribeiro C, Ribeiro AR, Maia AS, Goncalves VM, Tiritan ME (2014) New trends in sample preparation techniques for environmental analysis. Crit Rev Anal Chem 44:142–185

86. Parker WJ, Shi J, Fendinger NJ, Monteith HD, Chandra G (1999) Pilot plant study to assess the fate of two volatile Methylsiloxane compounds during municipal wastewater treatment. Environ Toxicol Chem 18:172–181

87. Wang DG, Aggarwal M, Tait T, Brimble S, Pacepavicius G, Kinsman L, Theocharides M, Smyth SA, Alaee M (2015) Fate of anthropogenic cyclic volatile methylsiloxanes in a wastewater treatment plant. Water Res 72:209–217

88. Wang DG, Alaee M, Steer H, Tait T, Williams Z, Brimble S, Svoboda L, Barresi E, Dejong M, Schachtschneider J, Kaminski E, Norwood W, Sverko E (2013) Determination of cyclic volatile methylsiloxanes in water, sediment, soil, biota, and biosolid using large-volume injection-gas chromatography-mass spectrometry. Chemosphere 93:741–748

89. Kaj L, Andersson J, Cousins AP, Remberger M, Brorström-Lundén E, Cato I (2004) Results from the Swedish national screening program - subreport 4: siloxanes. Swedish Environmental Research Institute, Stockholm

90. Kaj L, Schlabach M, Andersson J, Cousins AP, Schmidbauer N, Brorström-Lundén E (2005) Siloxanes in the Nordic environment. TemaNord/Nordic Council of Ministers, Copenhagen

91. Schlabach M, Andersen MS, Green N, Schøyen M, Kaj L (2007) Siloxanes in the environment of the inner Oslofjord. Norsk institutt for luftforskning (NILU), Kjeller

92. Souza Silva EA, Risticevic S, Pawliszyn J (2013) Recent trends in SPME concerning sorbent materials, configurations and in vivo applications. TrAC Trend Anal Chem 43:24–36

93. van Egmond R, Sparham C, Hastie C, Gore D, Chowdhury N (2013) Monitoring and modelling of siloxanes in a sewage treatment plant in the UK. Chemosphere 93:757–765

94. Oshita K, Omori K, Takaoka M, Mizuno T (2014) Removal of siloxanes in sewage sludge by thermal treatment with gas stripping. Energ Convers and Manage 81:290–297

95. Dewil R, Appels L, Baeyens J, Buczynska A, Van Vaeck L (2007) The analysis of volatile siloxanes in waste activated sludge. Talanta 74:14–19

96. Tavazzi S, Locoro G, Comero S, Sobiecka E, Loos R, Gans O, Ghiani M, Umlauf G, Suurkuusk G, Paracchini B, Cristache C, Fissiaux I, Riuz AA, Gawlik BM (2012) Occurrence and levels of selected compounds in European sewage sludge samples - results of a pan-European screening exercise (FATE SEES). J.R.C.o.t.E. Commission, Luxembourg

97. Evenset A, Leknes H, Christensen GN, Warner N, Remberger M, Gabrielsen GW (2009) Screening of new contaminants in samples from Norwegian Arctic - silver, platinum, sucralose, bisphenol A, Tetrabrombisphenol A, siloxanes, Phtalates (DEHP) and Phosphororganic flame retardants. Akvaplan-niva, Tromsø

98. Sanchez-Brunete C, Miguel E, Albero B, Tadeo JL (2010) Determination of cyclic and linear siloxanes in soil samples by ultrasonic-assisted extraction and gas chromatography-mass spectrometry. J Chromatogr A 1217:7024–7030

99. Meng F, Wu H (2015) Indoor air pollution by Methylsiloxane in household and automobile settings. PLoS One 10:e0135509

100. Genualdi S, Harner T, Cheng Y, Macleod M, Hansen KM, van Egmond R, Shoeib M, Lee SC (2011) Global distribution of linear and cyclic volatile methyl siloxanes in air. Environ Sci Technol 45:3349–3354

101. Wang XM, Lee SC, Sheng GY, Chan LY, Fu JM, Li XD, Min YS, Chan CY (2001) Cyclic organosilicon compounds in ambient air in Guangzhou, Macau and Nanhai, Pearl River Delta. J Appl Geochem 16:1447–1454

102. McLachlan MS, Kierkegaard A, Hansen KM, Egmond RV, Christensen JH, SkØth CA (2010) Concentrations and fate of Decamethylcyclopentasiloxane (D5) in the atmosphere. Environ Sci Technol 44:5365–5370

103. Buser AM, Kierkegaard A, Bogdal C, MacLeod M, Scheringer M, Hungerbuhler K (2013) Concentrations in ambient air and emissions of cyclic volatile methylsiloxanes in Zurich, Switzerland. Environ Sci Technol 47:7045–7051

104. Hodgson AT, Faulkner D, Sullivan DP, DiBartolomeo DL, Russell ML, Fisk WJ (2003) Effect of outside air ventilation rate on volatile organic compound concentrations in a call center. Atmos Environ 37:5517–5527

105. Katsoyiannis A, Anda EE, Cincinelli A, Martellini T, Leva P, Goetsch A, Sandanger TM, Huber S (2014) Indoor air characterization of various microenvironments in the Arctic. The case of Tromso, Norway. Environ Res 134:1–7

106. Cincinelli A, Martellini T, Amore A, Dei L, Marrazza G, Carretti E, Belosi F, Ravegnani F, Leva P (2016) Measurement of volatile organic compounds (VOCs) in libraries and archives in Florence (Italy). Sci Total Environ 572:333–339

107. de Hoffmann E, Stroobant V (2007) Mass spectrometry. Wiley, Hoboken

Fate of Volatile Methylsiloxanes in Wastewater Treatment Plants

De-Gao Wang and Mehran Alaee

Contents

Abstract Concentration, distribution, fate, removal efficiencies, daily and seasonal variations of volatile methylsiloxanes (VMS) in wastewater treatment plants (WWTPs) were reviewed in this chapter. Purge-and-trap, headspace, liquid-liquid extraction, liquid-solid extraction, membrane-assisted solvent extraction, and modified QuEChERS methods have been developed to analyze of VMS in samples from WWTPs. The different consumption quantities of commercial products containing siloxanes result in the difference of concentrations and proportion of VMS in the world. Daily fluctuations of VMS concentrations in water usage induce in flow variation to WWTP and VMS show seasonal variation in the WWTPs with different types of processes. In cold seasons, VMS prefer to stay in water phase rather than air or sludge because the air/water and organic carbon/water partition coefficients decrease with temperature. Although most WWTP remove siloxanes efficiently, long-term environmental monitoring of VMS is necessary in certain environments,

D.-G. Wang
College of Environmental Science and Engineering, Dalian Maritime University, Dalian, P. R. China

M. Alaee (✉)
Water Science and Technology Directorate, Environment and Climate Change Canada, Burlington, ON, Canada
e-mail: mehran.alaee@canada.ca

V. Homem and N. Ratola (eds.), *Volatile Methylsiloxanes in the Environment*,
Hdb Env Chem (2020) 89: 119–130, DOI 10.1007/698_2018_365,
© Springer Nature Switzerland AG 2018, Published online: 6 September 2018

considering the potential of VMS to bioaccumulate in biota and its toxicity to sensitive aquatic organisms.

Keywords Removal efficiencies, Seasonal variation, Siloxanes, Temperature-dependent, WWTP

1 Introduction

The beginnings of modern wastewater treatment can be traced back to the mid-nineteenth century with the industrial revolution. As cities grew larger, human and industrial waste production increased as did the outbreak of several diseases such as cholera [1]. Consequently, wastewater management became vital. In the early twentieth century, it was discovered that wastewater had high levels of suspended solids and biochemical oxygen demand (BOD) which resulted in deleterious and anoxic conditions for fish in receiving waters [2]. As a result, the treatment of solids and organic material from wastewater was added to the initial processes of sewage collection and disposal. In the following paragraphs, a very short description of the different types of wastewater treatment systems is presented; however, the reader, if interested in learning more about wastewater treatment systems, is encouraged to consult several text books including Metcalf and Eddy [3].

Lagoons are the earliest and simplest and cost-effective form of wastewater treatment. In this process, wastewater is held for a long period of time until dissolved organic matter has been decomposed by bacteria and suspended solids settle in the bottom of the lagoon. Wastewater from lagoons is released at regular intervals once the water meets discharge criteria. The efficiency of lagoons can be improved when several lagoons are joined together in serial or parallel arrangements or when mechanical aeration is added to enhance the otherwise facultative bacterial degradation processes. The main drawback of lagoons is that they occupy a large amount of land and are not amenable to urban centers. Environmental conditions and climate greatly influence the selection of lagoon type and treatment efficiency. In addition, volatilization of compounds such as siloxanes is halted in colder climates when lagoon surfaces are frozen surface during winter. Lagoons are a good option for rural and agricultural communities.

With population increases in urban centers, land mass has become less available, and as a result, different types of wastewater treatment processes have been developed. Primary wastewater treatment systems are another simple form of treating wastewater and consist of a settling tank where heavier solids are removed and sometimes a coagulant is added to assist with the precipitation and removal of organic matter. Synthetic polymer (polyetherimide and polyacrylamide) and natural polymer (chitosan) and chemical (alum) coagulants were often used to remove suspended solids and to reduce chemical oxygen demand. During this process, 40–60% of the suspended solids and oxygen demand is removed from wastewater. Secondary wastewater treatment involves an additional process that takes advantage of the bacterial digestion of organic matter in an aeration tank, which results in BOD reduction. In tertiary or advanced treatment plants, depending on the sensitivity of the receiving body of water, additional steps such as nitrification, denitrification, sand filtration, chemical precipitation, or advanced oxidation are used to remove additional contaminants, including phosphorus and/or nitrogen. After this stage, the effluent is usually discharge into the natural water. Depending on the operating condition and/or the receiving body is used for recreational purposes, additional disinfection treatment may be required; this is usually accomplished by chlorination, ozonation, and/or UV radiation.

Volatile methylsiloxanes (VMS) are high-volume chemicals that consist of repeating units of $[Me_2SiO]_n$, with Si–O atoms that are singly bonded forming a chain or a ring resulting in linear or cyclic VMS. Linear (L2, L3, L4, and L5) and cyclic (D3, D4, D5 and D6) volatile methylsiloxanes are used in personal care products such as shampoos and deodorants and in industrial applications such as the production of silicone-based polymers and dry-cleaning [4]. VMS are discharged "down-the-drain" or through industrial discharge into municipal wastewater treatment systems. As a result, wastewater treatment plants (WWTPs) have been identified as a conduit for entry of VMS into the environment. Due to their volatility and hydrophobicity, VMS have a tendency to partition into air and biosolids and to a lesser extent in effluent. VMS can enter the environment via volatilization into the atmosphere, biogas generated during anaerobic digestion of sludge, wastewater effluent discharges, and application of biosolids to agricultural lands [5–8].

As a result of all these factors, some studies of VMS emissions to the aquatic and terrestrial environment through wastewater treatment have been examined and are summarized in a review by Capela et al. [5]. Many studies have explored the partitioning of compounds within municipal WWTPs and their subsequent fate during sludge treatment. Analytical methods were developed to accurately detect and quantify VMS in challenging matrices such as municipal sludge, biogas, and wastewater. The goal of this chapter is to provide a comprehensive assessment of the current state of knowledge concerning the VMS in WWTPs.

2 Analysis

Reliable analytical methods are needed to accurately detect and quantify VMS in biogas, biosolids, and wastewater. These compounds are ubiquitous in the environment, and as such analytical determinations can be quite challenging. A number of methods for analysis of VMS have been developed in previous studies [9–12]. VMS are often separated by gas chromatography (GC), which facilitates the separation of simple or complex mixtures for qualitative or quantitative analysis. A variety of detection systems including flame ionization detectors and mass spectrometers have been used to detect the separated components as they elute from the column.

Atmospheric pressure chemical ionization-tandem mass spectrometry can be used in the direct analysis of VMS in air and biogas samples to avoid the contamination from GC systems and solvent extraction [13, 14].

Purge-and-trap, headspace, liquid-liquid extraction, liquid-solid extraction, membrane-assisted solvent extraction (MASE), and modified QuEChERS (quick, easy, cheap, effective, rugged, and safe) methods have been used in the analysis of VMS in samples. Usually, a purge-and-trap and headspace procedures are used for analysis of volatile organic compounds that have boiling points below 200°C [15]. Although VMS are very volatile, their strong affinity for dissolved organic carbon can reduce the transfer rate from water to air significantly [16]. The purge-and-trap and headspace methods may be limited by the properties of high affinity to organic matter, especially for wastewater with a high organic carbon content or sludge with a high humic acid content [17].

Liquid-liquid extraction and liquid-solid both are solvent extraction methods used in VMS sample preparation. Tetrahydrofuran and *n*-hexane are typically used to extract a variety of biological and biosolid matrices. VMS extraction using tetrahydrofuran has been shown to yield an efficiency of >90% in all biological matrices due to its water miscibility [18]. Solvent extraction using *n*-hexane of VMS in biosolids along with vortexing and centrifugation also showed excellent recoveries and repeatability due to the hydrophobicity of VMS [19].

In addition, a modified QuEChERS method was developed to detect VMS in biosolids, soil, sediment, and biota sample studies [11, 12]. The QuEChERS method was shown to have higher extraction efficiencies than MASE and headspace methods for samples with high lipid.

3 VMS in Influent and Effluent of WWTP

WWTP influent has been shown to contain a significant amount of VMS from intensive use of personal care and industrial products. VMS concentrations in influent ranged from a few nanograms to a hundred micrograms per liter in various regions with domestic and/or industrial origins [20]. As noted in previous reports, the source of wastewater plays a vital role on VMS levels in influent. WWTPs receiving a significant fraction of industrial effluent may contain higher levels of VMS since they are used in different industrial processes. These sites include industrial sectors producing cosmetic and personal care products, polymers, paper, etc. [8, 21]. Wang et al. [8] reported that one influent had concentrations of D5 at 135 µg L^{-1}. This WWTP received industrial wastewater from a facility that manufactures cosmetic and personal care products that incorporate VMS. Xu et al. [21] demonstrated that effluents from paper industries also contained high levels of VMS, which also resulted in high concentrations in the receiving WWTP influents.

VMS concentrations in 15 WWTPs in Ontario, Quebec, and British Columbia were determined (Fig. 1) [22]. Average total VMS concentrations in influent and effluent were 24.8 µg L^{-1} and 1.02 µg L^{-1}, varying in the ranges of

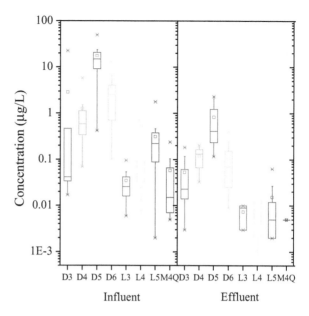

Fig. 1 Concentrations of VMS in influent and effluent from 15 Canadian WWTPs [22]

0.92–69.4 µg L^{-1} and <LOQ–2.87 µg L^{-1}, respectively. These values are similar to values reported by various studies and summarized by Capela et al. [5]. The total removal efficiency of VMS was about 96% through sorption on sludge and volatilization, indicating WWTPs can remove VMS effectively. Although concentrations in influent and effluent varied largely, the profiles in influent and effluent of various WWTPs were consistent with each other. Cyclic VMS were greater than linear VMS in influent and effluent and cyclic VMS accounted for 95% of the total VMS. For cyclic VMS, D5 was the dominant species in all influent and effluent, followed by D6, D4, and D3. Most of the D5 concentrations were >10 µg L^{-1} in influent samples and <1 µg L^{-1} in effluent samples. Most of the D6 concentrations ranged between 1 and 10 µg L^{-1} in influent and decreased to <0.1 µg L^{-1} in effluent. Most of D3 and D4 concentrations were <1 µg L^{-1} in influent samples and <0.1 µg L^{-1} in effluent samples. For linear VMS, L5 was the dominant species in all influent and effluent, followed by L4, L3, and M4Q. Most of L5 concentrations ranged between 0.1 and 1 µg L^{-1}, and the other linear VMS were <0.1 µg L^{-1} in influent. Most linear VMS were <0.01 µg L^{-1} in effluent samples. L3 and M4Q only were detected in three and one effluent samples.

4 VMS in Biosolid

During the wastewater treatment process, VMS with high K_{OW} have a tendency to partition into organic-rich sludge, which is further processed into biosolid. Mueller et al. [23] reported that concentrations of D4 ranged from 0.21 to 0.48 µg g^{-1} dw in biosolids collected by Ann Arbor Technical Services. Kaj et al. [7] determined cyclic VMS in biosolids collected from WWTPs in Sweden. D4 was detected in 37 out of 54 samples with a mean concentration of 0.39 µg g^{-1} dw. D5 and D6 were quantified in all 54 samples with mean concentrations of 9.5 and 1.3 µg g^{-1} dw, respectively. VMS showed a similar pattern of D5 > D6 > D4 in all biosolid samples [7]. Kaj et al. [7] also reported D4, D5, and D6 concentrations in all biosolid samples collected from Nordic countries. D5 was the dominant compound in all cases, making up 78–94% of the total.

Concentrations of VMS in biosolids are high because most of these compounds in influent water are removed by sorption onto the biosolid or volatilized into the atmosphere [24]. Zhang et al. [25] reported that the concentrations of D4, D5, and D6 ranged from 0.042 to 0.103 µg g^{-1} dw, 0.168 to 0.320 µg g^{-1} dw, and 0.088 to 0.569 µg g^{-1} dw in biosolid from one Chinese WWTP [25]. Liu et al. [26] reported 17 siloxanes in anaerobic digested sludge samples collected at the dewatering process from 42 WWTPs in China [26]. The mean concentrations of cyclic and linear VMS were 1.98 µg g^{-1} dw (in the range of LOQ–36.1 µg g^{-1} dw) and 0.937 µg g^{-1} dw (in the range of LOQ–13.2 µg g^{-1} dw). On average, cyclic siloxanes accounted for 68% of the total siloxanes [26]. Wang et al. [27] reported 1.69 µg g^{-1}, 67.8 µg g^{-1}, and 8.38 µg g^{-1} for D4, D5, and D6, respectively, for biosolids from a WWTP in Canada. However, high concentrations of linear VMS were found in Greece (54 µg g^{-1} dw), which is higher than those of cyclic VMS due to their high solid-liquid distribution coefficients [28]. The different consumption quantities of commercial products containing siloxanes explain the difference of concentrations and proportion among these regions. Overall, the concentrations in biosolid were in the range of <10 µg g^{-1} dw, <100 µg g^{-1} dw, and <10 µg g^{-1} dw, for D4, D5, and D6, respectively.

5 VMS in Biogas

Biogas is a by-product of the decomposition of organic matter during the anaerobic digestion of sludge. Biogas contains upward of 65% methane by volume which is a potent greenhouse gas and has the potential to explode. To eliminate these problems, biogas is collected and used as an energy source in WWTPs. During the combustion process, VMS are converted into SiO_2, a microcrystalline and abrasive material commonly known as sand which is known to damage the surfaces of combustion chambers and turbines [29].

As indicated in the previous section, a major portion of VMS entering WWTPs is partitioned into sludge, which is converted to biosolid. Due to the volatile nature of VMS, biogas generated during anaerobic digestion processes elevated levels of D4 and D5 concentrations ranging from 2,870 to 6,980 µg m^{-3}, and 2,750–9,650 µg m^{-3} have been observed [29]. Similar concentration ranges for D4 and D5 were also measured (6,300–8,200 µg m^{-3} and 9,400–15,000 µg m^{-3}, respectively) in biogas in a subsequent study [30]. Rasi et al. [31] found that D4 and

D5 levels were 30–870 µg m^{-3} and 100–1,270 µg m^{-3} in biogas. The amount of VMS in different studies varied largely with a range (100–10,000 µg m^{-3}) due to different raw materials and/or waste and wastewater treatment processes and process conditions at different sites [31]. Other VMS have been observed in biogas at lower concentrations; Tansel and Surita [32] reported 286, 260, 4,150, 1,800, and 90 µg m^{-3} for D3, L3, D4, D5, and D6, respectively. Based on free energy changes during the formation of VMS, they concluded that D4 and D5 were the most stable VMS under anaerobic digestion conditions.

6 Removal Efficiency and Mechanisms

The fate of VMS in WWTP is due to their very volatile ($\log K_{AW} > 2$) and hydrophobic ($\log K_{OW} > 6$) properties. Studies of the removal mechanisms of VMS during municipal wastewater treatment have been attributed to volatilization to air and adsorption onto sewage sludge in WWTPs. Biodegradation plays a minor role in the removal of cyclic VMS because of their low biological availability [33–36]. Volatilization is the important removal mechanism for D4, and partitioning to the particle phase was the dominant route for D5 and D6. Total suspended solid concentrations in effluent have an important influence on the emission of D5 and D6 in effluent, which can attribute to D5 and D6 having a greater organic carbon-water partition coefficient ($\log K_{OC} > 5$). The percentage of VMS in sludge increased with increasing molecular weight implying that more VMS with a higher hydrophobicity were adsorbed to the particle phase. WWTP treatment was also shown to have an influence on the removal efficiencies of cyclic VMS. The aeration process played a key role in removing VMS from the WWTPs. Taking into account temperature effects on K_{AW} and K_{OC}, VMS have a higher preference to remain in the water phase rather than in the air or sludge phases during the cold season. In addition, WWTP without aeration have lower removal efficiencies than those using aeration systems, and these are further reduced in the cold season. The average overall removal efficiencies were over 50% in most studies [37].

7 Daily and Weekly Variations

Daily fluctuation of VMS concentrations in water usage resulting in flow variation to WWTP has been reported, and similarly, the concentration and flux of VMS in influent varied with the progression of the day. Wang et al. [27] investigated the variability of VMS loading over the course of a day; mass input estimates were obtained from the influent concentrations and flow rates. Significant diurnal variability of VMS was observed to be caused by the influent flows. Morning and evening high mass flows contributed approximately equally to the total mass of VMS (80%) [27]. D5 can be considered an indicator of the manufacturing and consumption of cosmetic and personal care products, and, therefore, the influent mass flow of D5 may reflect the hygiene and household habits of the population. In this study, the morning and evening high mass flows of D5 accounted for 37% and 40% of total daily mass, respectively. The daily variation showed that the lowest concentrations were detected at night and the highest in the late morning. D5 can be considered an indicator of the manufacturing and consumption of cosmetic and personal care products; therefore, the influent mass flow of D5, corrected for industrial input, may reflect the hygiene habits of a population. This result implied that people frequently take showers in the morning and evening compared to the rest of the day. However, VMS showed no significant weekly pattern between weekends and working days [38].

8 Seasonal Variation

Wang et al. [22] investigated the seasonal behavior of VMS in WWTPs during warm and cold seasons in a 3-year sampling campaign in Canada. Concentrations in influent were seasonally dependent in these WWTPs and reached maximums in the WWTPs where the domestic wastewater was dominant in the warm season. The concentrations in influent ranged from 10.2 to 156 µg L^{-1} and from 4 to 6,510 µg L^{-1} in warm and cold seasons, respectively (Fig. 2). However, for the WWTP affected by industrial effluent with high VMS concentrations, the influent concentrations were high in cold seasons, when the domestic wastewater flow decreased. In comparison with the influents, VMS showed opposite seasonal variation patterns in the WWTP effluents: lower concentrations in the warm season and higher concentrations in the cold season. VMS concentrations notably increase in effluent due to lower partition coefficients of air/water and biomass/water resulting

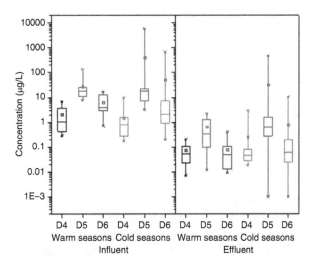

Fig. 2 Concentrations of VMS in all influent and effluent samples collected from 14 WWTPs in Southern Ontario and Southern Quebec in warm and cold seasons during 2010–2012 [22]

in lower volatilization and sludge adsorption. The VMS concentrations in effluent varied from 0.066 to 2.73 µg L^{-1} and 0.055 to 486 µg L^{-1} in warm and cold seasons, respectively.

Seasonal removal efficiencies were observed in various types of WWTPs including secondary activated sludge (SAS), aerated lagoon (ALA), facultative lagoon (FLA), and chemically assisted primary (CAP) (Fig. 3). Removal efficiencies were greater than 70% of cyclic VMS for the four treatment types during the warm season. During the cold season, the removal efficiencies decreased with decreasing water temperature. Temperature had greater effect on the FLA and CAP than SAS and ALA. The removal efficiencies of cyclic VMS in SAS and ALA were higher than in FLA and CAP in the cold season due to SAS and ALA, both having aeration processes. Aeration tanks play a key role in cyclic VMS removal.

In winter, VMS prefer to stay in the water phase rather than in the air or sludge phases because the K_{AW} and K_{OC} values decrease with decreasing the temperature. Temperature also has an effect on the partition processes of the air-water-biomass system. Therefore, the removal mechanism of VMS during municipal wastewater treatment, which is controlled by volatilization into the air and adsorption onto sewage sludge in WWTPs, is also affected [23, 24, 39]. VMS concentrations notably increase in effluent due to lower partition coefficients of air/water and biomass/water resulting in lower volatilization and sludge adsorption (Fig. 4). The values of K_{AW} and K_{OC} are a strong function of temperature (Fig. 4), with increases of 66–179 and 6–10 times for K_{AW} and K_{OC}, respectively, for a temperature increase from 0 to 40°C. These results suggest that VMS preferred to enter into air or the organic phase when the temperature increased and preferred to stay in the water phase in cold seasons, resulting in lower removal efficiency in cold seasons than in warm seasons. Therefore, FLA and CAP without aeration have lower removal efficiencies than the SAS and ALA in winter.

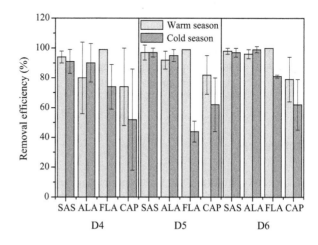

Fig. 3 Comparison of removal efficiencies in warm and cold seasons in four types of WWTPs: secondary activated sludge (SAS), aerated lagoon (ALA), facultative lagoon (FLA), and chemically assisted primary (CAP) [22]

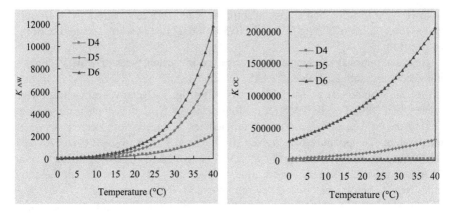

Fig. 4 Temperature dependences of the air/water partition coefficient (K_{AW}) and organic carbon/water partition coefficient (K_{OC}) for cyclic VMS from 0 to 40°C [22]

9 Conclusions

VMS can be found in biogas, biosolids, influent, and effluent from most WWTPs in different countries. A great variation in concentrations of VMS were found in biogas ($100-10,000$ μg m^{-3}), in biosolids ($0.1-100$ μg g^{-1} dw), in influent (<200 μg L^{-1}), and in effluent (<10 μg L^{-1}). It seems that the VMS in effluent have little influence on the aquatic organisms based on present information. The removal mechanism of VMS during municipal wastewater treatment is controlled by volatilization into the air and adsorption onto sewage sludge. Biodegradation plays a very small role in the removal because their low water solubility is thought to limit their biological availability. Temperature would also affect the partition processes of the air-water-biomass system. In general WWTPs have higher removal efficiencies of VMS in warm seasons, and seasonal variation of removal efficiencies occurred in various types of WWTPs in cold season, depending on the type of treatment types employed. However, considering the potential of VMS to bioaccumulate in biota and its toxicity to sensitive aquatic organisms, long-term environmental monitoring of VMS is necessary in certain environments.

References

1. Lofrano G, Brown J (2010) Wastewater management through the ages: a history of mankind. Sci Total Environ 408:5254–5264
2. Royal Commission on Sewage Disposal Eighth Report (1912) Standards and tests for sewage and sewage effluent discharging to rivers and streams. Cd 6464, London
3. Tchobanoglous G, Metcalf & Eddy (1991) Wastewater engineering treatment, disposal and reuse. Water resources and environmental engineering, vol 73. McGraw-Hill, New York, pp 50–51

4. Horii Y, Kannan K (2008) Survey of organosilicone compounds, including cyclic and linear siloxanes, in personal-care and household products. Arch Environ Contam Toxicol 55:701–710
5. Capela D, Ratola N, Alves A, Homem V (2017) Volatile methylsiloxanes through wastewater treatment plants - a review of levels and implications. Environ Int 102:9–29
6. Dewil R, Appels L, Baeyens J (2006) Energy use of biogas hampered by the presence of siloxanes. Energy Conver Manage 47:1711–1722
7. Kaj L, Andersson J, Palm Cousins A, Schmidbauer N, Brorstrom-Lunden E, Cato I (2005) Results from the Swedish national screening programme 2004. Subreport 4: siloxanes. IVL Swedish Environmental Research Institute, Stockholm
8. Wang D-G, Steer H, Tait T, Williams Z, Pacepavicius G, Young T, Ng T, Smyth SA, Kinsman L, Alaee M (2013) Concentrations of cyclic volatile methylsiloxanes in biosolid amended soil, influent, effluent, receiving water, and sediment of wastewater treatment plants in Canada. Chemosphere 93:766–773
9. Carpenter JC, Gerhards R (1997) Methods for the extraction and detection of trace organosilicon materials in environmental samples. In: Chandra G (ed) Organosilicon materials: the handbook of environmental chemistry. Springer, New York, pp 27–51
10. Varaprath S, Stutts DH, Kozerski GE (2006) A primer on the analytical aspects of silicones at trace levels-challenges and artifacts - a review. Silicon Chem 3:79–102
11. Wang D-G, Alaee M, Steer H, Tait T, Williams Z, Brimble S, Svoboda L, Barresi E, DeJong M, Schachtschneider J, Kaminski E, Norwood W, Sverko E (2013) Determination of cyclic volatile methylsiloxanes in water, sediment, soil, biota, and biosolid using large-volume injection-gas chromatography-mass spectrometry. Chemosphere 93:741–748
12. Wang D-G, Solla SRD, Lebeuf M, Bisbicos T, Barrett GC, Alaee M (2017) Determination of linear and cyclic volatile methylsiloxanes in blood of turtles, cormorants, and seals from Canada. Sci Total Environ 574:1254–1260
13. Badjagbo K, Furtos A, Alaee M, Moore S, Sauve S (2009) Direct analysis of volatile methylsiloxanes in gaseous matrixes using atmospheric pressure chemical ionization-tandem mass spectrometry. Anal Chem 81:7288–7293
14. Badjagbo K, Heroux M, Alaee M, Moore S, Sauve S (2009) Quantitative analysis of volatile methylsiloxanes in waste-to-energy landfill biogases using direct APCI-MS/MS. Environ Sci Technol 44:600–605
15. US EPA (2003) Method 5053C: purge-and-trap for aqueous samples
16. Whelan MJ, Sanders D, van Egmond R (2009) Effect of Aldrich humic acid on water-atmosphere transfer of decamethylcyclopentasiloxane. Chemosphere 74:1111–1116
17. David MD, Fendinger NJ, Hand VC (2000) Determination of Henry's law constants for organosilicones in actual and simulated wastewater. Environ Sci Technol 34:4554–4559
18. Varaprath S, Salyers KL, Plotzke KP, Nanavati S (1998) Extraction of octamethyl-cyclotetrasiloxane and its metabolites from biological matrices. Anal Biochem 256:14–22
19. Dewil R, Appels L, Baeyens J, Buczynska A, Van Vaeck L (2007) The analysis of volatile siloxanes in waste activated sludge. Talanta 74:14–19
20. Wang DG, Norwood W, Alaee M, Byer JD, Brimble S (2013) Review of recent advances in research on the toxicity, detection, occurrence and fate of cyclic volatile methyl siloxanes in the environment. Chemosphere 93:711–725
21. Xu L, He X, Zhi L, Zhang C, Zeng T, Cai Y (2016) Chlorinated methylsiloxanes generated in papermaking process and their fate in wastewater treatment processes. Environ Sci Technol 50:12732–12741
22. Wang D-G, Steer H, Pacepavicius G, Smyth SA, Kinsman L, Alaee M (2013) Seasonal variation and temperature-dependent removal efficiencies of cyclic volatile methylsiloxanes in fifteen wastewater treatment plants. Organohalogen Compd 75:1286–1290
23. Mueller JA, Di Toro DM, Maiello JA (1995) Fate of octamethylcyclotetrasiloxane (OMCTS) in the atmosphere and in sewage treatment plants as an estimation of aquatic exposure. Environ Toxicol Chem 14:1657–1666

24. Parker WJ, Shi J, Fendinger NJ, Monteith HD, Chandra G (1999) Pilot plant study to assess the fate of two volatile methyl siloxane compounds during municipal wastewater treatment. Environ Toxicol Chem 18:172–181
25. Zhang Z, Qi H, Ren N, Li Y, Gao D, Kannan K (2011) Survey of cyclic and linear siloxanes in sediment from the Songhua river and in sewage sludge from wastewater treatment plants, northeastern China. Arch Environ Contam Toxicol 60:204–211
26. Liu N, Shi Y, Li W, Xu L, Cai Y (2014) Concentrations and distribution of synthetic musks and siloxanes in sewage sludge of wastewater treatment plants in China. Sci Total Environ 476–477:65–72
27. Wang D-G, Aggarwal M, Tait T, Brimble S, Pacepavicius G, Kinsman L, Theocharides M, Smyth SA, Alaee M (2015) Fate of anthropogenic cyclic volatile methylsiloxanes in a wastewater treatment plant. Water Res 72:209–217
28. Bletsou AA, Asimakopoulos AG, Stasinakis AS, Thomaidis NS, Kannan K (2013) Mass loading and fate of linear and cyclic siloxanes in a wastewater treatment plant in Greece. Environ Sci Technol 47:1824–1832
29. Schweigkofler M, Niessner R (1999) Determination of siloxanes and VOC in landfill gas and sewage gas by canister sampling and GC-MS/AES analysis. Environ Sci Technol 33:3680–3685
30. Schweigkofler M, Niessner R (2001) Removal of siloxanes in biogases. J Hazard Mater 83:183–196
31. Rasi S, Lehtinen J, Rintala J (2010) Determination of organic silicon compounds in biogas from wastewater treatments plants, landfills, and co-digestion plants. Renew Energy 35:2666–2673
32. Tansel B, Surita SC (2014) Differences in volatile methyl siloxane (VMS) profiles in biogas from landfills and anaerobic digesters and energetics of VMS transformations. Waste Manag 34:2271–2277
33. Graiver D, Farminer KW, Narayan R (2003) A review of the fate and effects of silicones in the environment. J Polym Environ 11:129–136
34. Kent D, Fackler P, Hartley D, Hobson J (1996) Interpretation of data from nonstandard studies: the fate of octamethylcyclotetrasiloxane in a sediment/water microcosm system. Environ Toxicol Water Qual 11:145–149
35. Xu S (1999) Fate of cyclic methylsiloxanes in soils. 1. The degradation pathway. Environ Sci Technol 33:603–608
36. Xu S, Chandra G (1999) Fate of cyclic methylsiloxanes in soils. 2. Rates of degradation and volatilization. Environ Sci Technol 33:4034–4039
37. Wang D-G, Du J, Pei W, Liu Y, Guo M (2015) Modeling and monitoring cyclic and linear volatile methylsiloxanes in a wastewater treatment plant using constant water level sequencing batch reactors. Sci Total Environ 512–513:472–479
38. Sanchís J, Martínez E, Ginebreda A, Farré M, Barceló D (2013) Occurrence of linear and cyclic volatile methylsiloxanes in wastewater, surface water and sediments from Catalonia. Sci Total Environ 443:530–538
39. Stevens C, Annelin RB (1997) Ecotoxicity testing challenges of organosilicon materials. In: Chandra G (ed) Organosilicon materials: the handbook of environmental chemistry. Springer, New York, pp 83–103

Presence of Siloxanes in Sewage Biogas and Their Impact on Its Energetic Valorization

N. de Arespacochaga, J. Raich-Montiu, M. Crest, and J. L. Cortina

Contents

N. de Arespacochaga (✉) and J. Raich-Montiu
CETaqua, Barcelona, Spain
e-mail: narespacochaga@cetaqua.com

M. Crest
CIRSEE, Suez-Environnement, Le Pecq, France

J. L. Cortina
CETaqua, Barcelona, Spain

Department of Chemical Engineering, Universitat Politècnica de Catalunya-Barcelona
Tech (UPC), Barcelona, Spain

V. Homem and N. Ratola (eds.), *Volatile Methylsiloxanes in the Environment*,
Hdb Env Chem (2020) 89: 131–158, DOI 10.1007/698_2018_372,
© Springer Nature Switzerland AG 2019, Published online: 15 February 2019

Abstract Biogas produced in wastewater treatment plants (WWTPs) by microorganisms during the anaerobic degradation process of organic compounds is commonly used in energy production. Due to the increasing interest in renewable fuels, biogas has become a notable alternative to conventional fuels in the production of electricity and heat. Biomethane, upgraded from biogas, has also become an interesting alternative for vehicle fuel. Biogas contains mainly methane (from 40 to 60%) and carbon dioxide (40 to 55%), but it also contains trace compounds, such as hydrogen sulphide, halogenated compounds and volatile methyl siloxanes (VMS), which pose a risk on its energetic valorization.

It is reported that the concentrations of siloxanes in biogas are increasing in the recent years due to an increase in the use of silicon-containing compounds in personal care products, silicone oils and production of food, among others. This chapter reviews the presence of VMS in sewage biogas, depicting their concentrations and their speciation between linear and cyclic compounds depending on the wastewater treatment processes and operating conditions.

WWTP operators face therefore a choice between installing a gas purification equipment and controlling the problem with more frequent maintenance. Available technologies for siloxane removal are studied, and their impact on the performance of Energy Conversion Systems (ECS) is reported. The performance of adsorption systems using activated carbon, silica gel and zeolites is reviewed as it is a well-known and widespread used technology for siloxane abatement both at the scientific and industrial studies.

Keywords Biogas valorization, Siloxanes benchmarking, Siloxanes removal, Wastewater treatment plants

1 Introduction

Within the framework of sustainable development, energy on wastewater treatment plants (WWTPs) must be considered not only in terms of consumption reduction but also regarding the use as "green" energy. Self-sustainable wastewater treatment operations are a rapidly growing trend within the wastewater industry. Towards this goal, the production of biogas via anaerobic digestion represents the largest potential energy source in typical WWTP operations. Biogas produced from sludge treatment in sewage treatment plants, as well as from landfills, contains mainly methane and carbon dioxide. Other compounds found include nitrogen, oxygen, hydrogen sulphide, mercaptans, halogenated hydrocarbons and siloxanes [1]. The presence of contaminants in biogas varies with the source and, ultimately, with the treatment strategies used to clean it. Among the most critical compounds, volatile methyl siloxanes (VMS), composed of the repeating group $O-Si-(CH_3)_2$, have a significant adverse effect [2, 3], as their presence can greatly reduce the efficiency of energy recovery from the use of biogas [4].

Silicon-containing compound (silicone) consumption totalled around 0.85–2 million tons in 2002 [5, 6], and it increased by 24% from 2005 to 2009. They are widely used in oils/fluids, medicinal/pharmaceutical preparations and personal care products (cosmetics, toiletries and soaps) due to their low surface tension, smooth texture and high thermal stability. The organosilicon market has been growing around 6% each year in the 2010–2016 period [7] totalling eight million tons in 2015 [8]; and, differently from 2002, China is pulling the overall market instead of the USA (40 and 17% market share, respectively, in 2016). A similar market trend, 5% increase per year, is forecasted for the 2016–2021 period [9]. 10 to 20% of the consumed organosilicon materials may end up in wastewaters, from which 97% may adsorb on the sludge [10], and as a result they consequently end up in the biogas produced in WWTP operations.

There are different biogas recycling routes which include internal combustion engines, gas turbines, waste heat boilers, fuel cells and biogas upgrading (production of biomethane). Each of these end-use technologies have specifications regarding the quality of biogas (including siloxane content) that must be complied with to ensure an optimum efficiency and equipment lifetime guarantees. During combustion, siloxanes are converted into silicon dioxide deposits, leading to abrasion of engine parts or the buildup of layers that inhibit essential heat conduction or lubrication. The deposits may cause changes in the geometry of the combustion chamber. Also prone to deactivation by siloxanes are catalysts for both pre-combustion [11] and post-combustion gas purification, for example, to reduce formaldehyde concentrations in exhaust gas, possibly violating air emission regulations. Furthermore, parts of the deposited layers can break off and clog lines. Other undesired effects include the poisoning of catalysts employed in steam reforming [12] or fuel cells [13]. All these negative effects are associated with a rise in operating expenses (OPEX).

WWTP operators are therefore facing a choice of either installing gas purification equipment or controlling the problem with more maintenance. Moreover, engine warrantees linked to silicon levels were made significantly more stringent between 2002 and 2008 [14]. Growing importance is attributed to siloxane removal, as can be seen by the increase of related publications and patents filed over the last years. The siloxane problem has received increasing attention since the late 1990s, when older diesel engines were replaced by more efficient and less air-polluting lean burn engines. These were also more prone to silicon-related damage [15]. The extent of the problem became more visible as the use of biogas increased due to public subsidies [16] in countries like Germany. Biogas promotion policies spread since then to most of the European Union countries and elsewhere. Consequently, European biogas electricity has reached about 17,300 GWh$_e$ in 2006, with Germany alone producing approximately 7,300 GWh$_e$. Moreover, the production of silicon has risen, and so have the siloxane concentrations in sewage and landfill gas [17]. But siloxane removal from gas streams is not only relevant to biogas but also to off-air treatment in the silicone-producing industry or in mechanical-biological waste treatment. So, a reliable and robust analytical methodology is necessary (allowing a good quantification of siloxanes) in order to design, control and evaluate the removal efficiency of different systems of treatment that may be developed or used for the reduction of the concentration of these compounds.

This chapter focuses both on benchmarking the concentrations of siloxanes in sewage biogas (Sect. 2) and on evaluating the efficiency and economics of removal technologies depending on the final use of biogas to provide recommendations for WWTP operators (Sects. 3 and 4).

2 Benchmarking of Siloxane Levels in Sewage Biogas

Capela et al. [18] conducted an accurate review on the fate of VMS through wastewater treatment plants, showing how these compounds distribute between the sewage and sludge lines for different treatment processes. Due to their physico-chemical properties, rather than being biodegraded, siloxanes tend to transfer from the liquid to the solid phase, accumulating on the sludge and later, transferred to the biogas phase during the anaerobic digestion process. Several studies assessed the presence of siloxanes in sewage biogas [17, 19–25].

Raich-Montiu et al. [20] quantified siloxanes in biogas from five WWTPs from Spain, France and England and tried to correlate the anaerobic digestion process conditions with the siloxane profile and concentration. Biogas was sampled in triplicate on the conducting pipes after low pressure gasholders to reduce fluctuations on its chemical composition. The most relevant differences on the available processes and the operating conditions of the sludge line of each WWTP are summarized in Table 1.

Sludge type (primary, secondary or mixed sludge) has a significant effect on the biogas production and composition because the biodegradability (both rate and extent) of each sludge type is different. Wang et al. [26] observed higher siloxane levels in secondary sludge compared to primary sludge, indicating that their

Table 1 Description of the main sludge line processes in the selected WWTP (reprinted from Raich-Montiu et al. [20], Copyright 2014, with permission from Elsevier)

WWTP (biogas flow)	1 (5,750 m^3/day)	2 (29,000 m^3/day)	3 (5,850 m^3/day)	4 (29,500 m^3/day)	5 (1,560 m^3/day)
Sludge type	Mixed sludge	Indigenous and external sludge	Secondary sludge	Mixed sludge	Mixed sludge
Sludge pretreatment	None	Thermal hydrolysis	None	None	None
Temperature (and regime)	37°C (mesophilic)	42°C (mesophilic)	37°C (mesophilic)	55°C (thermophilic)	36°C (mesophilic)
Retention time (days)	19	26	21	21	25
Mixing system	Mechanical and sludge recirculation	Sludge recirculation	Sludge and biogas recirculation	Biogas recirculation	Mechanical and sludge recirculation

incidence in the biogas are expected to depend on the type of sludge digested. Sludge pretreatment technologies (i.e. thermal hydrolysis, as the case of WWTP-2) are used to break down the inlet organic matter before digestion, possibly favouring the degradation of silicon compounds by turning larger less-volatile silicon molecules into lighter and more volatile units. The temperature of the digester has a relevant impact both on the evaporation and on the stripping of silicon compounds from the liquid to the gas phase [17], while the retention time defines the extent of the degradation of organic matter in the reactor. Finally, the mixing system (mechanical, sludge recirculation or biogas recirculation) is used to homogenize hydraulically and thermally the content in the digester preventing stratification and can also affect the mechanism of siloxanes' transfer from the liquid to the gas phase.

In Raich-Montiu et al.'s [20] study, the macro-composition of the biogas showed the following patterns: CH_4 55–66%, CO_2 33–38% and O_2 0.2–1%. H_2S presented a wide range of concentration with values from 260 to 3,090 mg/Nm^3. Biogas composition regarding siloxane content is collected in Table 2.

As it can be observed, for all WWTPs, cyclic siloxanes and more specifically D_4 and D_5 were the major silicon compounds detected in sewage biogas. On the other hand, linear siloxanes, such as L_2, L_3, L_4 or L_5, were detected either at very low concentrations or not at all. The other two cyclic siloxanes, D_3 and D_6, were only quantified in WWTP-4 and WWTP-2, respectively, but with very low concentrations, indicating that D_4 and D_5 are the main congeners impacting the overall siloxane content in sewage biogas. Temperature was identified by authors as the most relevant operating condition impacting D_5 concentration as its presence on the thermophilic digester (124 mg/Nm^3 and 97% of total siloxanes in WWTP-4) largely exceeded the values observed for the other mesophilic digesters. Alternatively, the

Table 2 Mean value of siloxane concentrations in the different sites studied, sampled with adsorbent tubes

Concentration (mg/Nm^3)	WWTP-1	WWTP-2	WWTP-3	WWTP-4	WWTP-5
TMSOH	<0.02	0.9 (0.1)	<0.02	<0.02	<0.02
L_2	<0.01	<0.01	<0.01	<0.01	<0.01
L_3	0.1 (0.02)	0.1 (0.02)	0.1 (0.02)	0.3 (0.03)	0.1 (0.01)
L_4	0.1 (0.02)	<0.01	<0.01	0.3 (0.02)	<0.01
L_5	<0.01	<0.01	<0.01	0.8 (0.1)	<0.01
D_3	0.1 (0.02)	0.2 (0.03)	0.2 (0.03)	0.1 (0.02)	<0.01
D_4	6.8 (0.2)	10.1 (0.5)	1.5 (0.1)	1.9 (0.1)	2.5 (0.2)
D_5	12.9 (1.1)	43.8 (0.6)	12.5 (0.7)	124.0 (4.2)	13.1 (1.2)
D_6	0.5 (0.1)	–	–	–	0.4 (0.1)
Σ siloxanes	18.5	55.1	14.3	127.4	16.1
% D_4/Σ siloxanes	37	18	10	1	16
% D_5/Σ siloxanes	70	79	87	97	81

Standard deviation in brackets ($n = 3$) (Reprinted from Raich-Montiu et al. [20], Copyright 2014, with permission from Elsevier)

concentration of D_4 seemed to be more impacted by the pretreatment of the inlet sludge (WWTP-2), while retention time could also affect the D_4/D_5 ratio in sewage biogas (WWTP-1). Available information in this study is not sufficient to elucidate why the impact of thermal hydrolysis on siloxane content seems larger for D_4 rather than for D_5 or why lower retention times increase the contribution of D_4 on the overall siloxane concentration. On the other hand, the lower presence of siloxanes observed in WWTP-3 (both on D_4 and D_5) could be related to the fact that only secondary sludge was digested at that site. As the removal of silicon compounds from the water line of the WWTP was expected to occur through adsorption on the primary sludge flocks rather than due to biodegradation on the secondary treatment, its concentration on secondary sludge was expected to be lower.

Trimethylsilanol (TMSOH) was only detected in WWTP-2 at a concentration of 0.9 mg/Nm^3. consistent with Tansel and Surita [23], who concluded that this compound was more frequently observed on landfill gas rather than on sewage gas (0.075 vs 7.25 mg/Nm^3). Detection of TMSOH only in this sewage biogas indicates a slightly higher reduction potential in this digester (i.e. larger concentrations of H$^+$ and/or H$_2$ either dissolved or at the gas phase), which could be potentially explained by the presence of thermal hydrolysis pretreatment process upstream the anaerobic digester [23].

The temporal variability of siloxane concentration patterns was studied by Garcia et al. [25] over an entire year through a biogas characterization campaign (36 samples) on the same WWTP (Spain). While no linear siloxanes were detected, D_4 and D_5 were quantified in 75 and 100% of the samples (average concentrations of 1.6 and 7.7 mg/Nm^3; standard deviations 1.4 and 2.9 mg/Nm^3) and D_6 in 33% of the samples (around 0.1–0.2 mg/Nm^3). In addition, this study showed that biogas pressurization to 3 bar caused a reduction of around 15–20% on the total siloxane concentration due to their condensation.

Finally, a different geographical pattern of siloxane contamination levels has also been revealed through available literature. Rasi et al. [27] analysed the organic silicon compounds in WWTP-4 in Finland (and other biogas sources such as landfill and agricultural digesters, 48 samples) consistently quantifying siloxane concentrations in the range of 0.5 and 1 mg/Nm^3 for D_4 and D_5, respectively. Although no thorough comparative study assessing siloxane concentration patterns in different countries has been identified and relevant differences on siloxane content from plant to plant can be observed, values in Finland are significantly lower than those determined in other European biogases such as Germany, the UK, Spain and Italy, which are in the range of 3 and 16 mg/Nm^3.

The significance of D_4 and D_5 over global siloxane concentration in sewage biogas in the literature is compiled in Table 3. As it is depicted, these two cyclic siloxanes consistently contribute to over 95% of the total silicon contamination (25% D_4 and 70% D_5).

Table 3 Absolute and relative concentration of siloxanes D_4 and D_5 over global siloxanes in sewage biogas

WWTP	D_4 mg/Nm^3	%	D_5 mg/Nm^3	%	Sum $D_4 + D_5$ %	Reference
Augsburg (Germany)	2.95	50	2.78	48	98	Schweigkofler and Niessner [22]
Munich (Germany)	6.69	41	9.3	56	97	Schweigkofler and Niessner [22]
Severn Trent (UK)	3.8	12	8.2	88	100	Dewil et al. [17]
Jyväskylä (Finland)	0.2–0.9	30	0.4–1.2	57	88	Rasi et al. [27]
Rahola, Tampere (Finland)	0.1–0.5	34	0.2–0.8	56	90	Rasi et al. [27]
Viinikanlahti, Tampere (Finland)	0.1–0.3	46	0.2–0.35	48	94	Rasi et al. [27]
Espoo (Finland)	0.1–0.45	24	0.7–1	73	97	Rasi et al. [27]
Not specified	6.8	37	12.9	70	97	Raich-Montiu et al. [20]
Not specified	10.1	18	43.8	79	97	Raich-Montiu et al. [20]
Not specified	1.5	10	12.5	87	97	Raich-Montiu et al. [20]
Not specified	1.9	1	124	97	98	Raich-Montiu et al. [20]
Not specified	2.5	16	13.1	81	97	Raich-Montiu et al. [20]
Miami (Florida, USA)	4.15	62	1.8	27	89	Tansel and Surita [23]
Japan	Not available		Not available	86	>86	Oshita et al. [19]
Rincón de León (Alicante, Spain)	1.6	17	7.7	81.9	98.9	Garcia et al. [25]
Milan (Italy)	1.33	25	3.84	73	98	Paolini et al. [24]

3 Impact and Operating Concerns of Siloxane Content on the Performance of Energy Conversion Technologies

There are many Energy Conversion Systems (ECS) for biogas energy recovery employed in WWTPs [28–30]. Some have been long used, some are newly developed, and some are still in development. These energy recovery technologies include boilers, internal combustion engines, combustion turbines, micro-turbines, fuel cells and Stirling engines. Siloxanes, as well as the other biogas contaminants, pose a risk to the short- and long-term performance of ECS [17, 27].

During the combustion of biogas, siloxanes are converted into abrasive and microcrystalline silica (Eqs. 1 and 2), which deposits in the combustion chamber, turbine blades, heat exchangers, spark plugs, valves, cylinder heads, etc.

$$D_4 \text{ combustion}: \left[(CH_3)_2 SiO\right]_4 (g) + 16\, O_2 (g)$$
$$\rightarrow 4\, SiO_2 (s) + 8\, CO_2 (g) + 12\, H_2O (g) \tag{1}$$

$$D_5 \text{ combustion}: \left[(CH_3)_2 SiO\right]_5 (g) + 20\, O_2 (g)$$
$$\rightarrow 5\, SiO_2 (s) + 10\, CO_2 (g) + 15\, H_2O (g) \tag{2}$$

Silica has physicochemical properties similar to glass, hence causing abrasion on engine parts or the buildup of silica layers that hinder heat conduction and lubrication [31]. Different field tests with biogas-powered ECS [32] observed silica films (with a thickness of several millimetres) on the surfaces of the equipment, which was difficult to remove both chemically and mechanically. On the other hand, siloxanes can also affect the efficiency of the exhaust gas catalytic treatment [2, 33], poisoning Pt- and Pd-supported catalysts by blocking the surface of the metal with silicon atoms [34, 35]. Therefore, the combustion of siloxane-containing biogas causes serious damage to gas engines, micro-turbines and fuel cells, hampering the recovery of energy from biogas.

3.1 Internal Combustion Engines (ICE)

Reciprocating internal combustion engines are a well-established and proven power generation technology, and they have a long track record of use in biogas applications [29]. These reciprocating engines are somewhat less sensitive to contaminant levels in the biogas than some of the other engine technologies. However, serious breakdowns in internal combustion engines have been reported in the literature as a consequence of siloxane content in biogas [25, 36, 37]. Damage due to silica formation could basically take place in the combustion chamber (cylinder, valves, heads, piston), in the crankshaft, in the connecting rod bearings of the reciprocating mechanism and in the exhaust gas treatment system.

Alvarez-Florez and Egusquiza [36] reported siloxane-related damage on a 985 kW$_e$ reciprocating engine fuelled with landfill gas in Spain. Figure 1 shows the impact of siloxane combustion on the cylinder and the piston: (top) layers of deposited solids (0.1–0.3 mm thickness) on the engine head around the plate valves; (middle) soft-paste-like deposits on cylinder walls; and (down) piston surface with some fused material. According to the authors, the silica layers deposited on the cylinder head were brittle and could be easily broken off from the surface. On the other hand, the soft paste deposits in the ring grooves of the piston blocked the system, causing bad lubrication conditions and, consequently, an increase in the temperature of the piston. This eventually increased its size until

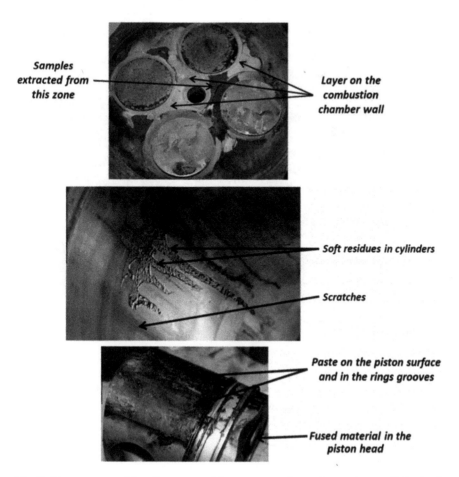

Fig. 1 Damages caused by siloxane-containing gas on reciprocating engines: cylinder head, cylinder walls and piston (reprinted from Alvarez-Florez and Egusquiza [36], Copyright 2015, with permission from Elsevier)

contacting with the cylinder surface and caused the scratches. Figure 2 shows silica deposits on the spark plugs of an internal combustion engine on a landfill site in Istanbul, Turkey [37].

Very few systematic scientific studies are available to elucidate the precise effect of siloxanes in internal combustion engines and to understand the relationship between siloxane concentration with deposited silica or with engine performance parameters. Nair et al. [38] assessed the impact of siloxane contamination on the exhaust gases' catalytic treatment in a pilot ICE unit of 250 W_e. The engine powered with contaminated biogas (D_4 and L_2 10 ppm_v; equimolar composition) showed a reduction on the efficiency of the catalytic CO reduction unit from 90 to 50%, and silica particles ranging from 10 to 180 nm were detected at the outlet of the exhaust gas treatment at a concentration of 10 $\mu g/m^3$. In addition, silica deposits were

Fig. 2 Silica deposit on spark plugs (reprinted from Sevimoglu and Tansel [37], Copyright 2013, with permission from Elsevier)

observed on piston heads, oxygen sensors and spark plugs after 96 h. These deposits consisted of a bottom layer, which was strongly bounded to the metal surfaces, and a porous and less dense top layer, which could be easily removed from the surface. Finally, silicon was also observed in the engine oil.

3.2 Micro-turbines

A micro-turbine is a smaller version of a combustion turbine, developed to be cost-effective at low output ranges. These systems generate power through a compressor, combustor and turbine using the Brayton power cycle. The available capacity ranging from 30 to 250 kW_e is well-suited to many biogas applications, and they have been installed at municipal WWTPs, landfills, and some dairy farms. The biggest technical challenge for micro-turbines in these applications has been assuring proper fuel treatment. Some early installations were shut down prematurely due to inadequate fuel moisture removal, gas compressor corrosion problems and the lack of siloxane filtering [39, 40]. Although inadequate moisture and hydrogen sulphide removal can cause corrosion issues in the components, siloxanes have been identified as the most important contaminant for micro-turbines, as silica causes significant erosion of the turbine nozzles, blades and bearings, which results in turbine failure. In this context, the siloxane limits are much more stringent than in ICEs. Figure 3 compiles siloxane damages on micro-turbines' blade wheels.

Fig. 3 Damages caused by the combustion of siloxane-rich biogas in the blade wheels of micro-turbines (left reprinted from ref. Urban et al. [11], Copyright 2009, with permission from Elsevier; right Iyer [41])

3.3 Fuel Cells

Fuel cells, through electrochemical reactions on the anode and cathode, convert chemical energy directly into electrical energy at a higher efficiency than combustion-based technologies. Five major types of fuel cells, characterized by the electrolyte and electrode materials used, are available: proton exchange membrane (PEMFC), phosphoric acid (PAFC), alkaline (AFC), molten carbonate (MCFC) and solid oxide (SOFC). Although all types of fuel cells can be potentially powered with biogas by extracting the hydrogen and removing all poisoning compounds, only a few types of fuel cells are effectively considered for biogas applications. Currently, there are a few industrial-scale PAFC and MCFC successfully operating with sewage biogas as well as some pilot-scale PEMFC and SOFC projects.

Differently from conventional electricity generation technologies, an extensive literature is published studying the effect of siloxane contamination in fuel cell systems [13, 42–46]. Both the anode and cathode materials within the fuel cell system and the catalyst used for biogas reforming into H_2 are deactivated by sulphur, siloxane and chlorine contamination [47, 48]. Regarding siloxanes, the formation of SiO_2 (s) according to Eqs. (3) and (4) [13] and its further deposition and accumulation in the porous cermet anodes are responsible for fuel cell performance degradation through catalyst deactivation. Table 4 compiles the poisoning effect of siloxanes in fuel cell units.

$$\left[(CH_3)_2SiO\right]_5 (g) + 25\,H_2O \rightarrow 5\,Si(OH)_4 (g) + 10\,CO (g) + 30\,H_2 (g) \quad (3)$$

$$Si(OH)_4 (g) \rightarrow SiO_2 (s) + 2\,H_2O \quad (4)$$

Table 4 Damage caused by siloxanes on fuel cell units

Fuel cell	Siloxane	Obtained results and conclusions	Reference
SOFC Ni-ScSZ	D_5 (10 ppm_v)	The open circuit voltage (OCV) dropped from 1 V to below 0.5 V in less than 30 h Impact faster at lower operating temperatures (800, 900 and 1,000°C) After 50 h of continuous operation, a fatal degradation was observed	Sasaki et al. [46]
SOFC Ni-YSC anode	D_4 (1–20 ppm_v)	At 1 ppm D_4, the cell voltage showed an average degradation rate of 0.25 mV/h (0.39 mV/h at 3 ppm D_4) A gradient in deposition of Si was observed: SiO_2 (s) is deposited at the fuel inlet regions (on the cell, on the current collector and on the interconnect plate)	Madi et al. [42]
SOFC Ni-YSC anode	D_4 (5 ppm_v)	5% reduction on cell voltage after 20 h Once D_4 contamination has been stopped, a progressive but slight (<1%) recovery is observed in 50 h	Madi et al. [43]
SOFC Ni-YSC anode	D4 (0.11–1.92 ppm_v)	10% reduction on cell voltage after 50 h Irreversible degradation	Papurello et al. [44]
SOFC Ni/Al_2O_3 steam reforming catalyst	D_5 (0.5–1 ppm_v)	CH_4 conversion rates reduced from 99 to 70% after 100 h SiO_2 (s) deposits on the catalyst surface accounted for almost 5% of the total silicon introduced	Papurello and Lanzini [45]

3.4 Stirling Engines

In the Stirling engine, the combustion occurs continuously in a combustion chamber that is separate from the working gas (usually helium) and engine moving parts, that is, it is an external process. Therefore, this engine may offer the advantage of being more tolerant to contaminants in the fuel gas [49]. At the time this chapter was written, no full-scale Stirling engine references were available for power generation [50]. However, because of its potential for applications using alternative fuels, such as biogas and waste heat, there have been significant developments in Stirling engine technologies in recent years. Some manufacturers (Stirling Biopower in the USA and Stirling Denmark and Clean Energy in Europe) claim that these engines can be powered without biogas treatment, and, therefore, no siloxane removal is required, and only regular cleaning of the heat exchanger and other engine parts is necessary.

3.5 Domestic Gas Boiler

The ECS in the previous sections are commonly used in WWTP for on-site electrical and thermal energy generation. In the recent years, biogas upgrading to biomethane and its further injection in the natural gas grid or its use in natural gas vehicles (NGVs) has become an interesting alternative for off-site biogas valorization in some countries [51–53]. But also here the major concern for gas supply companies is the potential impact of biomethane contaminants on the gas distribution network and pieces of equipment (instrumentation, compressors, deposits, etc.) and on domestic downstream equipment (i.e. gas boilers in households). Biomethane injection standards were defined in some European countries, and, more recently, two standards have been established by the CEN-EN 16723 for natural gas and biomethane for use in transport and biomethane injection in the natural gas network.

The Technical Committee 408 of the Comité Européen de Normalisation (CEN) assessed the effect of siloxane contamination on four 25-kW$_t$ domestic condensing boilers in the Netherlands with air-gas ratio control. The boilers had different configurations (materials and geometry) on their heat exchangers to represent the different pieces of equipment available: boiler 1, stainless steel tubular; boiler 2, aluminium lamellar; boiler 3, stainless steel concentric cylindrical wound tube; and boiler 4, aluminium finned tubes. Figure 4 compiles photographs taken after 850 h of continuous operation at a biogas contamination level of 7.5 mgSi/Nm^3 (D$_5$) on the four boilers.

Regardless of the appearance and chemical composition of the deposits, the thermal output of the boilers was reduced by about 35%. In addition, silica was also detected on the ionization probes typically employed as flame failure devices (FFD), which corresponded to an eventual shutdown of the boiler after only 50 h. As a result of the air-gas ratio control, the CO concentration in the flue gas in boilers was not affected by the siloxane concentration. However, if the boiler was operated without air-gas ratio control, an increase in the CO level would have been observed.

Nair et al. [55] also observed a detrimental effect of siloxanes on the flame sensors in a domestic pulse-combustion furnace exposed to L$_2$ and D$_4$ (equimolar composition: 2, 10 and 20 ppm$_v$ total). No clear trend in the shutdown time was observed, as the furnace exposed to 20 ppm$_v$ stopped later than that exposed to 10 ppm$_v$ but earlier than that exposed to 2 ppm$_v$. Similar to the experimentation with ICEs, silica particles with a mean diameter of 75 nm were detected in the flue gas.

3.6 Consolidation of Siloxane Concentration Limits in Energy Conversion Systems

Several technologies are available to recover energy from biogas, and different specific tolerance limits for siloxanes are requested. Unless siloxane content in the biogas is very low, gas purification systems should be installed to reduce ECS performance degradation and/or very frequent maintenance campaigns. Table 5 summarizes the limits for siloxanes and some other biogas contaminants for the each ECS. As it can be observed, despite the fact that each manufacturer requires a different concentration threshold, internal combustion engines are the ECS with less

Fig. 4 Silica deposits observed on four domestic boilers in the Netherlands after 850 h (D_5 1.5 mgSi/ Nm^3) (reprinted from van Essen et al. [54], Copyright 2013, with permission from TC408 CEN)

Table 5 Siloxane and other biogas contaminants concentration limits for different Energy Conversion Systems

| ECS | Manufacturer/model | Total siloxane limit (mg/m^3) | Limits on other biogas contaminants | | Reference |
			H$_2$S (mg/m^3)	NH$_3$ (mg/m^3)	
Internal combustion engine	Jenbacher	12	1,900	90	SEPA [56]
	Caterpillar	28	3,500	175	SEPA [56]
	Waukesha	30	1,200	Not available	SEPA [56]
	Tech 3 solution	5	Not available		Wheless and Pierce [57]
Micro-turbines	Capstone	0.03	10,600	Not available	Manufacturers
	Turbec	Not available	2,275		
Fuel cells	PAFC	0.05–0.1	1–5	10	Manufacturers
	MCFC		1–5	No limit (Fuel)	
	SOFC		0.5–1		
Stirling engine	Stirling biopower	No limit		Not available	Manufacturers
	Stirling Denmark				
	Clean Energy				
Biomethane injection into the grid	CEN-EN16723-1	0.3–1 mgSi/m^3	5	10	CEN-EN16723-1
Biomethane use in car fuel	CEN-EN16723-2	0.3 mgSi/m^3	5	10	CEN-EN16723-2

restrictive siloxane tolerance levels (from 5 to 30 mg/m^3); hence a thorough and deep cleaning system is usually not mandatory for most raw sewage biogases. On the other hand, fuel cells and micro-turbines show the most stringent cleaning requirements for silicon contamination (below 0.1 mg/m^3), thus gas cleaning units will have to be designed according to these specifications. Injection into the gas grid roots in the middle, with tolerance values in the range of 1 mg/m^3.

4 Removal of Siloxanes in the Energetic Valorization of Biogas

As it has been overviewed in the previous sections, typical siloxane concentrations in sewage biogas are in the range of 0.65–126 mg/Nm^3, while the ECS limits set by manufacturers range from 0.03 to 30 mg/Nm^3. As a result, siloxane abatement is a mandatory stage for the efficient energy recovery of biogas in WWTP. Several siloxane removal technologies are reported in literature, and some of them are currently commercialized at industrial level: based on adsorption, absorption, refrigeration/condensation, membrane separation and biological degradation. The number of publications on this field has been increasing in recent years. Ajhar et al. [58], de Arespacochaga et al. [4] and Shen et al. [59] conducted very complete reviews on siloxane removal from landfill and digester gas, including both experimental and commercial technologies. A summary of the main features, operating performance, energy consumption and costs for each technology can be found in Table 6.

4.1 Adsorption Processes

Currently, non-regenerative adsorption on porous solids represents the most common practice to abate siloxane compounds from sewage biogas, and there is a consensus that this technology provides the most efficient performance both in technical and economic points of view [61]. The most common adsorbents used to remove siloxanes include activated carbons (AC) [57, 60, 62, 63], silica gel [64–67] and zeolites [68–71] due to their excellent adsorption capacities. Other inorganic adsorbents, such as alumina [72], and polymeric adsorbents, such as polyurethane foams, have also been employed.

The accessibility of siloxane molecules to the internal adsorption surface depends on the adsorbent pore size distribution and on its textural properties. A positive correlation between the siloxane adsorption capacity and the specific surface area (BET area (m^2/g)) and the mesopore (2–50 nm) and micropore (0.7–2 nm) volumes was observed [12, 73, 74]. In contrast, no correlation with the narrow micropores (<0.7 nm) was observed [62], which is consistent with the siloxane cross-sectional molecular size (1.08–1.03 nm in the case of D_4; [75]). Siloxane adsorption mechanism is influenced by the presence of proton-donating groups (e.g. phenolic or carboxylic) on the activated carbon surface, which act as catalytic sites for the ring opening of cyclic siloxanes [69, 76]. This leads to the formation of α-ω-silanediols, which can further polymerize into longer siloxanes [77]. Detection of D_5, D_6 and D_7 on exhausted AC only exposed to D_4 as the silicon source revealed this polymerization mechanism on the surfaces of the carbons.

On the other hand, siloxane adsorption capacity is strongly dependent on biogas composition and siloxane concentration, as the presence of volatile organic compounds (VOCs) – including toluene, xylenes, limonene and halogenated hydrocarbons [67] – might compete with siloxanes in the adsorption process. A relevant competing compound is relative humidity, which at values of 50–70% can reduce the siloxane adsorption capacity both in activated carbons (a factor of 10; [32]) and in silica gels (90% reduction; [64]), probably as a result of water adsorption and formation of hydrogen bonds with oxygen functional groups on the adsorbent material surface. Notwithstanding, the presence of humidity up to 30% increased siloxane adsorption capacity for some activated carbons [77] due to an improvement in the open-ring and polymerization mechanism due to the higher accessibility of water. Table 7 collects siloxane adsorption capacities for different adsorbent

Table 6 Main features of siloxane removal technologies and conclusions on the operating performance and running costs (adapted from de Arespacochaga et al. [4])

Siloxane removal technology	Main features	Operating performance	Investment and operating costs[a]
Adsorption on carbon-based materials	Is the most widely used media to remove siloxanes from biogas, both in scientific studies and industrial full-scale facilities Impregnated (e.g. with H_3PO_4, KOH, KI) and non-impregnated activated carbon can be used The adsorption process is non-selective; hence adsorption capacity is reduced in the presence of humidity and other biogas contaminants (mainly sulphur and VOCs). Therefore, adsorption is usually installed downstream of a drying stage Regeneration of materials loaded with siloxanes is not commonly implemented due to their low efficiencies (deposition of amorphous silica and polymerization products); hence the most relevant operational expense is material replacement	Removal efficiencies greater than 99% can be achieved with adsorbent materials with meso- and microporous structures Adsorption capacities ranging 0.5–1.5% are experimentally observed in real biogas matrices	Investment costs: low 38 k€ for 190 Nm^3/h and 14 mg/Nm^3 siloxane [60] Operational cost: medium 4.5 k€/year for 190 Nm^3/h and 14 mg/Nm^3 siloxane [60] Adsorbent material approximately 0.5–1.5 €/kg and media disposal costs approximately 50–100 €/ton
Adsorption on inorganic materials	Silica gel and zeolites are other two adsorption materials commonly used to remove siloxanes from biogas; but they have been installed in full-scale systems to a lesser extent than activated carbon Siloxane adsorption on mesoporous silica gel is more selective than activated carbon. Nonetheless, gas humidity significantly reduces performance; thus, it is also installed downstream a drying stage Silica gel can be regenerated with hot air at 250°C		
Absorption	Chemical absorption with strong acids and bases destroys siloxanes. Only acidic reagents are suitable for biogas applications as strong bases produce carbonates due to the presence of CO_2; but concerns related to safety and corrosion arise Selexol™ process uses dimethyl ethers of polyethylene glycol in a physical absorption process	Removal efficiencies over 95% with concentrated and hot acids and Selexol™ Water is not efficient for siloxane scrubbing	Investment costs: medium Operational cost: high

(continued)

Table 6 (continued)

Siloxane removal technology	Main features	Operating performance	Investment and operating costs[a]
Refrigeration-condensation	Gas cooling to temperatures approximately 5°C is not very efficient as the removal efficiencies of D5 from landfill gas (0.6 mg/m^3) and sewage gas (9.7 mg/m^3) were 12% and 18%, respectively It is usually combined with adsorption to improve overall performance	Removal efficiencies approximately 5–20% (influence of gas temperature)	Investment costs: low Operational cost: medium
Deep chilling	80–90% removal efficiency at -30°C for siloxane concentrations 7–15 mgSi/m^3 It can only provide economic feasibility at large-scale facilities	Removal efficiencies around 80–90%	Investment and operational costs: high
Membranes	Selective siloxane permeation by solution and diffusion through a dense polymeric membrane Less reported than other removal technologies	Up to 80%	Investment costs: high Operational cost: medium
Biological	Limited by mass transfer limitations and reduced siloxane biodegradability Studied some years ago; but provided very low removal efficiencies at high residence time; even with the use of a second organic phase to enhance availability Current investigations are assessing the combination of adsorption and biological degradation	Removal efficiency of 10–40% Elimination capacity limited to 100 mg/m^3/h (very small compared to biological desulphurisation)	Investment costs: medium Operational costs: low

[a]Investment cost (k€/Nm^3/h): <0.5 (low), 0.5–1 (medium), >1 (high). Operating cost (c€/Nm^3): <1.5 (low), 1.5–3 (medium), >3 (high)

materials reported in recent literature, including the conditions of the test. Most of the experiments are conducted in N_2 dry matrices and with siloxane concentrations much larger than those commonly reported for sewage biogas, thus adsorption capacities reported usually outstand those observed in real operation conditions.

Moreover, a well-reported problem related to adsorption technologies is concentration roll-up, which consists of the desorption of compounds previously adsorbed in downstream zones of the filter bed, being replaced by more strongly adsorbed compounds. In the context of siloxanes and sewage biogas, it is assumed that D_5 can displace previously adsorbed D_4 and eventually result in outlet concentrations higher than the inlet ones [60, 73]. Due to its smaller size, the early breakthrough of D_4 may be prevented or delayed using materials with higher percentages of micropores.

Finally, according to both landfill gas treatment practices and laboratory experiments, adsorbent materials can only be partially regenerated from siloxanes after

Table 7 Adsorption capacities for siloxane abatement of different adsorbent materials under different conditions

Test matrix	Targeted siloxane	Siloxane concentration (mg/m^3)	Humidity (%)	Other impurities	Adsorbent material	Adsorption capacity (mg/g)	Reference
Real landfill gas	D_4, D_5	30	30–70	Not available	Activated carbon	5–15	Wheless and Pierce [57]
N_2	D_4	4,500	0	None	Various activated carbons	55–180	Matsui and Imamura [73]
					Silica gel	100	
50:50 CH$_4$:CO$_2$	D_4	1,440	0	None	Activated carbon	22–225	Yu et al. [74]
N_2	L_2	725–1,450	0	None	Activated carbon	100	Gislon et al. [78]
N_2	L_2, D_4, D_5	20,000	0	None	Activated carbon	242–307	Nam et al. [72]
					Silica gel	202	
N_2	D_4	400	0	None	Silica gel	250	Sigot et al. [64]
N_2 55:45 CH$_4$:CO$_2$	D_4	13,200	0 and 20	None	Various activated carbons	249–1,732	Cabrera-Codony et al. [62]
Real sewage biogas	D_4, D_5	10	50.5	Linear alkanes, aromatics	Activated carbon	5	de Arespacochaga et al. [60]
N_2	D_5, D_6	192	0	Limonene, toluene	Activated carbon	120	Kajolinna et al. [67]
					Silica gel	131	
					Molecular sieve	53	
N_2	D_4	2,000	0	None	Mesoporous synthetic zeolite	105	Jiang et al. [69]
N_2	D_4	6,600	0	None	Mesoporous silica gel	353–642	Jafari et al. [65]
N_2	D_4	3,000	0	None	Natural zeolite (clinoptilolite)	11	Cabrera-Codony et al. [68]
					Various synthetic zeolites	28–143	
N_2	L_2, D_4, D_5	2	0 and 30	Limonene, toluene	Steam activated carbon and phosphoric activated carbon	Not provided	Cabrera-Codony et al. [77]

use [79], even with advanced oxidation processes. Cabrera-Codony et al. [80] studied the regeneration of AC previously exposed to D_4 with O_3, H_2O_2 and Fenton processes showing regeneration efficiencies below 50%. Induced changes on the activated carbon surface, rather than the formation of insoluble SiO_2 (s) or failure to oxidize adsorbed siloxanes, was responsible for these low values. Although several patents have been issued claiming the development of regenerable adsorbents for siloxanes (see the review by Ajhar et al. [58]), it still seems to be an unsolved problem in practice. As a result, to increase the lifetime of the adsorbent bed, choosing the most efficient AC for siloxane removal becomes crucial to reduce operational costs.

4.2 Absorption Processes

Due to the chemical nature of siloxanes, the most suitable absorbents are polar organic solvents, such as Selexol™ (polyethylene glycol or dimethyl ethers) [58]. Reactive absorption methods with concentrated solutions of acids (which cleave Si-O bonds) achieved moderate siloxane removal efficiencies, but their use complicated plant design due to safety concerns, resulting in higher capital and operational costs. The application of alkaline solutions is unpractical, as the presence of carbon dioxide results in high caustic consumption and precipitation of carbonates in the absorption column [81]. Finally, because many organic silicon compounds are at least partially water-soluble, water could also be a physical absorbent, but it has not been a very effective medium for siloxane removal [21]. Thus, more comprehensive studies on this field are required [82]. In practice, gas scrubbing systems often combine other pre- and post-treatment methods. Because the absorption process operates better at lower temperatures, gas is often cooled prior to the absorption stage. AC adsorption can also be used as a polishing step after gas scrubbing [83].

4.3 Refrigeration/Condensation and Deep Chilling Processes

The condensation of water contained in the biogas also leads to the removal (with the condensed moisture) of many pollutants such as H_2S, siloxanes, SO_2 and halogens. Water condensation allows some siloxanes to be removed and solubilized in the water stream. In addition, low temperatures can condense volatile siloxanes (apart from water) allowing further removal. The performance of a refrigeration/condensation system can vary from low to moderate (15–50% siloxane removal) depending on the refrigeration temperature employed and the initial concentration of siloxanes [57]. Therefore, the refrigeration/condensation process is used as a gas pretreatment prior to the use of AC (increasing significantly the AC lifetime), but it cannot achieve the overall siloxane removal requirements as a stand-alone system [83].

In deep chilling conditions, the theoretical removal efficiency depends on saturation partial pressure (P_{sat} (T)) of each siloxane. Although the theory predicts removal efficiencies of 26% at $-25°C$ [17, 58], experimentation with deep chilling processes clearly exceeds the theoretical performance [63]. In any case, the energetic requirements for the application of deep chilling systems are so high that such a process is not profitable unless with very high biogas flows and/or siloxane concentrations [11].

4.4   Membrane Separation Processes

These technologies consist of selective siloxane permeation by dissolution and diffusion through dense polymeric or inorganic membrane material [58, 84], while methane is retained on the other side of the membrane. Membranes are characterized by large surface areas available for separation while occupying small volumes, which makes the technology very compact [85–87]. A selection of elastomeric membranes was assessed to determine the permeability and selectivity of some siloxanes (L_2, L_3, L_4, D_3, D_4 and D_5) towards methane in a real landfill gas matrix at 40°C and ambient pressure. A higher siloxane/methane selectivity will result in lower methane losses over the membrane. Membrane processes seem especially suitable when biogas upgrading is required, as CO_2 can also be removed.

4.5   Biological Degradation Processes

The biological removal of pollutants from gaseous phases presents certain advantages over physicochemical technologies, particularly their reduced operating costs [58, 59]. Bio-scrubbers (BS), bio-filters (BF) and bio-trickling filters (BTF) are typical reactor configurations suitable for these applications. The removal efficiencies of gas contaminants in biological systems depend both on the gas-liquid mass transfer rate and biodegradability of the target contaminants [88, 89]. Aerobic and anaerobic batch studies for the biodegradation of siloxanes did not show significant cell growth, indicating that these chemicals are not a suitable carbon source for microorganisms with limited biodegradation potential. In this context, Popat and Deshusses [90] conducted a thorough investigation to assess D_4 abatement in bio-trickling filters, concluding that the removal efficiency was limited to 40% even at very long retention times and the elimination capacity to 30–100 mg/m^3/h, which is 1,000 times smaller than biological H_2S removal [91]. These results introduced considerable doubts about the industrial implementation of biological processes for siloxane removal, and the investigations on this alternative slowed down in recent years.

More recently, Soreanu [92] conducted lab-scale experimentation with BTF for the removal of D4 and D5 from biogas obtaining very similar results than those from

8 years before (60% removal efficiency and 29 mg/m^3/h). At the time this chapter is written, the University of Girona (Spain) is working on a hybrid system combining adsorption and biological degradation of siloxanes (results not yet published). The rationale of this design is based on the capability of activated carbon to convert non-biodegradable D_4 and D_5 into more soluble and more biodegradable organic silicon compounds (silanediols) which can be further biodegraded. In this way, the most relevant limitations to improve the efficiencies of biological processes for siloxane removal are targeted.

The prospects for biological biogas treatment must also take into consideration the presence of hydrogen sulphide in biogas. BTF have already been tested for the removal of hydrogen sulphide from biogas under anoxic conditions. Simultaneous removal of both hydrogen sulphide and siloxane could represent an attractive cost-effective alternative to more expensive conventional biogas treatment technologies.

5 Conclusions

Siloxanes D_4 and D_5 are the most relevant volatile methyl siloxanes (VMS) present in sewage biogas, contributing consistently to over 95% of the total silicon contamination (25% D_4 and 70% D_5). Even though relevant differences have been observed on siloxane content from plant to plant, it can be concluded that average D_4 and D_5 concentrations are in the range of 3 and 16 mg/Nm^3, respectively. The anaerobic digestion temperature seems the most relevant operating condition affecting siloxane content in sewage biogas. Thermophilic digestion showed D_5 contents up to 127 mg D_5/Nm^3, around 15 times higher than average concentrations on mesophilic digesters. The contribution of sludge thermal pretreatments should be further studied, but a positive relationship is also envisaged. On the other hand, the link between mixing, retention time and type of sludge would probably contribute to a lesser extent to the contents of siloxanes.

Together with the other biogas contaminants, siloxanes pose a risk to the short- and long-term performance of Energy Conversion Systems (ECS). Hence, a purification stage should be installed in most of WWTP for reliable biogas valorization. The limits established for siloxanes differ from one technology to the other and even from one manufacturer to the other. Nonetheless, the available literature demonstrates that siloxane concentrations comparable to those observed in sewage biogas affect negatively the performance of energy generation units. While the impact on conventional technologies (internal combustion engines, boilers) can be observed in around 800–1,000 h, on innovative technologies (fuel cells), it starts to be relevant much earlier, i.e. 20–50 h.

Siloxane removal technologies are commercially available and can comply with the technical specifications of the different ECS. Refrigeration followed by adsorption (basically on activated carbon) is the most implemented technology in full-scale facilities, with operating costs in the range of 1.5–3 c€/Nm^3 depending on siloxane concentrations and humidity conditions. The use of adsorbent materials of increased lifetime would allow a reduction in costs related to adsorbent replacement. Adsorbent materials with large micro- and mesoporosities and textural properties with proton-donating groups showed the largest adsorption capacities. Within this context, alternative adsorbent materials with enhanced properties are being studied, and a combination of adsorption and biological degradation can deliver a breakthrough in siloxane removal technologies in the near future.

References

1. Eklund B, Anderson EP, Walker BL, Burrows DB (1998) Characterization of landfill gas composition at the fresh kills municipal solid-waste landfill. Environ Sci Technol 32:2233–2237
2. Ohannessian A, Desjardin V, Chatain V, Germain P (2008) Volatile organic silicon compounds: the most undesirable contaminants in biogases. Water Sci Technol 58:1775–1781
3. VDI (2008) In: K.d.L.i.V.u.D.-N. KRdL (eds) Measurement of landfill gases – measurements in the gas collection system. Beuth Verlag, Berlin
4. de Arespacochaga N, Valderrama C, Raich-Montiu J, Crest M, Mehta S, Cortina JL (2015) Understanding the effects of the origin, occurrence, monitoring, control, fate and removal of siloxanes on the energetic valorization of sewage biogas – a review. Renew Sust Energ Rev 52:366–381
5. Lassen C, Hansen CL, Mikkelsen SH, Maag J (2005) Siloxanes-consumption, toxicity and alternatives. Environmental Project No. 1031, Danish Environmental Protection Agency, Environmental Protection Agency. http://www.miljoestyrelsen.dk/udgiv/publications/2005/87-7614-756-8/pdf/87-7614-757-6.pdf. Accessed 17 Oct 2014
6. Tran TM, Abualnaja KO, Asimakopoulos AG, Covaci A, Gevao B, Johnson-Restrepo B et al (2015) A survey of cyclic and linear siloxanes in indoor dust and their implications for human exposures in twelve countries. Environ Int 78:39–44
7. Chemical Economics Handbook (2010) In: Will RK, Fink U, Kishi A (eds) Chemical economics handbook marketing research report silicones. IHS Publication, Denver
8. Mojsiewicz-Pieńkowska K, Jamrógiewicz M, Szymkowska K, Krenczkowska D (2016) Direct human contact with siloxanes (silicones) – safety or risk part 1. Characteristics of siloxanes (silicones). Front Pharmacol 7:132
9. Chemical Economics Handbook (2017) Chemical economics handbook marketing research report silicones. IHS Publication, Denver
10. Graiver D, Farminer KW, Narayan R (2003) A review of the fate and effects of silicones in the environment. J Polym Environ 11:129–136
11. Urban W, Lohmann H, Salazar Gómez JI (2009) Catalytically upgraded landfill gas as a cost-effective alternative for fuel cells. J Power Sources 193:359–366
12. Finocchio E, Montanari T, Garuti G, Pistarino C, Federici F, Cugino M, Busca G (2009) Purification of biogases from siloxanes by adsorption: on the regenerability of activated carbon sorbents. Energy Fuel 23:4156–4159
13. Haga K, Adachi S, Shiratori Y, Itoh K, Sasaki K (2008) Poisoning of SOFC anodes by various fuel impurities. Solid State Ionics 179:1427–1431
14. McBean EA (2008) Siloxanes in biogases from landfills and wastewater digesters. Can J Civ Eng 35:431–436
15. Appels L, Baeyens J, Degreve J, Dewil R (2008) Principles and potential of the anaerobic digestion of waste-activated sludge. Prog Energy Combust Sci 34:755–781

16. EEG (2009) In: N.C.a.N.S. Federal Ministry for the Environment (eds) Act revising the legislation on renewable energy sources in the electricity sector and amending related provisions – Renewable Energy Sources Act – EEG

17. Dewil R, Appels L, Baeyens J (2006) Energy use of biogas hampered by the presence of siloxanes. Energy Convers Manag 47:1711–1722

18. Capela D, Ratola N, Alves A, Homem V (2017) Volatile methylsiloxanes through wastewater treatment plants – a review of levels and implications. Environ Int 102:9–29

19. Oshita K, Omori K, Takaoka M, Mizuno T (2014) Removal of siloxanes in sewage sludge by thermal treatment with gas stripping. Energy Convers Manag 81:290–297

20. Raich-Montiu J, Ribas-Font C, de Arespacochaga N, Roig-Torres E, Broto-Puig F, Crest M, Bouchy L, Cortina JL (2014) Analytical methodology for sampling and analysing eight siloxanes and trimethylsilanol in biogas from different wastewater treatment plants in Europe. Anal Chim Acta 812:83–91

21. Rasi S, Läntelä J, Veijanen A, Rintala J (2008) Landfill gas upgrading with countercurrent water wash. Waste Manag 28:1528–1534

22. Schweigkofler M, Niessner R (1999) Determination of siloxanes and VOC in landfill gas and sewage gas by canister sampling and GC-MS/AES analysis. Environ Sci Technol 33:3680–3685

23. Tansel B, Surita SC (2014) Differences in volatile methyl siloxane (VMS) profiles in biogas from landfills and anaerobic digesters and energetics of VMS transformations. Waste Manag 34:2271–2277

24. Paolini V, Petracchini F, Carnevale M, Gallucci F, Perilli M, Esposito G, Segreto M, Galanti L, Scaglione D, Ianniello A, Frattoni M (2018) Characterisation and cleaning of biogas from sewage sludge for biomethane production. J Environ Manag 217:288–296

25. García M, Prats D, Trapote A (2015) Presence of siloxanes in the biogas of a wastewater treatment plant separation in condensates and influence of the dose of iron chloride on its elimination. Int J Waste Resour 6:1

26. Wang DG, Aggarwal M, Tait T, Brimble S, Pacepavicius G, Kinsman L, Theocharides M, Smyth SA, Alaee M (2015) Fate of anthropogenic cyclic volatile methylsiloxanes in a wastewater treatment plant. Water Res 72:209–217

27. Rasi S, Lehtinen J, Rintala J (2010) Determination of organic silicon compounds in biogas from wastewater treatments plants, landfills, and co-digestion plants. Renew Energy 35:2666–2673

28. Bensaid S, Russo N, Fino D (2010) Power and hydrogen co-generation from biogas. Energy Fuel 24:4743–4747

29. Björklund J, Geber U, Rydberg T (2001) Energy analysis of municipal wastewater treatment and generation of electricity by digestion of sewage sludge. Resour Conserv Recycl 31:293–316

30. Pöschl M, Ward S, Owende P (2010) Evaluation of energy efficiency of various biogas production and utilization pathways. Appl Energy 87:3305–3321

31. Mokhov AV (2011) Silica formation from siloxanes in biogas: novelty or nuisance. In: International Gas Union research conference, October19–21, Seoul, Korea

32. Schweigkofler M, Niessner R (2011) Removal of siloxanes in biogases. J Hazard Mater 83(3):183–196

33. Badjagbo K, Heroux M, Alaee M, Moore S, Sauve S (2010) Quantitative analysis of volatile methylsiloxanes in waste-to-energy landfill biogases using direct APCI-MS/MS. Environ Sci Technol 44:600–605

34. Libanati C, Ullenius DA, Pereira CJ (1998) Silica deactivation of bead VOC catalyst. Appl Catal B Environ 43:21–28

35. Pirnie M (2003) Retrofit digester gas engine with fuel gas cleanup and exhaust emission control technology – pilot testing of emission control system plant 1 engine 1. Orange County Sanitation District, Irvine

36. Alvarez-Florez J, Egusquiza E (2015) Analysis of damage caused by siloxanes in stationary reciprocating internal combustion engines operating with landfill gas. Eng Fail Anal 50:29–38

37. Sevimoglu O, Tansel B (2013) Composition and source identification of deposits forming in landfill gas (LFG) engines and effect of activated carbon treatment on deposit composition. J Environ Manag 128:300–305

38. Nair N, Zhang X, Gutierrez J, Chen J, Egolfopoulos F, Tsotsis T (2012) Impact of siloxane impurities on the performance on an engine operating on renewable natural gas. Ind Eng Chem Res 51:15786–15795
39. Bruno JC, Ortega-López V, Coronas A (2009) Integration of absorption cooling systems into micro gas turbine trigeneration systems using biogas: case study of a sewage treatment plant. Appl Energy 86:837–847
40. Somehsaraei HN, Majoumerd MM, Breuhaus P, Assadi M (2014) Performance analysis of a biogas-fueled micro gas turbine using a validated thermodynamic model. Appl Therm Eng 66:181–190
41. Iyer S (2011) Gas treatment systems. Nrgtek, Orange http://nrgtekusa.com/technology/gas_separation_membranes
42. Madi H, Lanzini A, Diethelm S, Papurello D, van Herle J, Lualdi M, Larsen JG, Santarelli M (2015a) Solid oxide fuel cell anode degradation by the effect of siloxanes. J Power Sources 279:460–471
43. Madi H, Diethelm S, Poitel S, Ludwig C, van Herle J (2015b) Damage of siloxanes on Ni-YSZ anode supported SOFC operated on hydrogen and bio-syngas. Fuel Cells 15:718–727
44. Papurello D, Chiodo V, Maisano S, Lanzini A, Santarelli M (2018) Catalytic stability of a Ni-catalyst towards biogas reforming in the presence of deactivating trace compounds. Renew Energy 127:481–494
45. Papurello D, Lanzini A (2018) SOFC single cells fed by biogas: experimental tests with trace contaminants. Waste Manag 72:306–312
46. Sasaki K, Haga K, Yoshizumi T, Minematsu D, Yuki E, Liu RR, Uryu C, Oshima T, Ogura T, Shiratori Y, Ito K, Koyama M, Yokomoto K (2011) Chemical durability of Solid Oxide Fuel Cells: influence of impurities on long-term performance. J Power Sources 196:9130–9140
47. Erekson EJ, Bartholomrw CH (1983) Sulfur poisoning of nickel methanation catalysts. II. Effects of H2S concentration, CO and H2O partial pressures and temperature on reactivation rates. Appl Catal 5:323–336
48. Papurello D, Lanzini A, Drago D, Leone P, Santarelli M (2016) Limiting factors for planar solid oxide fuel cells under different trace compound concentrations. Energy 95:67–78
49. Aschmann V, Kissel R, Gronauer A (2007) Untersuchungen zum Leistungs- und Emissionsverhalten biogasbetriebener Blockheizkraftwerke an Praxisbiogas-anlagen, 8th Conference Bau, Technik und Umwelt in der landwirtschaftlichen Nutztierhaltung
50. Thomas B, Bekker M, Kelm T, Oechsner H, Wyndorps A (2009) Gekoppelte Produktion von Kraft und Wärme aus Bio-, Klär- und Deponiegas in kleinen, dezentralen Stirling-Motor-Blockheizkraftwerken. Final report, BWPLUS Projekt No. 25008-25010, 2009. http://www.fachdokumente.lubw.baden-wuerttemberg.de
51. Petersson A, Wellinger A (2009) Biogas upgrading technologies-developments and innovations. IEA Bioenergy 20:1–19
52. Bekkering J, Broekhuis AA, van Gemert WJT (2010) Optimisation of a green gas supply chain: a review. Bioresour Technol 101:450–456
53. Starr K, Talens Peiro L, Lombardi L, Gabarrell X, Villalba G (2014) Optimization of environmental benefits of carbon mineralization technologies for biogas upgrading. J Clean Prod 76:32–41
54. van Essen M, Visser P, Gersen S, Levinsky H, Vainchtein D, Dutka M, Mokhov A (2013) Regarding specifications for siloxanes in biomethane for domestic equipment. Fifth Research Day of the Energy Delta Gas Research, Nunspeet
55. Nair N, Vas A, Zhu T, Sun W, Gutierrez J, Chen J, Egolfopoulos F, Tsotsis T (2013) Effect of siloxanes contained in natural gas on the operation of a residential furnace. Ind Eng Chem Res 52:6253–6261
56. SEPA (2004) Guidance on gas treatment technologies for landfill gas engines. Environment Agency, Bristol
57. Wheless E, Pierce J (2004) Siloxanes in landfill and digester gas update. http://www.scsengineers.com/Papers/Pierce_2004Siloxanes_Update_Paper.pdf. Accessed 17 Oct 2014

58. Ajhar M, Travesset M, Yuce S, Melin T (2010) Siloxane removal from landfill and digester gas – a technology overview. Bioresour Technol 101:2913–2923

59. Shen M, Zhang Y, Hu D, Fan J, Zeng G (2018) A review on removal of siloxanes from biogas: with a special focus on volatile methylsiloxanes. Environ Sci Pollut Res. https://doi.org/10.1007/s11356-018-3000-4

60. de Arespacochaga N, Valderrama C, Mesa C, Bouchy L, Cortina JL (2014) Biogas deep clean-up based on adsorption technologies for Solid Oxide Fuel Cell applications. Chem Eng J 255:593–603

61. Kuhn JN, Elwell AC, Elsayed NH, Joseph B (2017) Requirements, techniques, and costs for contaminant removal from landfill gas. Waste Manag 63:246–256

62. Cabrera-Codony A, Montes-Moran MA, Sanchez-Polo M, Martín MJ, Gonzalez-Olmos R (2014) Biogas upgrading: optimal activated carbon properties for siloxane removal. Environ Sci Technol 48:7187–7195

63. Rossol D, Schmelz KG, Hohmann R (2003) Siloxane im Faulgas. Abwasser Abfall 8:8

64. Sigot L, Ducom G, Benadda B, Labouré C (2014) Adsorption of octamethylcyclotetrasiloxane on silica gel for biogas purification. Fuel 135:205–209

65. Jafari T, Jiang T, Zhong W, Khakpash N, Deljoo B, Aindow M, Singh P, Suib SL (2016) Modified mesoporous silica for efficient siloxane capture. Langmuir 32:2369–2377

66. Sigot L, Ducom G, Germain P (2015) Adsorption of octamethylcyclotetrasiloxane (D4) on silica gel (SG): retention mechanism. Microporous Mesoporous Mater 213:118–124

67. Kajolinna T, Aakko-Saksa P, Roine J, Kåll L (2015) Efficiency testing of three biogas siloxane removal systems in the presence of D5, D6, limonene and toluene. Fuel Process Technol 139:242–247

68. Cabrera-Codony A, Georgi A, Gonzalez-Olmos R, Valdés H, Martín MJ (2017) Zeolites as recyclable adsorbents/catalysts for biogas upgrading: removal of octamethylcyclotetrasiloxane. Chem Eng J 307:820–827

69. Jiang S, Qiu T, Li X (2010) Kinetic study on the ring-opening polymerization of octamethylcyclotetrasiloxane (D4) in miniemulsion. Polymer 51(18):4087–4094

70. Oshita K, Ishihara Y, Takaoka M, Takeda N, Matsumoto T, Morisawa S, Kitayama A (2010) Behaviour and adsorptive removal of siloxanes in sewage sludge biogas. Water Sci Technol 61:2003

71. Ricaurte Ortega D, Subrenat A (2009) Siloxane treatment by adsorption into porous materials. Environ Technol 30:1073–1083

72. Nam S, Namkoong W, Kang JH, Park JK, Lee N (2013) Adsorption characteristics of siloxanes in landfill gas by the adsorption equilibrium test. Waste Manag 33:2091–2098

73. Matsui T, Imamura S (2010) Removal of siloxane from digestion gas of sewage sludge. Bioresour Technol 101:S29–S32

74. Yu M, Gong H, Chen Z, Zhang M (2013) Adsorption characteristics of activated carbon for siloxanes. J Environ Chem Eng 1:1182–1187

75. Hamelink JL, Simon PB, Silberhorn EM (1996) Henry's law constant volatilization rate, and aquatic half-life of octamethylcyclotetrasiloxane. Environ Sci Technol 30(6):1946–1952

76. Yashiro T, Kricheldorf HR, Schwarz G (2010) Polymerization of cyclosiloxanes by means of triflic acid and metal triflates. Macromol Chem Phys 211(12):1311–1321

77. Cabrera-Codony A, Santos-Clotas E, Ania CO, Martín MJ (2018) Competitive siloxane adsorption in multicomponent gas streams for biogas upgrading. Chem Eng J 344:565–573

78. Gislon P, Galli S, Monteleone G (2013) Siloxanes removal from biogas by high surface area adsorbents. Waste Manag 33(12):2687–2693

79. Boulinguiez B, Le Cloirec P (2009) Biogas pre-upgrading by adsorption of trace compounds onto granular activated carbons and an activated carbon fibercloth. Water Sci Technol 59:935–944

80. Cabrera-Codony A, Gonzalez-Olmosa R, Martin MJ (2015) Regeneration of siloxane-exhausted activated carbon by advanced oxidation processes. J Hazard Mater 285:501–508

81. Soreanu G, Beland M, Falletta P, Edmonson K, Svoboda L, Al-Jamal M, Seto P (2011) Approaches concerning siloxane removal from biogas, a review. Can Biosyst Eng 53:8–18

82. Läntelä J, Rasi S, Lehtinen J, Rintala J (2012) Landfill gas upgrading with pilot-scale water scrubber: performance assessment with absorption water recycling. Appl Energy 92:307–314

83. Abatzoglou N, Boivin S (2009) A review of biogas purification processes. Biofuels Bioproducts Biorefining 3:42–71

84. Ajhar M, Melin T (2006) Siloxane removal with gas permeation membranes. In: Conference of the European-Membrane-Society (EUROMEMBRANE 2006) Giardini Naxos, Italy, 24-28 September 2006. Elsevier Science Bv, Amsterdam, pp 234–235

85. Meinema HA, Dirrix RWJ, Terpstra RA, Jekerle J, Kösters PH (2005) Ceramic membranes for gas separation-recent developments and state of the art. Interceram 54:8691

86. Pandey P, Chauhan RS (2001) Membranes for gas separation. Prog Polym Sci 26:853–893

87. Strathman H, Bell CM, Kimmerle K (1986) Development of synthetic membranes for gas and vapor separation. Pure Appl Chem 58:1663–1668

88. Gabriel D, Deshusses MA (2003) Retrofitting existing chemical scrubbers to biotrick-ling filters for H2S emission control. Proc Natl Acad Sci U S A 100:6308–6312

89. Li Y, Zhang W, Xu J (2014) Siloxanes removal from biogas by a lab-scale biotrickling filter inoculated with Pseudomonas aeruginosa S240. J Hazard Mater 275:175–184

90. Popat SC, Deshusses MA (2008) Biological removal of siloxanes from landfill and digester gases: opportunities and challenges. Environ Sci Technol 42:8510–8515

91. Soreanu G, Beland M, Falletta P, Edmonson K, Seto P (2008) Laboratory pilot scale study for H2S removal from biogas in an anoxic biotrickling filter. Water Sci Technol 57:201–207

92. Soreanu G (2016) Insights into siloxane removal from biogas in biotrickling filters via process mapping-based analysis. Chemosphere 146:539–546

Volatile Dimethylsiloxanes in Aquatic Systems

Josep Sanchís and Marinella Farré

Contents

Abstract Volatile methylsiloxanes are high-volume synthetic chemicals that are included in a plethora of domestic and industrial formulations. Because of their widespread use, these organosilicon molecules are emitted to the environment and reach the aquatic systems, where they may cause potential adverse effects to some aquatic organisms. The study of the occurrence and fate of volatile methylsiloxanes in the aquatic media has progressed considerably during the last years thanks to the development of new analytical methods, which decrease the limits of detection substantially while minimising and stabilising the contamination levels of the blank assays. The present chapter briefly reviews the most relevant analytical strategies that have been developed for the analysis of volatile methylsiloxanes in the aquatic environment, with a focus in water matrices, sediments and biota. The behaviour and fate of cyclic and linear methylsiloxanes in the aquatic environment are summarised, as well as the levels at which these compounds have been reported.

J. Sanchís and M. Farré (✉)
Institute of Environmental Assessment and Water Research (IDAEA-CSIC), Barcelona, Catalonia, Spain
e-mail: mfuqam@cid.csic.es

V. Homem and N. Ratola (eds.), *Volatile Methylsiloxanes in the Environment*, Hdb Env Chem (2020) 89: 159–180, DOI 10.1007/698_2018_363,
© Springer Nature Switzerland AG 2018, Published online: 6 September 2018

Keywords Analytical methods, Aquatic systems, Occurrence, Volatile methylsiloxanes

1 Introduction

Siloxanes are anthropogenic organosilicon compounds of polymeric nature with outstanding properties such as chemical inertness, thermal stability, electrical resistivity, gas permeability, hydrophobicity and fluidity, among others [1, 2]. Siloxanes are characterised by the alternation of organosilane groups and oxygen atoms, although their structures may vary considerably concerning the molecular weight, the degree of functionalisation and embranchment. Those that are mainly formed by the repetition of dimethylsiloxy groups ($-SiMe_2-O-$) are named *polydimethylsiloxanes* or *dimethicones*, and, among them, those with a low molecular weight – typically containing seven or less silicon atoms – are called *volatile methylsiloxanes* (*VMS*).

VMS are commonly classified according to their structure. Those VMS with a circular structure (empirical formula $(CH_3)_2Si_nO_n$, $n \geq 2$) are named *cyclic volatile methylsiloxanes* (*cVMS*). Those consisting in a linear chain (formulae $(CH_3)_{2n+2}Si_nO_{n-1}$, $n \geq 2$) are named *linear volatile methylsiloxanes* (*lVMS*).

VMS are mass-produced chemicals. Their global production in 2002 accounted for 2×10^6 tons/year – 34% in North America, 33% in Europe and 28% in Asia [3–6] – and it is estimated that more than 8×10^6 tons/year were produced in 2016 [7]. VMS are employed in a wide range of applications, including paintings formulations, detergents, cosmetics, lubricants and other industrial and household products [8].

The number of articles dealing with the environmental occurrence, behaviour and fate of siloxanes has increased exponentially during the last years. Figure 1 illustrates how, from the early 1990s, VMS have awakened a sustained interest among the scientific community, environmental agencies and regulatory organisms. According to Mojsiewicz-Pieńkowska et al. [7], the awareness about the potential environmental issues of VMS has increased given the emergence of some reports warning about the long-term exposure of living organisms – including humans – to siloxanes, allied to the increased accessibility to scientific knowledge, reports and adequate analytical instrumentation and methods. All these factors have contributed to better understanding the behaviour and fate of siloxanes in the environment, particularly in the case of cVMS, which have gathered most of the attention and are produced at higher amounts than their linear homologues.

A consensus exists in observing cVMS as non-harmful chemicals when considering the current emission levels and their limited environmental persistence [9–11]. However, some studies have warned about the potential toxic effects to some aquatic organisms. For example, octamethylcyclotetrasiloxane (D4) exhibited LC_{50} of 10 µg/L to rainbow trouts (*Oncorhynchus mykiss*) after 14 days of exposure [12] and its no-observed-effect concentration (NOEC) after 93 days was 4.4 µg/L. Similarly, the NOEC of decamethylcyclopentasiloxane (D5) and dodecamethylcyclohexasiloxane

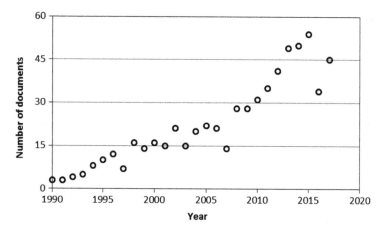

Fig. 1 Number of published SCI documents related to siloxanes with an environmental scope, including articles (81%), conference papers (14%), reviews (3%) and others (~2%: book chapters, letters, editorials and an erratum). Search performed in Scopus® database, searching the terms "siloxane" *or* "cVMS" *or* "lVMS" *and* "environment" *or* "lake" *or* "river" *or* "air" *or* "dust" *or* "wastewater" *or* "soil" *or* sediment" in the title *or* the abstract *or* the keywords of the document. Results filtered to the subject area "environmental science"

(D6) for fathead minnow (*Pimephales promelas*) was 8.7 µg/L after 65 days of exposure [13] and 4.4 µg/L after 49 days of exposure [14], respectively. The bioaccumulation potential of D5 was demonstrated in field studies conducted in rivers, lakes and marine environments [15–18]. Also, some recent models challenge the non-persistent character of cVMS and lVMS in some particular cases of the aquatic environment [19]. Because of all these, several authors [20–23] include cVMS and lVMS in the category of emerging contaminant, together with other families of well-known pollutants such as perfluoroalkyl substances, pharmaceuticals or nanomaterials, among others.

The present chapter reviews the analytical methods commonly employed for the analysis of VMS in aquatic samples as well as their fate and occurrence in this environmental compartment.

2 Analysis of VMS in the Aquatic Environment

The need to assess trace levels of VMS in complex environmental samples is one of the main challenges of current instrumentation, both regarding sensitivity (in particular for lVMS) and concerning contamination issues (in particular, in the case of cVMS). To overcome these challenges, the study of VMS in the aquatic environment has required the development of analytical methods that try to maximise the preconcentration factors while maintaining a strict control of the potential sources of contamination along all the sampling and analytical procedure. The following subsections briefly review the main analytical methods for the analysis of VMS in samples from the aquatic environment.

2.1 Extraction Techniques

2.1.1 Extraction and Microextraction of VMS from Water Matrices

Liquid-liquid extraction is a common methodology that can be universally implemented in analytical laboratories with cheap instrumentation. In the case of VMS, this approach reaches good recovery percentages in wastewater and freshwater samples [24–27] with common solvents such as hexane, dichloromethane, ethyl acetate or a combination of these. After phase separation, the organic extract is collected and concentrated under gentle rotary evaporation or N_2 blowdown. Cortada et al. [28] published a liquid-liquid microextraction method assisted by ultrasounds (UA-DLLME) which did not require the evaporation of the extract, minimising its manipulation and the subsequent risk of contamination.

Other approaches for the extraction of VMS from water matrices include headspace extraction [29–31], headspace solid-phase microextraction [32–34] and purge and trap [35–37].

2.1.2 Extraction of VMS from Sediments

Solid-liquid extraction, eventually assisted by sonication [26], is the most common technique for the extraction of VMS from sediment samples. Dried sediments can be extracted with solvents such as hexane [38], pentane [39] or ethyl acetate:hexane [26, 40]. Lee et al. [41] proposed a sequential extraction with hexane, hexane: dichloromethane (1:1) and hexane:ethyl acetate (1:1). The cited methods present adequate recovery percentages for the most relevant compounds (typically D4–D6 and L3–L5) and limits of detection below (or close to) 1 ng/g using sediment:solvent that range from 0.2 ± 0.04 to 0.5 g/mL.

Alternatively, Kaj et al. [36] employed the purge and trap technique. River sediments were homogenised with ultrapure water obtaining a slurry that was analysed similarly than surface water samples.

2.1.3 Extraction of VMS from Aquatic Biota

Extraction of VMS from biota can also be obtained with straightforward methods based on solid-liquid extraction, with organic solvents like hexane [37, 38, 42, 43], pentane [42], tetrahydrofuran [42, 44, 45] and methanol [46] and mixtures of hexane-ethyl acetate [25], purge and trap [47–49], and, more recently, using QuEChERS ("quick, easy, cheap, effective, rugged and safe") technology [50].

2.2   Instrumental Analysis of VMS

Gas chromatography coupled to mass spectrometry (GC-MS) is the optimal technique for the determination of cVMS and lVMS, taking into account the physico-chemical properties of these compounds: relatively high 1-octanol-water partition coefficients (K_{ow}) and vapour pressures (Table 1), the lack of polar functional groups in their structures and their affinity towards GC-friendly organic solvents.

Well-resolved Gaussian peaks of cVMS and lVMS can be easily achieved employing fused-silica capillary columns with common stationary phases based on dimethylsiloxane/methylphenylsiloxane [24, 26, 32, 59, 60] or cyanopropylphenyl/ methylphenylsiloxane copolymers [28, 61, 62]. The main limitation of this approach is the release of cVMSs from the column, particularly after prolonged use. Capillary columns with polyethylene glycol stationary phases have been employed as contamination-free alternatives [30, 31, 38, 63]. Silicon-based septa are another potential source of contamination. Silicon-free septa, such as Merlin® MicroSeal, are commonly used as alternatives [32, 61].

GC-MS provides good ionisation performance of VMSs by standard electron ionisation sources (IE), which result in outstandingly sensitive and selective analyses. The acquisition is typically performed by single ion monitoring (SIM), with simple quadrupoles or ion-trap analysers. Alternatively, selected reaction monitoring (SRM) acquisition with triple quadrupole analysers has been purposed. SRM offers a lower instrumental response than SIM, but in the analysis of complex water extracts, this loss in sensitivity is compensated by an increase in its selectivity and neater chromatograms [26, 41].

Both GC-IE-MS and GC-EI-MS/MS offer better sensitivity and selectivity than other detection techniques, such as GC with flame ionisation detectors (GC-FID) [45, 64–66], atmospheric pressure chemical ionisation sources (APCI-MS) [67, 68], selected ion flow tube mass spectrometry (SIFT-MS) [69] or liquid scintillation counter (LSC) [45, 70, 71].

2.3   Analysis of Degradation Products

Hydroxyl-terminated siloxanes (*silanols*) are common degradation products of cVMS and lVMS (also in aqueous media), which have short lives in the environment [72]. The analytical methods employed for the analysis of precursor cVMS and lVMS are less adequate for the analysis of these molecules because of their higher water affinity and polarity. In fact, the K_{OW} of silanols is drastically different than that of cVMS/lVMS [72]. Because of this, substantially different extraction procedures are required for these compounds.

In the case of water matrices, dimethylsilanediol has been extracted by LLE, with 1-pentanol and methyl isobutyl ketone [73], and by SPE, with ENV-Carb cartridges

Table 1 Relevant physicochemical properties of selected IVMS and VMS

Name	Abbreviation	CAS number	Formula	Vapour pressure (Pa)	log K_{OW}	Water solubility
Octamethyltrisiloxane	L3	107-51-7	$C_8H_{24}O_2Si_3$	521 ± 5[a]	6.87 ± 0.24[b]	34.49 ± 1.0 ppb[c]
Decamethyltetrasiloxane	L4	141-62-8	$C_{10}H_{30}O_3Si_4$	58.8 ± 0.7[a]	8.09 ± 0.13[b]	6.74 ± 0.8 ppb[c]
Dodecamethylpentasiloxane	L5	141-63-9	$C_{12}H_{36}O_4Si_5$	6.0 ± 0.5[d]	6.0[e]	70.41 ± 8.3 ppb[c]
Tetradecamethylhexasiloxane	L6	107-52-8	$C_{14}H_{42}O_5Si_6$	0.66 ± 0.16[d]	6.6[e]	1.3×10^{-5} mg/L[f]
Hexamethylcyclotrisiloxane	D3	541-45-9	$C_6H_{18}O_3Si_3$	1,140 ± 8[a]	3.85[g]	1.56 ± 0.03 ppm[c]
Octamethylcyclotetrasiloxane	D4	556-67-2	$C_8H_{24}O_4Si_4$	124.5 ± 6.2[d]	6.98 ± 0.13[h]	56.20 ± 2.50 ppb[c]
Decamethylcyclopentasiloxane	D5	541-02-6	$C_{10}H_{30}O_5Si_5$	20.4 ± 1.1[d]	8.07 ± 0.22[h]	17.03 ± 0.72 ppb[c]
Dodecamethylcyclohexasiloxane	D6	540-97-6	$C_{12}H_{36}O_6Si_6$	2.26 ± 1.44[d]	8.87 ± 0.14[h]	5.13 ± 0.48 ppb[c]

[a]Calculated at $T = 298.15$ K from constants for the AIChE DIPPR equation determined in Flaningam [51]
[b]Calculated at 293.95°C and at 292.85°C for L3 and L4, respectively [52]
[c]Experimentally calculated at 296.15 K [53]
[d]Liquid state vapour pressure at 298.15 K [54]
[e]From [55]
[f]Estimated using the model "Water Solubility for Organic Compounds Program for Microsoft Windows v.1.41" [56]
[g]From [57]
[h]Calculated at 21.7°C, 24.6°C and 23.6°C for D4, D5 and D6, respectively [58]

eluted with tetrahydrofuran [74]. Recently, Xu et al. [27] analysed the occurrence of hydroxylated-D4 and hydroxylated-D5 in several matrices, including landfill leachate. The extraction of these compounds was conducted by LLE with (1) 25 mL of *n*-hexane and (2) 20 mL of *n*-hexane/ethyl acetate (1:1).

Varaprath and Lehmann [75] were the first to show that some low-molecular-weight silanols could be easily analysed by GC with the same methodologies employed for the analysis of cVMS and lVMS and with no derivatisation. THF standards of hydroxyl-diterminated linear dimethylsiloxanes with two to five silicon atoms (namely, *di-*, *tri-*, *tetra-*, *penta-* and *hexamerdiol*) were analysed with a commercial HP-5 capillary column (Agilent), previously deactivated with a silylating agent, bis(trimethylsilyl)acetamide.

This procedure has been applied in more recent studies employing state-of-the-art instrumentation and with no reported previous column silylation [74]. The analysis of hydroxylated-D4 and hydroxylated-D5 can also be conducted by conventional GC-MS [27]. It should be mentioned that some authors opted for derivatisation protocols, such as the acid-catalysed trimethylsilylation with hexamethyldisiloxane (HMDS) [76], in order to reduce peak tailing and to analyse compounds with a higher number of hydroxyls.

When GC is coupled to mass spectrometry, electron ionisation (EI) offers a proper ionisation of merdiols [75]. However, the EI-MS spectra of those merdiols and cVMS with the same number of silicon atoms share the same radical ions, with only slightly different relative abundances. Positive chemical ionisation (PCI) was proposed as an alternative to avoid the potential misidentification of merdiols [27, 75].

Alternatively, silanols have been fractionated or separated chromatographically in aqueous matrices by reversed-phase liquid chromatography (LC), using water-acetonitrile gradients [77]. LC systems may be coupled to an optical emission spectroscope with an inductively coupled plasma atomic source (LC-ICP-OES) [75, 78], to a radiometric detector [58] or to mass spectrometry.

3 Occurrence and Behaviour of VMS in the Aquatic Environment

3.1 Sources of Siloxanes to the Aquatic Environment

VMS present high vapour pressure values. Therefore, a significant fraction of VMS from personal care product formulations is expected to volatilise during its application and the few following hours. Jovanovic et al. [79] carried out in vivo dermal exposure experiments and observed that <1.0% of D4 and only 0.2% of D5 were absorbed by organisms after 24 h [79]. Despite this general scheme, siloxanes are produced and used in high amounts, and a significant loading of cVMS still reaches wastewater treatment plants (WWTPs). The wastewater treatments significantly reduce the concentrations of polydimethylsiloxanes [80] and VMS in the treated effluents, but some concentrations are still there. For example, Kaj et al. [37] reported concentrations of cVMS and lVMS in the influents and the effluents of WWTPs from Nordic countries. cVMS were the dominant compounds, particularly D5. The prevalence of cyclic VMS over linear VMS is, in addition, in good agreement with the profiles in personal care products [8] and their production rates, as it has been reported. The concentrations of cVMS in the influents were <MLOD–3.7 µg/L, 0.33–26 µg/L and 0.12–3.8 µg/L, for D4, D5 and D6, respectively, while in the effluents, the concentration levels were about one order of magnitude lower: <MLOD–0.11 µg/L, 0.063–5.2 µg/L and <MLOD–0.33 µg/L, for D4, D5 and D6, respectively. Regarding lVMS, concentrations in the influent ranged from 0.0034 to 0.014 µg/L for L3, from <MLOD to 0.078 µg/L for L4 and from <MLOD to 0.23 µg/L for L5 and decreased significantly after their treatment. In Table 2, the levels reported in different surveys from Canada [81], Greece [24], Catalonia [26] and China [33] are presented.

The elimination of cVMS and lVMS is explained by sorption to wastewater sludge, volatilisation and degradation (i.e. to dimethylsilanediol) [24, 59]. This process was observed to be compound dependant and dependent on the amount of dissolved organic matter [81]. In any case, the removal of VMS is incomplete, and trace concentrations of lVMS and, particularly, cVMS are emitted by WWTP discharges, thus being an important source of siloxanes to the aquatic environment [81] (see Fig. 2).

The introduction of VMS into the aquatic environment due to atmospheric deposition has generally been considered as negligible because of the combination of physicochemical properties: high volatility and short half-lives in the atmosphere [83, 84]. However, some recent works pointed out that snow scavenging could favour the introduction of VMS into the aquatic environment. Concentrations of cVMS in the air of Toronto (Canada) were observed to decrease from 121 to 73 ng/m^3 in snow events [85]. Also, Sanchís et al. [86] justified the occurrence of

Table 2 Occurrence of cVMS in selected surveys

Reference	Location	Type of sample	D4 (µg/L)	D5 (µg/L)	D6 (µg/L)
Kaj et al. [37]	DNK, FIN, FRO, NOR	Influent	<MLOD–3.7	0.33–26	0.12–3.8
		Effluent	<MLOD–0.11	0.063–5.2	<MLOD–0.33
Wang et al. [39]	CAN	Influent	0.282–6.69	7.75–691	31.2–338
		Effluent	<MLOD–0.045	<MLOD–1.56	<MLOD–0.093
Sanchís et al. [26]	ESP	Influent	<MLOQ–1.09	<MLOD–24.4	Not analysed
		Effluent	<MLOD–0.476	0.0421–3.59	Not analysed
Bletsou et al. [24]	GRE	Influent	0.099–0.187	0.544–5.36	1.16–3.19
		Effluent	0.103–0.197	0.125–6.02	0.026–0.020
Xu et al. [33]	CHN	Influent	2.42–2.89	3.04–3.29	2.20–2.56
		Effluent	0.25–0.55	0.50–1.00	0.52–0.96

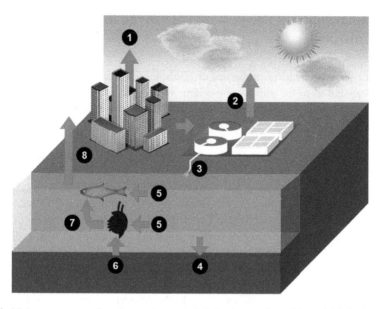

Fig. 2 Main processes ruling the environmental behaviour and partition of VMS, including emission to the atmosphere from urban sources (1) or wastewater treatment plants (2), release into the aquatic environment (3), partition to sediments (4), introduction to the food web from water (5) or sediments (6), transfer through the trophic chain (7) and water-air partition (8) (adapted from [82])

VMS in remote areas with the action of snow scavenging and ice melting, leading to unexpected nonequilibrium concentrations of cVMS in soils and vegetation from remote areas [86, 87]. On the other hand, models have shown that hydroxylated cVMS, including the main (photo)degradation product, dimethylsilanediol, do undergo wet deposition because of their higher water affinity [88, 89] at warmer temperatures. Thus, the role of atmospheric deposition as an additional source of siloxanes to the aquatic environment should not be dismissed.

The degradation of polydimethylsiloxanes by hydrolysis in the environment also produces VMS, but it cannot be considered a major source of cVMS and lVMS to the aquatic environment. The hydrolysis of oligo- and large polydimethylsiloxanes mainly results in dimethylsilanediol ($HO-Si(CH_3)_2-O-Si(CH_3)_2-OH$), the main product, and minor amounts of trimethylsilanol from the hydrolysis of terminal groups [78, 90, 91]. These reactions have been mainly studied in soils [92], and cVMS and lVMS are formed only at trace level because of the degradation of low-life intermediates and low-molecular-weight silanols [78, 93]. In sediments (which is the relevant compartment for virtually insoluble high-molecular weight polydimethylsiloxanes), slow hydrolysis rates were determined [94]: for D5, half-lives in sediments ranged from 800 days to 3,100 days at 24°C, depending on the experimental conditions (sterile/biotic and aerobic/anaerobic exposures) [39]. It should be mentioned that in one study, the concentrations of some cVMS were higher in effluents than in influents, which was justified by the degradation of polydimethylsiloxane (PDMS) during the WWTP treatment [24].

3.2 Behaviour and Partition of VMS in the Aquatic Environment

Given the physicochemical properties of VMS, these compounds tend to be significantly transferred to the vapour phase once released into the aquatic systems. For example, regarding cVMS, the log K_{AW} of D4 and D5 was experimentally measured as 2.69 ± 0.13 (21.7°C) and 3.13 ± 0.13 (24.6°C), respectively [58], while for lVMS, it was 2.32 ± 0.03 for L2 (19.7°C), 3.04 ± 0.20 for L3 (20.8°C) and 3.22 ± 0.13 for L4 (19.7°C) [52]. Also, VMS will tend to be adsorbed to suspended material and sediments. log K_{OC} for D4, D5 and D6 was experimentally assessed as 5.17 ± 0.23, 6.22 ± 0.09 and 7.36 ± 0.10 [95], although it should be taken into account that these values may vary substantially with temperature [95] and salinity [96]. K_{OC} correlates inversely with T and directly with the ionic strength of the aqueous medium, which is relevant when assessing the fate of cVMS in different seasons and water media [19].

In the water phase, hydrolysis processes mitigate the concentrations of cVMS. Hydrolyses happen as pseudo-first order reactions, with degradation rates that are system dependent, as they vary according to the physicochemical properties of the medium (i.e. pH and temperature). Half-lives of D4 and D5 in freshwater at pH = 7 and 12°C were 16.7 and 315 days, respectively, while they were 2.9 and 64 days in seawater at pH = 8 and 9°C [39, 97, 98].

Table 3 indicates the distribution of cVMS in the different compartments according to a Level III fugacity model published in the corresponding monographs conducted by Canadian environmental and health agencies [99–101]. As it can be seen, these fugacity models estimate that around 72.2% of D4, 30.9% of D5 and 16.0% of D6 emitted to the aquatic environment are expected to remain in this compartment after equilibrium. Later, an updated version of the model was applied to a single cVMS (D5). The new model concluded that even a lower amount of siloxanes remains in the water compartment (5.2% in water, 94.0% in sediment) [102]. The modelled scenario presented a water compartment of 10^{10} m^2 and 20 m depth, containing suspended particulate (1:5 × 10^{-6} water:particulate volumes, $\rho = 1{,}500$ kg/m^3); a sediment compartment of 10^{10} m^2 and 5 cm depth containing 80% of pore water ($\rho = 1{,}280$ kg/m^3); an atmosphere compartment of 10^{11} m^2 and 1,000 m depth containing aerosol (2 × 10^{-11} air:aerosol volumes, $\rho = 2{,}000$ kg/m^3); and a soil compartment of 9 × 10^{10} and 20 cm depth (0.2:0.3:0.5 air:water:solid). In another scenario, where D5 was emitted simultaneously to all the environmental compartments at realistic rates (94.5% to air, 0.8% to water and 4.7% to soil) resulted

Table 3 Estimated distribution of cVMS in the different environmental compartments when emitted directly to the water system

	Water (%)	Air (%)	Sediment (%)
D4	72.2	13.6	14.2
D5	30.9	4.4	64.7
D6	16.0	0.9	83.0

in 1.84% of D5 remaining in the water. In that case, 33.2% of D5 remained in sediment in equilibrium.

Panagopoulos and MacLeod [19] modelled the persistence of cVMS in several environmental compartments of a particular location (Longyearbyen fjord, Adventjorden, Norwegian Arctic), simulating the seasonality of K_{OC} values and its dependence with environmental factors (temperature, amount of suspended particles, sediment deposition rates, etc.). The overall residence time of D4, D5, D6, L4 and L5 was 287 h, 5,496 h, 9,889 h, 2,872 h and 8,143 h, respectively. The authors of this study concluded that the evaluated VMS meet several criteria for being considered persistent compounds, at least in the particular scenario of this study (freshwater and cold environment): the REACH seawater criterion (only from October to March), the REACH marine sediment criterion and the 100-day overall residence time criterion [103].

3.3 Levels of Siloxanes in River, Lake and Coastal Environments

The presence of organosilicon compounds in environmental waters attracted the attention of scientists already in the late 1970s. Pellberg [104] reported average concentrations of organosilicon compounds close to ~30 µg/L in surface microlayer samples from Potomac River and Delaware Bay, USA. Also, levels up to 54 µg/L were found in five of the nine water samples analysed in the Nagara River (Japan) [105], while no silicones were detected above the limit of detection (2.5 µg/L) in the Tokyo, Ise and Osaka bays [106]. However, these reports were unable to discriminate between VMS and large molecular weight dimethylsiloxanes, and most of the volatile compounds may have been lost during the extraction and preconcentration steps. Since then, several surveys have been conducted in coastal water and surface freshwater bodies covering different regions of the Northern Hemisphere. Kaj et al. [37] analysed the concentrations of VMS in a large number of water samples from the Nordic countries. cVMS (D4–D6) and lVMS (L2–L5) were found below the method limit of detection in all the surface water samples, including coastal locations of Denmark, Iceland and Norway, the lakes Bergsjøen and Røgden (Norway) and the Swedish river Nissan. In the same survey, detectable concentrations of VMS were reported in WWTP influents and effluents and, eventually, in some landfill leachates and runoffs, which suggest that these emissions are considerably attenuated when they reach surface water bodies and coastal areas.

Later, Sparham et al. [30] analysed D5 in the waters of two English rivers, the Great Ouse River ($n = 9$, including a sample from the tributary Ouzel) and River Nene ($n = 5$). The concentrations in the Ouse River basin ranged from the MLOD (10 ng/L) to 29.4 ng/L. In the Nene River, concentrations ranged from 12.9 ± 0.4 to 26.8 ± 3.3 ng/L. These concentrations were significantly lower than those of the Great Billing STW treated effluent, which discharged 400 ± 12 ng/L of D5 into the waters of Nene River.

In other works, a much higher influence of the WWTPs in receiving waters was reported. For example, Wang et al. [81] reported concentrations of D4, D5 and D6 of <9–23, 27–1,480 and <22–151 ng/L. In that study, the concentrations of D5 in surface water almost matched the same concentrations found in the treated effluents discharged in some sampling points.

In Catalonia, NE of Spain, two surveys were conducted. Companioni-Damas et al. [32] analysed seven samples from the Llobregat River and five samples from the Besòs River. In that study, D3 and D4 were not detected, but D5 and D6 were found in three and one water samples, at concentrations of 22.2–58.5 ng/L and 21.2 ng/L, respectively. These concentrations are in agreement with those previously detected in Nene and Great Ouse Rivers [30]. lVMS (L2–L5) were the most frequently detected compounds but at lower levels, below or close to 1 ng/L. In another later work, three samples from the Llobregat River and three samples from its tributary, the Riera de Rubí, were analysed. In this study, significantly higher concentrations of cVMS were detected (27.1–76.0 ng/L of D3, 61.2–987 ng/L of D4 and 87.2–468 ng/L of D5). The highest concentrations were found in the Riera de Rubí, which is a low-flow brook located in a highly industrialised area with influence from large WWTPs plants.

Horii et al. [35] analysed water samples from the Tokyo Bay and several discharging rivers. cVMS were detected in more than 88 of samples at mean concentrations of 10, 13, 180, 18 ng/L for D3, D4, D5 and D6, respectively. Again the concentrations were particularly high in the areas close to WWTP discharges, such as in Ara and Motokoyama rivers. In these cases, the concentration of D5 reached the μg/L order, while the reference points located upstream were significantly lower. Regarding L3, L4 and L5, these lVMS were detected less frequently (in 0, 8.3 and 63% of samples) and at lower levels (<0.6 ng/L, <0.6–2.3 ng/L and 0.7–7.1 ng/L, respectively) than cVMS.

Recently, three works studied the occurrence of VMS in surface waters from China [25, 74, 107]. Zhi et al. [74] analysed 41 surface water samples from different lakes located in Daqing (an oilfield area in Heilongjiang Province, China). In this study, three locations were investigated:

1. An area with a recent presence of oilfields (*new oilfields*, sum of cVMS ranged from the limit of quantification to 92.3 ng/L, median concentration 39.6 ng/L)
2. An area with prolonged oilfield activity (*old oilfields*, sum of cVMS ranged from the limit of quantification to 242 ng/L, median concentration 37.4 ng/L)
3. A reference location, far from oilfields (sum of cVMS up to 149 ng/L, median concentration, 5.1 ng/L)

No significant differences were found among the new and the old oilfield areas, which was justified by the differences in the lake's pH (higher pH values in old oilfield lakes accelerated the degradation of the compounds). Concentrations of lVMS (from L7 to L16) were similar in the three locations (median concentrations, 41.1, 67.8 and 52.3 ng/L), although the maximum concentrations in the reference point were significantly lower than the extreme points of the other groups. In the second work, slightly higher concentrations were observed in the Dongting Lake

(Hunan Province, China) [107]. The concentrations of D4, D5 and D6 were 6.07–85.59 ng/L, 16.35–313.10 ng/L and 14.24–309.35 ng/L, respectively, and all the cVMS were detected in the 13 samples. These concentrations were justified because of the direct input of VMS from nearby WWTPs. Finally, the presence of VMS was investigated in open seawaters and coastal locations [25]. The levels were found notably lower, as expected. Average levels of D4, D5, D6 and D7 were 6.93 ± 9.48 ng/L, 12.2 ± 10.3 ng/L, 16.5 ± 13.9 ng/L and 5.66 ± 4.52 ng/L, respectively, while the average concentrations of lVMS (L8–L12) were around 1 ng/L and never exceeded the 5 ng/L. A clear difference could be observed between samples located close (65.6 ± 32.1 ng/L of total siloxanes) and relatively far (32.7 ± 6.26 ng/L) from urban areas.

To summarise, the predominance of cVMS over lVMS was observed in all the reported works. The levels of VMS are in general at the ng/L order, and they are highly dependent on the presence of nearby local emission sources, such as urban WWTPs and silicone facilities. The characteristics of the locations in terms of salinity, suspended material, organic material content, climate and seasonality also play important roles in the fate of the detected VMS [19].

3.4   Levels of Siloxanes in Sediments

Once emitted to the aquatic environment, VMS tend to be adsorbed readily to sediment where they are relatively persistent. For this reason, many studies have studied the levels of these compounds in river, lake and marine sediments. The main results of these studies are summarised in Table 4.

As can be seen, the levels of cVMS (typically, D4, D5 and/or D6) range from the low ng/g$_{dw}$ levels to, occasionally, the low μg/g$_{dw}$, with D5 being universally the most frequently detected siloxane and the one quantified at higher concentrations. In general, the highest concentrations of cVMS were observed in sediments from freshwater bodies that received the direct input of local contamination sources, such as in Kaj et al. [37], Sparham et al. [108], Sanchís et al. [26], Wang et al. [81] and Lee et al. [41]. In these studies, sediment samples were taken in freshwater bodies from highly industrialised areas or near WWTP discharges, and concentrations of D5 were observed to be higher than 1 μg/g$_{dw}$.

Significantly lower concentrations have been determined in marine sediments. For instance, the concentrations of D5 in sediments from Huanghai Sea ranged from MLOD to only 22.7 ng/g$_{dw}$. The highest concentrations were observed close to Dalian (China). Warner et al. [38] also detected quantifiable concentrations of D5, 0.4–2.1 ng/g$_{ww}$, in superficial sediments from Adventfjorden (Svalbard, Norway), in the Arctic Ocean. These concentrations were directly related to the inputs from Longyearbyen, the largest settlement of the island. D4 and D6 were not detected in any sediment sample from the Arctic.

In general, the presence of lVMS in sediments has received much less attention than cVMS, as it happens with other environmental matrices. Sanchís et al. [26]

Table 4 Selected works reporting the occurrence of siloxanes in sediments in ng/g$_{dw}$

Reference	Location	Type of sediment	D4	D5	D6	L2	L3	L4	L5	L6
Kaj et al. [37]	DNK, FIN, FRO, ISL, NOR, SWE	Reference sites	<MLOD	<MLOD	<MLOD	<MLOD	<MLOD	<MLOD	<MLOD	N.A.
		Site with diffuse inputs	<MLOD–160				MLOD–160			N.A.
		Site with local inputs	<MLOD–2,300				MLOD–2,300			N.A.
Warner et al. [38]	SJM	Coastal sediments in a high latitude and local WWTP	<0.9[a]	0.4–2.1[a]	<0.6[a]	N.A.	N.A.	N.A.	N.A.	N.A.
Zhang et al. [40]	CHN	Sediments from Songhua River (local inputs)	0.98–33.0	3.40–155	1.52–527	N.A.	N.A.	1.14–79.9 (sum of L4–L16)		N.A.
Sparham et al. [108]	GBR	Sediments from River Great Ouse (local inputs)	12 ± 2–24 ± 3	820 ± 130–1,450 ± 71	N.A.	N.A.	N.A.	N.A.	N.A.	N.A.
		Sediment from Humber estuary (local inputs)	N.A.	49 ± 7–256 ± 44	N.A.	N.A.	N.A.	N.A.	N.A.	N.A.
Kierkegaard et al. [48]	GBR	Estuary sediments from Humber estuary	<MLOD	59–272	30–109	N.A.	N.A.	N.A.	N.A.	N.A.
Wang et al. [81]	CAN	Miscellaneous sediments from S. Ontario and S. Quebec	<3–49	11–5,840	4–371	N.A.	N.A.	N.A.	N.A.	N.A.
Sanchís et al. [26]	CAT	Sediments from Llobregat River basin (many industrial inputs)	<1.8–679	3.39–1,270	N.A.	N.A.	0.4–4.35	0.3–11.1	0.9–75.3	N.A.
Hong et al. [25]	CHN	Sediments from the Huanghai Sea, near Dalian	<MLOD–1.81	<MLOD–22.7	<MLOD–5.10	N.A.	N.A.	N.A.	N.A.	N.A.
Schøyen et al. [109]	NOR	Coastal sediment from inner Oslofjord	N.I.	412[a, b]	N.I.	N.A.	N.A.	N.A.	N.A.	N.A.

Lee et al. [41]	KOR	Superficial sediments from Ulsan and Onsan Bays and related rivers	N.A.	8.29–1,000	<MLOQ–601	N.A.	N.A.	<LOQ–2.00	<LOQ–891	<LOQ–330
		Core sediments from Ulsan Bay	N.A.	2.21–98.4 (D5–D7)	N.A.	N.A.	N.A.	<MLOD–9.7 (L4–L17)		
Zhang et al. [107]	CHN	Dongting Lake (intensive agricultural activity and industry)	3.98–360.29	13.57–304.52	4.63–332.14	N.A.	N.A.	N.A.	N.A.	N.A.

N.A. not analysed, *N.I.* not indicated
[a]Values in ng/g_{ww}
[b]Average value

detected concentrations of L3, L4 and L5 two orders of magnitude lower than those of cVMS in the Llobregat River basin, and Zhang et al. [40] observed levels of lVMS (L4 to L17) up to 79.9 ng/g$_{dw}$ in the Songhua River. The highest concentrations of lVMS in sediments were recorded in Korean coastal areas (Ulsan and Onsan Bays), where concentrations of L4, L5 and L5 reached 2.00, 891 and 330 ng/g$_{dw}$, respectively.

4   Conclusions and Final Remarks

During the last decades, a myriad of analytical methods was developed, with appropriate recovery rates and limits of detection for the investigation of VMS in the environment, including aqueous media. The levels of VMS in the surface waters are normally in the ng/L order, and they are highly dependent on the presence of nearby local emission sources (such as urban WWTPs and silicone facilities). Matrix characteristics (such as salinity, suspended material and organic matter) and environmental parameters (such as climate and seasonality) play relevant roles in the fate and behaviour of VMS. Levels of sediments have also been reviewed. D5 is commonly the most frequently detected compound and also the one presenting the highest concentrations. D5 can reach the 1 µg/g$_{dw}$ order in rivers and lakes with direct inputs from local WWTPs. In general, marine sediments register significantly lower concentrations than sediments from freshwater bodies, and levels of lVMS are lower than those of cVMS in all the sediments samples reported in the literature.

According to the current knowledge about the ecotoxicology of VMS, the reported concentrations are unlikely to be any imminent problem to the aquatic fauna. However, since some recent studies estimates that these pollutants may be persistence in certain fractions and conditions, the scientific community and regulatory agencies should still include these compounds in their lists and check their occurrence periodically in the aquatic environment. This is particularly true for sediments, where cVMS are likely to accumulate because of the high sedimentation rates and low hydrolysis constants.

Finally, it should be mentioned that the behaviour and toxicity of some other VMS remains largely unknown. For instance, chlorinated cVMS, may be generated as by-products in conventional papermaking processes where cVMS and element chlorine co-occur [110]. Also, there are many potentially bioaccumulable and/or persistent siloxanes, such as fluorinated siloxanes, diphenylsiloxanes (e.g. octaphenylcyclotetrasiloxane), tetradecanoic acid methyl(3,3,4,4,5,5,6,6,6-nonafluorohexyl)silylene ester, triphenylsilanol, chloro-diphenyl-hydroxylsilane, tetrakis(trimethylsilyloxy)silane, etc. [111]. As some of these *emerging siloxanes* have little or no publicly available information about their behaviour and occurrence [112], there is still a lot of work to do with them.

Acknowledgements This work was supported by the Spanish Ministry of Economy and Competitiveness through the project NANO-transfer (ERA-NET SIINN PCIN-2015-182-CO2-02) and by the Catalan Government (Consolidated Research Group "2017 SGR 1404 – Water and Soil Quality Unit").

References

1. Colas A (2005) Silicones: preparation, properties and performance. Dow corning, life sciences. Dow Corning Corporation, Midland
2. Hunter M, Hyde J, Warrick E, Fletcher H (1946) Organo-silicon polymers. The cyclic dimethyl siloxanes. J Am Chem Soc 68:667–672
3. Brooke DN, Crookes MJ, Gray D, Robertson S (2009) Environmental risk assessment report: decamethylcyclopentasiloxane. Environment Agency, Bristol
4. Brooke DN, Crookes MJ, Gray D, Robertson S (2009) Environmental risk assessment report: dodecamethylcyclohexasiloxane. Environment Agency, Bristol
5. Brooke DN, Crookes MJ, Gray D, Robertson S (2009) Environmental risk assessment report: octamethylcyclotetrasiloxane. Environment Agency, Bristol
6. Tran TM, Abualnaja KO, Asimakopoulos AG, Covaci A, Gevao B, Johnson-Restrepo B et al (2015) A survey of cyclic and linear siloxanes in indoor dust and their implications for human exposures in twelve countries. Environ Int 78:39–44
7. Mojsiewicz-Pieńkowska K, Jamrógiewicz M, Szymkowska K, Krenczkowska D (2016) Direct human contact with siloxanes (silicones)–safety or risk part 1. Characteristics of siloxanes (silicones). Front Pharmacol 7:132
8. Horii Y, Kannan K (2008) Survey of organosilicone compounds, including cyclic and linear siloxanes, in personal-care and household products. Arch Environ Contam Toxicol 55:701–710
9. Fairbrother A, Woodburn KB (2016) Assessing the aquatic risks of the cyclic volatile methyl siloxane D4. Environ Sci Technol Lett 3:359–363
10. Gobas FA, Xu S, Kozerski G, Powell DE, Woodburn KB, Mackay D et al (2015) Fugacity and activity analysis of the bioaccumulation and environmental risks of decamethylcyclopentasiloxane (D5). Environ Toxicol Chem 34:2723–2731
11. Redman AD, Mihaich E, Woodburn K, Paquin P, Powell D, McGrath JA et al (2012) Tissue-based risk assessment of cyclic volatile methyl siloxanes. Environ Toxicol Chem 31:1911–1919
12. Sousa JV, McNamara PC, Putt AE, Machado MW, Surprenant DC, Hamelink JL et al (1995) Effects of octamethylcyclotetrasiloxane (OMCTS) on freshwater and marine organisms. Environ Toxicol Chem 14:1639–1647
13. Parrott J, Alaee M, Wang D, Sverko E (2013) Fathead minnow (*Pimephales promelas*) embryo to adult exposure to decamethylcyclopentasiloxane (D5). Chemosphere 93:813–818
14. Drottar KR (2005) 14C-dodecamethylcyclohexasiloxane (^{14}C–D6): bioconcentration in the fathead minnow (*Pimephales promelas*) under flow-through test conditions. Dow Corning Corporation, Silicones Environment, Health and Safety Council (SEHSC) (cited from the Report of the Assessment for D6 by Environment Canada and Health Canada)
15. Kierkegaard A, Bignert A, McLachlan MS (2012) Cyclic volatile methylsiloxanes in fish from the Baltic Sea. Chemosphere 93:774–778
16. Kierkegaard A, Bignert A, McLachlan MS (2013) Bioaccumulation of decamethylcyclopentasiloxane in perch in Swedish lakes. Chemosphere 93:789–793
17. Sanchís J, Llorca M, Picó Y, Farré M, Barceló D (2016) Volatile dimethylsiloxanes in market seafood and freshwater fish from the Xúquer River, Spain. Sci Total Environ 545:236–243

18. Zhi L, Xu L, He X, Zhang C, Cai Y (2018) Distribution of methylsiloxanes in benthic mollusks from the Chinese Bohai Sea. J Environ Sci
19. Panagopoulos D, MacLeod M (2018) A critical assessment of the environmental fate of linear and cyclic volatile methylsiloxanes using multimedia fugacity models. Environ Sci: Processes Impacts 20:183–194
20. Álvarez-Muñoz D, Llorca M, Blasco J, Barceló D (2017) Contaminants in the marine environment. Marine ecotoxicology. Elsevier, Amsterdam, pp 1–34
21. Ginebreda A, Pérez S, Rivas D, Kuzmanovic M, Barceló D (2015) Pollutants of emerging concern in rivers of Catalonia: occurrence, fate, and risk. Experiences from surface water quality monitoring. Springer, Cham, pp 283–320
22. Richardson SD, Kimura SY (2015) Water analysis: emerging contaminants and current issues. Anal Chem 88:546–582
23. Thomaidis NS, Asimakopoulos AG, Bletsou A (2012) Emerging contaminants: a tutorial mini-review. Global NEST J 14:72–79
24. Bletsou AA, Asimakopoulos AG, Stasinakis AS, Thomaidis NS, Kannan K (2013) Mass loading and fate of linear and cyclic siloxanes in a wastewater treatment plant in Greece. Environ Sci Technol 47:1824–1832
25. Hong W-J, Jia H, Liu C, Zhang Z, Sun Y, Li Y-F (2014) Distribution, source, fate and bioaccumulation of methyl siloxanes in marine environment. Environ Pollut 191:175–181
26. Sanchís J, Martínez E, Ginebreda A, Farré M, Barceló D (2013) Occurrence of linear and cyclic volatile methylsiloxanes in wastewater, surface water and sediments from Catalonia. Sci Total Environ 443:530–538
27. Xu L, Xu S, Zhi L, He X, Zhang C, Cai Y (2017) Methylsiloxanes release from one landfill through yearly cycle and their removal mechanisms (especially hydroxylation) in leachates. Environ Sci Technol 51:12337–12346
28. Cortada C, dos Reis LC, Vidal L, Llorca J, Canals A (2014) Determination of cyclic and linear siloxanes in wastewater samples by ultrasound-assisted dispersive liquid-liquid microextraction followed by gas chromatography-mass spectrometry. Talanta 120:191–197
29. Krogseth IS, Whelan MJ, Christensen GN, Breivik K, Evenset A, Warner NA (2016) Understanding of cyclic volatile methyl siloxane fate in a high latitude lake is constrained by uncertainty in organic carbon–water partitioning. Environ Sci Technol 51:401–409
30. Sparham C, Van Egmond R, O'Connor S, Hastie C, Whelan M, Kanda R et al (2008) Determination of decamethylcyclopentasiloxane in river water and final effluent by headspace gas chromatography/mass spectrometry. J Chromatogr A 1212:124–129
31. van Egmond R, Sparham C, Hastie C, Gore D, Chowdhury N (2013) Monitoring and modelling of siloxanes in a sewage treatment plant in the UK. Chemosphere 93:757–765
32. Companioni-Damas EY, Santos FJ, Galceran MT (2012) Analysis of linear and cyclic methylsiloxanes in water by headspace-solid phase microextraction and gas chromatography–mass spectrometry. Talanta 89:63–69
33. Xu L, Shi Y, Cai Y (2013) Occurrence and fate of volatile siloxanes in a municipal wastewater treatment plant of Beijing, China. Water Res 47:715–724
34. Zhang Y (2014) Analysis of octamethylcyclotetrasiloxane and decamethylcyclopentasiloxane in wastewater, sludge and river samples by headspace gas chromatography/mass spectrometry. Doctoral dissertation, Colorado State University
35. Horii Y, Minomo K, Ohtsuka N, Motegi M, Nojiri K, Kannan K (2017) Distribution characteristics of volatile methylsiloxanes in Tokyo Bay watershed in Japan: analysis of surface waters by purge and trap method. Sci Total Environ 586:56–65
36. Kaj L, Andersson J, Cousins AP, Remberger M, Ekheden Y, Dusan B et al (2004) Subreport 4: siloxanes. IVL Swedish Environmental Research Institute, Stockholm Results from the Swedish National Screening Programme
37. Kaj L, Schlabach M, Andersson J, Cousins AP, Schmidbauer N, Brorström-Lundén E (2005) Siloxanes in the nordic environment. Nordic Council of Ministers

38. Warner NA, Evenset A, Christensen G, Gabrielsen GW, Borgå K, Leknes H (2010) Volatile siloxanes in the European arctic: assessment of sources and spatial distribution. Environ Sci Technol 44:7705–7710

39. Wang D-G, Norwood W, Alaee M, Byer JD, Brimble S (2012) Review of recent advances in research on the toxicity, detection, occurrence and fate of cyclic volatile methyl siloxanes in the environment. Chemosphere 93:711–725

40. Zhang Z, Qi H, Ren N, Li Y, Gao D, Kannan K (2011) Survey of cyclic and linear siloxanes in sediment from the Songhua River and in sewage sludge from wastewater treatment plants, Northeastern China. Arch Environ Contam Toxicol 60:204–211

41. Lee S, Lee S, Choi M, Kannan K, Moon H (2018) An optimized method for the analysis of cyclic and linear siloxanes and their distribution in surface and core sediments from industrialized bays in Korea. Environ Pollut 236:111–118

42. Warner NA, Kozerski G, Durham J, Koerner M, Gerhards R, Campbell R et al (2012) Positive vs. false detection: a comparison of analytical methods and performance for analysis of cyclic volatile methylsiloxanes (cVMS) in environmental samples from remote regions. Chemosphere 93:749–756

43. Warner NA, Nøst TH, Andrade H, Christensen G (2014) Allometric relationships to liver tissue concentrations of cyclic volatile methyl siloxanes in Atlantic cod. Environ Pollut 190:109–114

44. Varaprath S, Seaton M, McNett D, Cao L, Plotzke KP (2000) Quantitative determination of octamethylcyclotetrasiloxane (D4) in extracts of biological matrices by gas chromatography-mass spectrometry. Int J Environ Anal Chem 77:203–219

45. Woodburn K, Drottar K, Domoradzki J, Durham J, McNett D, Jezowski R (2013) Determination of the dietary biomagnification of octamethylcyclotetrasiloxane and decamethylcyclopentasiloxane with the rainbow trout (Oncorhynchus mykiss). Chemosphere 93:779–788

46. McGoldrick DJ, Letcher RJ, Barresi E, Keir MJ, Small J, Clark MG et al (2014) Organophosphate flame retardants and organosiloxanes in predatory freshwater fish from locations across Canada. Environ Pollut 193:254–261

47. Kierkegaard A, Adolfsson-Erici M, McLachlan MS (2010) Determination of cyclic volatile methylsiloxanes in biota with a purge and trap method. Anal Chem 82:9573–9578

48. Kierkegaard A, van Egmond R, McLachlan MS (2011) Cyclic volatile methylsiloxane bioaccumulation in flounder and ragworm in the Humber Estuary. Environ Sci Technol 45:5936–5942

49. McGoldrick DJ, Chan C, Drouillard KG, Keir MJ, Clark MG, Backus SM (2014) Concentrations and trophic magnification of cyclic siloxanes in aquatic biota from the Western Basin of Lake Erie, Canada. Environ Pollut 186:141–148

50. Wang D-G, de Solla SR, Lebeuf M, Bisbicos T, Barrett GC, Alaee M (2017) Determination of linear and cyclic volatile methylsiloxanes in blood of turtles, cormorants, and seals from Canada. Sci Total Environ 574:1254–1260

51. Flaningam OL (1986) Vapor pressures of poly (dimethylsiloxane) oligomers. J Chem Eng Data 31:266–272

52. Xu S, Kropscott B (2014) Evaluation of the three-phase equilibrium method for measuring temperature dependence of internally consistent partition coefficients (K_{OW}, K_{OA}, and K_{AW}) for volatile methylsiloxanes and trimethylsilanol. Environ Toxicol Chem 33:2702–2710

53. Varaprath S, Frye CL, Hamelink J (1996) Aqueous solubility of permethylsiloxanes (silicones). Environ Toxicol Chem 15:1263–1265

54. Lei YD, Wania F, Mathers D (2010) Temperature-dependent vapor pressure of selected cyclic and linear polydimethylsiloxane oligomers. J Chem Eng Data 55:5868–5873

55. Bruggeman W, Weber-Fung D, Opperhuizen A, Van Der Steen J, Wijbenga A, Hutzinger O (1984) Absorption and retention of polydimethylsiloxanes (silicones) in fish: preliminary experiments. Toxicol Environ Chem 7:287–296

56. Government of Canada (2011) Screening assessment for the challenge. Siloxanes and silicones, di-Me, hydrogen-terminated. Chemical Abstracts Service Registry Number 70900-21-9

57. Mazzoni S, Roy S, Grigoras S (1997) Eco-relevant properties of selected organosilicon materials. Organosilicon materials. Springer, Berlin, pp 53–81

58. Xu S, Kropscott B (2012) Method for simultaneous determination of partition coefficients for cyclic volatile methylsiloxanes and dimethylsilanediol. Anal Chem 84:1948–1955

59. Lee S, Moon H-B, Song G-J, Ra K, Lee W-C, Kannan K (2014) A nationwide survey and emission estimates of cyclic and linear siloxanes through sludge from wastewater treatment plants in Korea. Sci Total Environ 497:106–112

60. Lu Z, Martin PA, Burgess NM, Champoux L, Elliott JE, Baressi E et al (2017) Volatile methylsiloxanes and organophosphate esters in the eggs of European starlings (*Sturnus vulgaris*) and congeneric gull species from locations across Canada. Environ Sci Technol 51:9836–9845

61. Pieri F, Katsoyiannis A, Martellini T, Hughes D, Jones KC, Cincinelli A (2013) Occurrence of linear and cyclic volatile methyl siloxanes in indoor air samples (UK and Italy) and their isotopic characterization. Environ Int 59:363–371

62. Raich-Montiu J, Ribas-Font C, De Arespacochaga N, Roig-Torres E, Broto-Puig F, Crest M et al (2014) Analytical methodology for sampling and analysing eight siloxanes and trimethylsilanol in biogas from different wastewater treatment plants in Europe. Anal Chim Acta 812:83–91

63. Hanssen L, Warner NA, Braathen T, Odland JØ, Lund E, Nieboer E et al (2013) Plasma concentrations of cyclic volatile methylsiloxanes (cVMS) in pregnant and postmenopausal Norwegian women and self-reported use of personal care products (PCPs). Environ Int 51:82–87

64. Dewil R, Appels L, Baeyens J, Buczynska A, Van Vaeck L (2007) The analysis of volatile siloxanes in waste activated sludge. Talanta 74:14–19

65. Huppmann R, Lohoff HW, Schröder HF (1996) Cyclic siloxanes in the biological waste water treatment process – determination, quantification and possibilities of elimination. Fresenius J Anal Chem 354:66–71

66. Kochetkov A, Smith JS, Ravikrishna R, Valsaraj KT, Thibodeaux LJ (2001) Air-water partition constants for volatile methyl siloxanes. Environ Toxicol Chem 20:2184–2188

67. Badjagbo K, Furtos A, Alaee M, Moore S, Sauvé S (2009) Direct analysis of volatile methylsiloxanes in gaseous matrixes using atmospheric pressure chemical ionization-tandem mass spectrometry. Anal Chem 81:7288–7293

68. Badjagbo K, Héroux M, Alaee M, Moore S, Sauvé S (2009) Quantitative analysis of volatile methylsiloxanes in waste-to-energy landfill biogases using direct APCI-MS/MS. Environ Sci Technol 44:600–605

69. Langford V, Gray J, Maclagan R, McEwan MJ (2013) Detection of siloxanes in landfill gas and biogas using SIFT-MS. Curr Anal Chem 9:558–564

70. Fackler PH, Dionne E, Hartley DA, Hamelink JL (1995) Bioconcentration by fish of a highly volatile silicone compound in a totally enclosed aquatic exposure system. Environ Toxicol Chem 14:1649–1656

71. Kent DJ, McNamara PC, Putt AE, Hobson JF, Silberhorn EM (1994) Octamethylcyclotetrasiloxane in aquatic sediments: toxicity and risk assessment. Ecotoxicol Environ Saf 29:372–389

72. Rücker C, Kümmerer K (2015) Environmental chemistry of organosiloxanes. Chem Rev 115:466–524

73. Parker R (1978) Determination of organosilicon compounds in water by atomic absorption spectroscopy Bestimmung von siliciumorganischen Verbindungen in Wasser mit Hilfe der Atomabsorptions-Spektralphotometrie. Fresenius Z Anal Chem 292:362–364

74. Zhi L, Xu L, He X, Zhang C, Cai Y (2018) Occurrence and profiles of methylsiloxanes and their hydrolysis product in aqueous matrices from the Daqing oilfield in China. Sci Total Environ 631:879–886

75. Varaprath S, Lehmann RG (1997) Speciation and quantitation of degradation products of silicones (Silane/Siloxane Diols) by gas chromatography – mass spectrometry and stability of dimethylsilanediol. J Environ Polym Degrad 5:17–31

76. Mahone L, Garner P, Buch R, Lane T, Tatera J, Smith R, Frye CL (1983) A method for the qualitative and quantitative characterization of waterborne organosilicon substances. Environ Toxicol Chem 2:307–313

77. Dorn SB, Skelly Frame SB (1994) Development of a high-performance liquid chromatographic–inductively coupled plasma method for speciation and quantification of silicones: from silanols to polysiloxanes. Analyst 119:1687–1694

78. Grümping R, Hirner A (1999) HPLC/ICP-OES determination of water-soluble silicone (PDMS) degradation products in leachates. Fresenius J Anal Chem 363:347–352

79. Jovanovic ML, McMahon JM, McNett DA, Tobin JM, Plotzke KP (2008) In vitro and in vivo percutaneous absorption of 14C-octamethylcyclotetrasiloxane (14C-D4) and 14C-decamethylcyclopentasiloxane (14C-D5). Regul Toxicol Pharmacol 50:239–248

80. Fendinger N, McAvoy D, Eckhoff W, Price B (1997) Environmental occurrence of polydimethylsiloxane. Environ Sci Technol 31:1555–1563

81. Wang D-G, Steer H, Tait T, Williams Z, Pacepavicius G, Young T et al (2013) Concentrations of cyclic volatile methylsiloxanes in biosolid amended soil, influent, effluent, receiving water, and sediment of wastewater treatment plants in Canada. Chemosphere 93:766–773

82. Sanchís JÀ (2015) Occurrence and toxicity of nanomaterials and nanostructures in the environment. Doctoral dissertation, Universitat de Barcelona, Barcelona

83. Genualdi S, Harner T, Cheng Y, MacLeod M, Hansen KM, van Egmond R et al (2011) Global distribution of linear and cyclic volatile methyl siloxanes in air. Environ Sci Technol 45:3349–3354

84. Xu S, Wania F (2013) Chemical fate, latitudinal distribution and long-range transport of cyclic volatile methylsiloxanes in the global environment: a modeling assessment. Chemosphere 93:835–843

85. Ahrens L, Harner T, Shoeib M (2014) Temporal variations of cyclic and linear volatile methylsiloxanes in the atmosphere using passive samplers and high-volume air samplers. Environ Sci Technol 48:9374–9381

86. Sanchís J, Cabrerizo A, Galbán-Malagón C, Barceló D, Farré M, Dachs J (2015) Response to comments on "unexpected occurrence of volatile dimethylsiloxanes in Antarctic soils, vegetation, phytoplankton and krill". Environ Sci Technol 49:7510–7512

87. Sanchís J, Cabrerizo A, Galbán-Malagón C, Barceló D, Farré M, Dachs J (2015) Unexpected occurrence of volatile dimethylsiloxanes in Antarctic soils, vegetation, phytoplankton, and krill. Environ Sci Technol 49:4415–4424

88. Janechek NJ, Hansen KM, Stanier CO (2017) Comprehensive atmospheric modeling of reactive cyclic siloxanes and their oxidation products. Atmos Chem Phys 17:8357

89. Whelan MJ, Estrada E, Van Egmond R (2004) A modelling assessment of the atmospheric fate of volatile methyl siloxanes and their reaction products. Chemosphere 57:1427–1437

90. Lehmann RG, Varaprath S, Frye CL (1994) Fate of silicone degradation products (silanols) in soil. Environ Toxicol Chem 13:1753–1759

91. Stevens C (1998) Environmental degradation pathways for the breakdown of polydimethylsiloxanes. J Inorg Biochem 69:203–207

92. Xu S, Lehmann RG, Miller JR, Chandra G (1998) Degradation of polydimethylsiloxanes (silicones) as influenced by clay minerals. Environ Sci Technol 32:1199–1206

93. Carpenter JC, Cella JA, Dorn SB (1995) Study of the degradation of polydimethylsiloxanes on soil. Environ Sci Technol 29:864–868

94. Xu S (2010) Aerobic and anaerobic transformation of decamethylcyclopentasiloxane (^{14}C-D5) in aquatic sediment systems. Centre Européen des Silicones (CES). CES report 27 Jan 2010

95. Panagopoulos D, Jahnke A, Kierkegaard A, MacLeod M (2017) Temperature dependence of the organic carbon/water partition ratios (K_{OC}) of volatile methylsiloxanes. Environ Sci Technol Lett 4:240–245

96. Panagopoulos D, Kierkegaard A, Jahnke A, MacLeod M (2016) Evaluating the salting-out effect on the organic carbon/water partition ratios (K_{OC} and K_{DOC}) of linear and cyclic volatile

methylsiloxanes: measurements and polyparameter linear free energy relationships. J Chem Eng Data 61:3098–3108

97. Durham J, Kozerski G (2005) Hydrolysis of octamethylcyclotetrasiloxane (D4). Silicones Environmental Health and Safety Council Study

98. Durham J, Kozerski G (2006) Hydrolysis of decamethylcyclopentasiloxane (D5). Silicones Environmental Health and Safety Council Study

99. Environment Canada, Health Canada (2008) Screening assessment for the challenge - decamethylcyclopentasiloxane (D5)

100. Environment Canada, Health Canada (2008) Screening assessment for the challenge - dodecamethylcyclohexasiloxane (D6)

101. Environment Canada, Health Canada (2008) Screening assessment for the challenge - octamethylcyclotetrasiloxane (D4)

102. Hughes L, Mackay D, Powell DE, Kim J (2012) An updated state of the science EQC model for evaluating chemical fate in the environment: application to D5 (decamethylcyclopentasiloxane). Chemosphere 87:118–124

103. Webster E, Mackay D, Wania F (1998) Evaluating environmental persistence. Environ Toxicol Chem 17:2148–2158

104. Pellenbarg R (1979) Environmental poly (organosiloxanes)(silicones). Environ Sci Technol 13(5):565–569

105. Watanabe N, Nakamura T, Watanabe E, Sato E, Ose Y (1984) Distribution of organosiloxanes (silicones) in water, sediments and fish from the Nagara River watershed, Japan. Sci Total Environ 35:91–97

106. Watanabe N, Nagase H, Ose Y (1988) Distribution of silicones in water, sediment and fish in Japanese rivers. Sci Total Environ 73:1–9

107. Zhang Y, Shen M, Tian Y, Zeng G (2018) Cyclic volatile methylsiloxanes in sediment, soil, and surface water from Dongting Lake, China. J Soils Sediments 18:2063–2071

108. Sparham C, van Egmond R, Hastie C, O'Connor S, Gore D, Chowdhury N (2011) Determination of decamethylcyclopentasiloxane in river and estuarine sediments in the UK. J Chromatogr A 1218:817–823

109. Schøyen M, Øxnevad S, Hjermann D, Mund C, Böhmer T, Beckmann K et al (2016) Levels of siloxanes (D4, D5, D6) in biota and sediments from the Inner Oslofjord, Norway, 2011-2014 (Poster presentation). In: The SETAC Europe 26th annual meeting, Nantes

110. Xu L, He X, Zhi L, Zhang C, Zeng T, Cai Y (2016) Chlorinated methylsiloxanes generated in the papermaking process and their fate in wastewater treatment processes. Environ Sci Technol 50:12732–12741

111. Howard PH, Muir DCG (2010) Identifying new persistent and bioaccumulative organics among chemicals in commerce. Environ Sci Technol 44:2277–2285

112. Thomas KV, Schlabach M, Langford K, Fjeld E, Øxnevad S, Rundberget T et al (2014) Screening program 2013. New bisphenols, organic peroxides, fluorinated siloxanes, organic UV filters and selected PBT substances. Norwegian Environment Agency, Oslo

Cyclic and Linear Siloxanes in Indoor Environments: Occurrence and Human Exposure

A. Cincinelli, T. Martellini, R. Scodellini, C. Scopetani, C. Guerranti, and A. Katsoyiannis

Contents

Abstract Methylsiloxanes (MSs) are an important class of additive chemicals that due to their physicochemical properties have been broadly used in several industrial applications and consumer products. The purpose of this chapter is to provide a literature review on the current state of knowledge on the occurrence and distribution of MSs in air samples from different indoor environments, including, for example, residential houses, offices, public buildings, cars, industries or hair salons. Literature studies on the levels of cyclic and linear siloxanes in indoor dust, which is a major

A. Cincinelli (✉)
Department of Chemistry "Ugo Schiff", University of Florence, Florence, Italy

Consorzio Interuniversitario per lo Sviluppo dei Sistemi a Grande Interfase (CSGI), University of Florence, Florence, Italy
e-mail: alessandra.cincinelli@unifi.it

T. Martellini, R. Scodellini, and C. Scopetani
Department of Chemistry, University of Florence, Florence, Italy

C. Guerranti
Consorzio Interuniversitario per lo Sviluppo dei Sistemi a Grande Interfase (CSGI), University of Florence, Florence, Italy

A. Katsoyiannis
Norwegian Institute for Air Research (NILU) at FRAM – High North Research Centre on Climate and the Environment, Tromsø, Norway

V. Homem and N. Ratola (eds.), *Volatile Methylsiloxanes in the Environment*,
Hdb Env Chem (2020) 89: 181–200, DOI 10.1007/698_2019_410,
© Springer Nature Switzerland AG 2019, Published online: 21 January 2020

source of MS due their particle-binding affinity, are discussed. A wide range of MS concentrations in air and dust samples has been reported together with an evident different level in indoor air samples from building of different classification. Among cyclic methylsiloxanes, D5 was usually the dominant congener in the investigated samples. In general, the levels from industrial facilities were one or more orders of magnitude higher than those in residential buildings. The mean inhalation exposure doses to total siloxanes for infants, toddlers, children, teenagers and adults are also presented. Recent investigations on human exposure to MSs through dust ingestion were also included.

Keywords Bioaccumulative, Transdermal permeation, Endocrine disruptors, Immunologic responses, Ubiquitous, Particle-binding affinity

1 Introduction

Siloxanes, a portmanteau word deriving from the words silicon, oxygen and alkanes, are a large group of organosilicone chemicals with molecular weight ranging from a few hundred to several hundred thousand [1]. Siloxanes are characterized by structural units of $(R)_2SiO$, where R is a methyl group, and they can be linked into cyclic and linear structures, with the linear siloxanes terminated by additional methyl groups.

Methylsiloxanes (MSs) are an important class of additive chemicals that due to their physicochemical properties have been broadly used in several industrial applications and consumer products [2, 3]. Low molecular weight cyclic volatile methylsiloxanes (cVMSs) include hexamethylcyclotrisiloxane (D_3), octamethylcyclotetrasiloxane (D_4), decamethylcyclopentasiloxane (D_5) and dodecamethylcyclohexasiloxane (D_6), which are the most widely studied. Similarly, the most common linear siloxanes (lVMS) are hexamethyldisiloxane (L_2), octamethyltrisiloxane (L_3), decamethyltetrasiloxane (L_4) and dodecamethylpentasiloxane (L_5).

The thermal stability (150–200°C), good low temperature performance (<70°C), strong hydrophobicity, antifriction, good dielectric character and good radiation resistance have made them preferred solvents or main ingredients for a variety of products including cosmetic and personal care products (PCPs) for hair and skin care products, i.e. shampoos, conditioners, deodorants [4]; antiperspirants; pharmaceuticals; surfactants and mould release agents; and lubricants, paints, polishes and coatings on a range of substrates including textiles, carpeting and paper [5]. Some MSs were identified as high production volume (HPV) chemicals by the Organization for Economic Co-operation and Development [6] and by the US Environmental

Protection Agency [7]; it was estimated that the total annual volume of cVMSs in the USA and China, including D4, D5 and D6, reached 470 and 800 million kg, respectively.

In the last few years, several protection agencies have expressed their concern about the possible toxic and hazard effects of siloxanes to humans and the environment [8]. In general, organosiloxanes are related to potential adverse effects [2], such as direct or indirect estrogen mimicry, connective tissue disorders, toxic effects on human reproductive, immune and nervous system [9], adverse immunologic responses and fatal liver and lung damage [9–11].

The limited number of studies present in the literature on the health effects of MSs focused their attention on cVMSs. Estrogenic and androgenic activities of D4 and D5 were reported in rats [12, 13], and some studies have also indicated that some MSs are potential endocrine disruptors and show neurotoxic effects in addition to reproductive toxicity [9].

The European Commission classified D5 in category 3 for reproductive toxicity (with the risk phrase R62 "possible risk of impaired fertility" in Annex 1 to the substance Directive 67/548/EEC) and based this classification on reproductive effects observed in rats. D4 and D5 are also included in the list of potential PBT (persistent, bioaccumulative and toxic) and vPvB (very persistent and very bioaccumulative) substances selected on the basis of screening criteria in the EU [14].

In addition, MSs are also characterized by high vapour pressure, low water solubility and high Henry's law constant [15] that make them to be present and persist in outdoor and indoor environments. As a result of the widespread use of MSs in the aforementioned products and their physicochemical properties, these chemicals are now ubiquitous in the environment through various emission pathways or as a result of leakage. In particular, cVMSs and lVMSs have been detected in almost all environmental compartments, including outdoor air [16], wastewater [15], sediments [17], soils [1] and sewage sludge [18]. Their presence in the environment and tendency to bioaccumulate and to persist to degradation processes together with the suspected adverse human and ecological health effects [1, 3] have led to a continuously increasing scientific interest and public health concern.

Siloxanes are thus considered a new group of emerging contaminants, and especially the cVMSs are under consideration for regulation in different countries in order to restrict their marketing or use in PCPs, because of concerns regarding their potential toxicity [9].

This chapter provides a literature review of the current state of knowledge on the primary environmental sources, occurrence and levels of MSs in indoor air and dust and estimations of human exposures to this class of organic compounds through inhalation and dust ingestion.

2 Siloxanes in Indoor Environments

2.1 Siloxanes in Indoor Air

Despite their wide use in common products and their volatility, limited studies have investigated the occurrence of siloxanes in indoor environments [19–21].

However, in view of their toxic properties, human exposure to MSs has aroused increasing concern [22, 23]. Indoor air quality (IAQ) refers to the health and wellbeing of the occupants of a building environment. In modern societies, people spend on average more than 90% of their time in indoor environments [24]. Thus, individuals may be exposed to a variety of contaminants (in the form of gases and particles) from cleaning products, carpets and furnishings, perfumes, tobacco smoke, construction materials and activities, office machines, microbial growth (fungal, mould and bacterial), insects and outdoor pollutants. In addition, other factors such as indoor temperatures, relative humidity and ventilation levels may also affect how building occupants respond to the IAQ. Understanding the levels and sources of indoor contaminants and how to control them could help to prevent or resolve building-related worker/occupants symptoms and the health and comfort of building occupants [25]. In the last decade, numerous studies have focused the attention on the occurrence of a large number of volatile organic compounds (VOCs), such as carbon monoxide, radon, particulate matter, polycyclic aromatic hydrocarbons, polybrominated diphenyl ethers, perfluorinated alkyl substances, spray insecticides and more recent pollutants like cVMSs for evaluating the IAQ, which influences the human's wellbeing [26–30].

There are only few studies in the literature regarding the levels and distribution of cVMS and lVMS in indoor environments (see Table 1). A first study in 1992 [31] reported the concentrations of cVMSs, together with the levels of other VOCs, measured at a telephone switching centre at Neenah, Wisconsin (USA). They found maximum concentration levels of 0.9 $\mu g\ m^{-3}$ for D3 and 3.7 $\mu g\ m^{-3}$ for D5, while D4 co-eluted with limonene and was not possible to measure its single concentration. A few years later, Shields et al. [32] reported the comparison of VOCs in indoor atmospheric samples from rooms of three types of US commercial buildings (telecommunications and administrative offices, data centres) with different occupant densities. The authors found that D4 and D5 were almost always observed indoors (69 of the 70 building samples), with concentrations ranging from 2.5 to 10 and from 7.0 to 39.6 $\mu g\ m^{-3}$, respectively. D4 and D5 showed higher concentrations when there were more occupants, indicating that they were good markers of occupant density. The results also suggested that personal care products were the dominant source of D5 within these buildings. Similar conclusions were reported by Yucuis et al. [33], who determined high levels of cVMSs in indoor air samples collected from a laboratory ($n = 10$) and two offices ($n = 4$) in the USA. In particular, the median concentration for the sum of D4–D6 was 2,200 ng m^{-3}, with D5 being the dominant chemical in all indoor air samples, accounting for the 96% of the total cVMSs. The maximum value determined was 56,000 ng m^{-3} in one office.

Table 1 Mean air concentrations (ng/m³) of MSs in indoor air samples

Site	Indoor air G = Gas phase P = Particulate phase	Country	Number of samples	D3, ng/m³	D4, ng/m³	D5, ng/m³	D6, ng/m³	D7, ng/m³	ΣL4-L16, ng/m³	ΣL4-L9, ng/m³	References
Telco offices	G	USA	50		2.5	7.0					[32]
Data centers	G	USA	9		9.4	26.1					
Administration offices	G	USA	11		10.2	39.6					
Bathroom	G	United Kingdom	9	11,000	68,000	120,000	26,000				[19]
Boy bedroom	G		5	13,000	8,400	150,000	24,000				
Girl bedroom	G		5	58,000	15,000	97,000	8,100				
Living room	G		5	160,000	43,000	84,000	12,000				
Adult room	G		1	1,200	1,900	45,000	5,400				
Kindergarten	G		1	14,000	17,000	270,000	43,000				
Offices	G		4	6,600	9,800	54,000	4,600				
Supermarket	G		3	15,000	6,900	230,000	8,400				
Bathroom	G	Italy	15	69,000	42,000	98,000	16,000				
Boy bedroom	G		6	26,000	8,000	110,000	1,000				
Girl bedroom	G		6	27,000	19,000	37,000	5,400				
Living room	G		5	3,500	8,200	38,000	45,000				
Adult room	G		9	36,000	11,000	170,000	19,000				
Supermarket	G		2	3,900	5,200	54,000	2,200				
Offices	G		5	3,300	2,200	7,500	ND				
Lab 1	G	University of Iowa (USA)		46	1,300	<59					[33]
Lab 2	G			51	1,700	<59					
Lab 3	G			57	970	<59					

(continued)

Table 1 (continued)

Site	Indoor air G = Gas phase P = Particulate phase	Country	Number of samples	D3, ng/m³	D4, ng/m³	D5, ng/m³	D6, ng/m³	D7, ng/m³	ΣL4-L16, ng/m³	ΣL4-L9, ng/m³	References
Lab 4	G				63	1,900	<59				
Office A	G				73	2,300	<59				
Lab 5	G				23	8,100	<59				
Lab 6	G				47	970	70				
Lab 7	G				58	14,000	200				
Lab 8	G				31	1,100	<59				
Lab 9	G				75	39,000	220				
Lab 10	G				27	1,000	ND				
Office B1	G				340	53,000	1,600				
Office B2	G				260	7,600	314				
Office B3	G				500	56,000	2,800				
Work – Offices	G	Tromso (Norway)		2,230	984	3,419	162				[35]
Work – Lab	G			1,120	427	2,816	179				
Work – Storage room	G			1,180	962	3,224	182				
Work – Gym	G			1,425	1,071	3,824	244				
Work – Bathroom	G			1,864	1,500	5,168	334				
Home – Bedroom	G			1,136	926	4,885	200				
Home – Living rooms	G			1,081	937	7,634	235				
Home – Bathrooms	G			1,149	960	14,977	344				
Homes	G + P	Albany (USA)	20	24.6	56.7	446	75.4	52.9			[40]
Offices	G + P		7	12.2	26.5	150	50.8	555			

			13	20.4	38	87.6	28.9	178		
Lab	G + P		13							
Schools	G + P		6	15.9	98.9	649	107	46.6		
Salons	G + P		6	13	495	3,200	444	48.4		
Public places	G + P		8	22.6	199	1,140	151	73.8		
Homes undergoing decoration	G	Beihang University (China)	40		1,500	6,670	1,700		1,300	[36]
Ordinary homes	G		30		577	4,985	645		547	
Cars	G		20		165	1,052	899		181	
Historical library – Reading room (RR)	G	Florence (Italy)			1,880	5,880	8,700	7,300		[43]
Historical library – Repository (R)	G				1,870	4,470	10,540	5,150		
Modern university library (RR)	G				1,680	5,450	32,470	9,740		
Modern university library (R)	G				2,350	7,000	10,390	7,140		
Modern university library (RR)	G				930	10,150	28,310	11,170		
Modern university library (RR)	G				1,480	5,990	42,250	5,330		
Modern university library (R)	G				1,280	4,680	72,500	3,510		
Historical archives (R)	G				3,540	13,260	13,980	7,130		
Historical archives (R)	G				2,810	20,390	24,860	13,400		
Historical library (RR)	G				2,980	26,000	5,840	2,380		
Historical library (RR)	G				3,330	3,370	5,310	4,680		

(continued)

Table 1 (continued)

Site	Indoor air G = Gas phase P = Particulate phase	Country	Number of samples	D3, ng/m³	D4, ng/m³	D5, ng/m³	D6, ng/m³	D7, ng/m³	ΣL4-L16, ng/m³	ΣL4-L9, ng/m³	References
Historical library (RR)	G				4,120	12,720	32,350	5,670			
Historical library (R)	G				10,440	7,820	14,350	10,920			[57]
Homes	G + P	Hanai (Vietnam)	19	8.95	71.4	151	93.3			226	
Cars	G + P		8	10.6	15.6	34.7	63.2			341	
Kindergartens	G + P		7	4.05	9.35	30.7	20.8			224	
Laboratories	G + P		19	3.64	8.63	20.8	36.5			208	
Offices	G + P		9	5.41	9.64	57.8	7.96			178	
Hair salon	G + P		13	16.5	227	385	249			444	
Homes	G + P	Bachin	8	5.13	15.4	23.1	19.3			228	
Homes	G + P	Thaibinh	6	3.19	10.9	19.6	17			106	
Homes	G + P	Tuyenquang	8	2.56	6.60	8.48	13.7			102	

Indoor air measurements were also conducted by Zhou et al. [34] and Pieri et al. [19]. Zhou et al. [34] reported siloxane concentrations in nonoccupational environments in Canada, where D4 and D5 were the two most frequently detected VMSs. The maximum concentrations for cyclic siloxanes were 2,773 µg m^{-3} for D4 in a school and 495 µg m^{-3} and 168 µg m^{-3} for D5 and D6, respectively, in a general office. Pieri et al. [19] evaluated the occurrence of lVMSs and cVMSs in various indoor environments, occupational and domestic, in Italy and the UK. The results showed that the cVMSs were the most abundant, with average concentrations ranging from 7.5 to 170 µg m^{-3} in samples from Italy and from 45 to 270 µg m^{-3} from the UK. Linear VMSs were found at low levels, and only in a few cases, average concentrations exceeded 10 µg m^{-3}. Results highlighted the differences that can be observed between various indoor environments (e.g. domestic-like bathrooms, bedrooms or occupational) and between two countries. In most cases, the concentrations found in the UK were higher than in the respective indoor environments in Italy. The exposure assessment of these two countries for adults and children revealed significant differences both in the levels of exposure and in the respective patterns. In Italy, the biggest part of the exposure to VMSs takes place domestically, whereas in the UK, it is observed for occupational environments.

To this date there are no studies on MS pollution in the indoor atmosphere of remote regions, except for one [35] conducted in Tromsø (Norway) aiming at a better understanding the IAQ in Arctic regions and its impact on the overall population exposed to chemicals. In fact, in these latitudes the winter is longer and colder than in other regions, and thus people spend more time indoors and are supposedly exposed to a higher level of pollutants due to heating equipment that facilitate the evaporation of chemicals from household products and building materials, VOCs emissions from fireplaces or emissions from new materials used for frequent refurbishment works in wood houses. Katsoyiannis et al. [35] focused the study on various microenvironments, including professional and domestic environments and vehicles, and found that D5 was the dominant cVMS in all samples, ranging from 2.8 to 15 µg m^{-3}, followed by D3 (range 1.1–2.2 µg m^{-3}) and D4 (range 0.4–1.5 µg m^{-3}). The authors concluded that the targeted microenvironments shown good IAQ for VOCs even though the levels of aerosols exceeded in several cases the proposed regulations and guidelines.

An evident different level of MSs was observed between renovated/redecorated indoor environments and ordinary family homes not undergoing renovation. Meng and Wu [36] found concentrations of D4–D6 slightly higher (mean 9.6 µg m^{-3}, range 0.20–169 µg m^{-3}) than that in ordinary family homes (mean 6.27 µg m^{-3}, range 0.09–23.0 µg m^{-3}) and explained this trend considering that MSs from ordinary family homes originate from furniture, household electrical goods, personal care products and various kind of consumables, whereas in redecorated rooms, the sources are not only those but also various redecoration products (i.e. paint, latex paint), showing a strong correlation with the duration of the construction work. The authors also determined MSs inside private cars evidencing a lower mean indoor air concentration of 1.73 µg m^{-3}. In conclusion, Meng and Wu [36] demonstrated that interior renovation and decoration work, and even traveling in cars, can expose humans to siloxane contamination during daily ordinary activities.

The human exposure to MSs was also studied in three different industries (textile, automobile and building) and residential areas in China [37]. Results showed that cyclic compounds (D4–D6) in industrial facilities ranged from 58.6 to 572 $\mu g\ m^{-3}$ and were 1–3 orders of magnitude higher than those determined in residential houses (45.4–139 ng m^{-3}). Although lVMSs had higher concentrations than the cyclic compounds in most industrial additives/products, their total concentrations ranged from 0.05 to 11.2 $\mu g\ m^{-3}$ in indoor air samples, and their levels were much lower than those measured for cyclic siloxanes. This evidence should be due to their lower vapour pressures [38, 39].

Tran and Kannan [40] investigated the levels of five cyclic and nine linear siloxanes in both particulate and vapour phases of indoor air in Albany, New York, USA, and compared airborne particle concentrations with those reported in indoor dust from China [41]. MS concentrations in the vapour phase of indoor air showed D4–D6 at higher concentrations than in the particulate phase. In the latter, D3–D7 were found in all samples; L5 to L9 were frequently found in the samples (range 75–95%), whereas L3, L4 and L11 were rarely detected. D5 and L8 were the most abundant congeners, with concentrations ranging from 29.3 to 34,300 $\mu g\ g^{-1}$ (mean 2,420 $\mu g\ g^{-1}$) and from below the method quantification limit (MQL) to 12,700 $\mu g\ g^{-1}$ (mean 1,320 $\mu g\ g^{-1}$), respectively.

The concentrations of siloxanes in airborne particles were found to be four times higher than those determined in indoor dust samples from China [41]. In particular, the sum of mean concentration of five cyclic and nine linear siloxanes in the particulate phase of indoor air samples was 6,000 $\mu g\ g^{-1}$, and, as expected, the highest levels of siloxanes were found in beauty salons, as personal care products are the major sources of siloxanes in the indoor environment [2, 22].

The strong correlation between high MS concentrations and the extensive use of PCPs was evident in the study conducted by Tran and Kannan [40], who analysed bulk indoor air (particulate plus vapour phases) from six hair and nail salons. The mean level of siloxanes found in hair salons was 6.21 $\mu g\ m^{-3}$, which represented the highest concentration found in the air samples collected from Albany (New York, USA). The mean total concentration of siloxanes was 25 times higher in hair salon than the lowest value of 249 ng m^{-3} found in laboratories and 4 times higher than the total mean value for all samples (1,470 ng m^{-3}), indicating that high siloxane concentrations can be explained by the extensive use of PCPs (i.e. conditioners, shampoo) in hair salon environments. Moreover, poor ventilation conditions, high stability of MSs and long service hours may expose hairdressers to MSs to a higher degree than the general public [42].

Over the past few years, IAQ in non-residential buildings, in particular museum and school buildings, has received particular attention. In a recent paper [43], cyclic MSs together with other VOCs were assessed in indoor air samples from libraries and archives in Florence (Italy) in order to evaluate not only the IAQ for visitors, librarians and curators but also the impact that indoor air quality can have on exposed and/or preserved objects. D5 was the predominant congener (range

7.5–170 µg m^{-3}), followed by D6 and D4, according to the siloxane profiles reported in literature and known composition of PCPs [19].

Regarding school buildings, it should be taken into consideration that students spend over 30% of their life in schools and about 70% of their time inside a classroom during school days. Thus, a poor IAQ in their classrooms may have a marked impact on their health, leading to reduced school attendance, respiratory infections, asthma-related symptoms, allergies and compromised performance [44, 45]. Unfortunately, compared to adults, children are more vulnerable, and risks associated with indoor air pollution are twofold [46]. In fact, children's lungs are exposed to more air in any given time period because they breathe more rapidly than adults; children's lungs are smaller and less developed, so they are more likely to experience complications from the exposure to indoor air pollutants. A recent interesting study was conducted by Villanueva et al. [47], who presented for the first time the levels of VOCs at 18 primary schools located in 3 different areas (rural, urban and industrial) of the province of Ciudad Real (Spain). Among them, D4 and D5 were found in all investigated areas, with D5 at higher concentrations reflecting the formula commonly used in personal care products, such as antiperspirants, cosmetics, hair care products and household products [2, 22].

2.2 Siloxanes in Indoor Dust

Analysis of siloxanes in house dust has been suggested as a possible measurement for indoor contamination because it is a complex mixture of biological material, matter from indoor aerosols and soil particles and can be considered a sink, reservoir and repository for semi-volatile organic compounds and particle bound matter [48]. House dust analysis could also provide a valuable information towards the assessment of human indoor exposure [48]. House dust and the compounds adsorbed therein may enter the human body by inhalation, through non-dietary ingestion, ingestion of particles adhering to food, surfaces in the home and on the skin as well as by adsorption through the skin [49]. If compared to the number of studies on persistent organic pollutants (POPs), such as polycyclic aromatic hydrocarbons (PAHs), polychlorinated biphenyls (PCBs) and polybromodiphenylethers (PBDEs) in indoor environments, only a few studies have reported the occurrence of MSs in indoor dust samples from residential houses, offices, schools and laboratories and even less in dust from occupational areas. However, MSs are a class of organic compounds that should be considered taking into account that indoor dust is a major reservoir of MSs due to their particle-binding affinity [50].

The first study reporting the occurrence and distribution of siloxanes in indoor dust samples was conducted in China in 2009 [41]. Samples were collected in household vacuum cleaners in living rooms of several houses, student dormitories, offices and chemical laboratories from China, and all showed the presence of all cyclic (D4–D7) and linear (L4–L14) siloxanes investigated. Total siloxane concentrations ranged from 21.5 to 21,000 ng g^{-1} (mean 1,540 ± 2,850 ng g^{-1}) with the linear isomers being the predominant compounds, at 1–2 orders of magnitude higher than concentrations of total cVMSs, in all indoor dust samples (see Table 2). Siloxane concentrations in dust were associated with the number of electrical/electronic appliances and number of occupants and smokers living in the house.

An extensive sampling survey of siloxanes in indoor dust samples was conducted by Tran et al. [21], who analysed 310 samples collected from various environments (offices, labs, cars) in 12 countries. The highest concentrations of total cVMSs (D3–D7) were determined in Greece with a median value of 1,380 ng g^{-1} (range 118–25,100 ng g^{-1}) whereas a median value of 772 ng g^{-1} for total lVMSs (L4–L14) (ranging from 129 to 4,990 ng g^{-1}). The median total concentrations of siloxanes in the indoor dust samples from the investigated countries were as follows: Greece (2,970 ng g^{-1}) ($n = 28$), Kuwait (2,400 ng g^{-1}) ($n = 28$), South Korea (1,810 ng g^{-1}) ($n = 28$), Japan (1,500 ng g^{-1}) ($n = 13$), the USA (1,220 ng g^{-1}) ($n = 22$), China (1,070 ng g^{-1}) ($n = 18$), Romania (538 ng g^{-1}) ($n = 23$), Colombia (230 ng g^{-1}) ($n = 28$), Vietnam (206 ng g^{-1}) ($n = 38$), Saudi Arabia (132 ng g^{-1}) ($n = 28$), India (116 ng g^{-1}) ($n = 28$) and Pakistan (68.3 ng g^{-1}) ($n = 28$). As observed in other studies, D5 was the predominant siloxane and found with 100% frequency in dust samples, except for Vietnam (82%), India (79%), Saudi Arabia (75%) and Pakistan (50%). Regarding lVMSs, L8–L10 were predominant, with 100% frequency in samples from Greece, Japan, Kuwait and South Korea. Even if the limited number of samples does not allow to discern the geographical distribution of siloxanes, authors were able to attribute the variation in concentrations to the different consumption and use of products, in particular PCPs [22], main sources of siloxanes among the investigated countries.

Data on occupational exposure to MSs are also limited to the studies of Xu et al. [37] and Liu et al. [50]. Xu et al. [37] determined MSs in dust samples from urban areas and industrial plants. In residential houses, the authors found cVMSs (D4–D6) ranging from 23.3 to 73.5 ng g^{-1} and lVMSs (L8–L12) ranging from 18.0 to 156 ng g^{-1}. A significant correlation between daily used PCPs and cyclic and linear siloxanes suggested that PCPs could also be an important source of MSs in indoor environment of residential houses [41]. In industrial facilities (such as paint production, building automobile, engine and textile industries), the levels were higher, with cVMSs ranging from 2.74 to 50.4 μg g^{-1} and lVMSs varying from 184 to 6,700 μg g^{-1}.

Taking into account that air samples from hair salons showed the highest siloxane concentrations, the determination of MSs in indoor dust from these and similar locations and the evaluation of occupational exposure are imperative to enable risk assessment studies. Recently, Liu et al. [50] investigated the occurrence of MSs in indoor dust samples collected from barbershops, university dormitories, urban households and bathhouses in Tianjin (China) and evaluated the exposure of hairdressers through indoor dust ingestion. The authors found the total concentration of MSs (D4–D6 and L4–L16) ranging from 5,290 to 133,000 ng g^{-1} (mean 43,100 ng g^{-1}) in the barbershop dusts and from 250 to 8,680 ng g^{-1} (mean 2,500 ng g^{-1}) in the other nonoccupational dust samples (bathhouses 5,680 ng g^{-1} > houses 2,450 ng g^{-1} > dormitories 2,090 ng g^{-1}). Linear VMSs

Table 2 Mean air concentrations (ng/g) of MSs in indoor dust samples

Site	Country	Number of samples	D4, ng/g	D5, ng/g	D6, ng/g	D7, ng/g	ΣD3-D7, ng/g	ΣL4-L14, ng/g	ΣL4-L16, ng/g	References
House	Tianjin (China)	56	13.8	33.3	23.9	14.6		1,420		[41]
Dormitory		17	5.15	24.4	48.9	17.0		719		
Office		9	46.3	71.1	33.4	30.7		1,650		
Laboratory		6	44.9	25.6	23.2	58.4		3,380		
PC		5	6.69	2.34	8.77	2.67		628		
Air conditioners		7	10.9	24.4	26.7	38.2		2,310		
Homes	Beihang University (China)	10	80	100	75				10,001	[36]
Homes/laboratory	Shangai (China)	18					458			[21]
Homes	Cartagena (Colombia)	28					304			
Homes	Athens, Erateni, Komatri (Greece)	28					4,100			
Homes	Patna (India)	28					90.4			
Homes, offices	Kumamoto, Nagasaki, Fuknova, Saitama, Saga (Japan)	13					296			
Homes, cars	Kuwait (Kuwait)	28					847			
Homes, cars, offices	Faisalabad (Pakistan)	28					118			
Homes	Iasi (Romania)	23					317			
Homes, cars, air conditioners	Jeddah (Saudi Arabia)	28					194			
Homes, Labs, offices	Ansau, Any aug (South Korea)	28					430			
Homes, Labs, offices	Albany (USA)	22					587			

(continued)

Table 2 (continued)

Site	Country	Number of samples	D4, ng/g	D5, ng/g	D6, ng/g	D7, ng/g	ΣD3-D7, ng/g	ΣL4-L14, ng/g	ΣL4-L16, ng/g	References
Homes, Labs, offices	Vietnam	38					111			
Barbershops	China	36	3,644	8,626	1,848				29,009	[50]
Bathhouses		6	375	1,150	402				3,746	
Dormitory		42	55.7	171	87.9				1,776	
House		114	1,212	2,901	658				10,560	

represented 66–85% of all MSs in the studied dusts, with concentrations 2.2 times higher than those measured for cVMSs. These trends reflected the levels and distribution of MSs in personal care products in China [23], which showed higher concentrations of high molecular weight lVMSs than low molecular weight lVMSs in most of the products. Moreover, the higher level of lVMS (in particular congeners with Si atoms above six) in dust samples with respect to cVMSs could be related to their less volatility and higher tendency to adsorb to dusts. D5 was the dominant congener followed by D4 and D6, showing a pattern similar to those of PCPs. The same authors estimated the daily intakes of total MSs through indoor dust ingestion using the model of worst-case exposure (95th percentile concentration) and high dust ingestion. Human exposure matters are discussed in detail below.

3 Occurrence of MSs in Indoor Environments and Human Exposure

In indoor environments (houses, offices, public buildings, etc.) humans are constantly exposed to siloxanes which are used as additives in many of the common consumer products that they use or come into contact with every day. The exposed population include not only consumers and general public but also occupational workers who are involved in the production of MSs or of products that contain MSs in their formulations and/or use these products in professional settings, such as beauticians and barbers/hair dressers [42].

The routes of exposure to consumers can be dermal, inhalation and ingestion. As suggested by Anderson et al. [51], inhalation is a significant route for human exposure because high amounts of silicone are released to the indoor environment during and directly after the use of consumer products containing silicone products. Utell et al. [52] evidenced the potential for human exposure to cVMS by the respiratory route. They determined and compared the respiratory intake and uptake of D4 in 12 healthy volunteers (25–49 years) in 2 different experimental set ups (nasal and mouthpiece) and estimated a mean total intake and uptake of 11.6 and 1.1 mg day^{-1}, for nasal, and 14.9 and 2 mg day^{-1}, for mouthpiece, respectively.

Several approaches were used to estimate consumer dermal exposure to cVMSs. A wide range of exposure estimates were derived as a result of the uncertainty in adsorption efficiency and penetration rates of cVMSs in the human skin as well as in the degree of evaporative loss during product use. A recent study indicated that inhalation may be a more significant contributor to VMSs intake than transdermal permeation [53]. It should be considered that besides the workers involved in siloxane production, also people working in some general industries, such as building, textile and automobile industries (where siloxane products are largely used), could have potential exposure to MSs [4]. Xu et al. [54] determined plasma concentrations of MSs in workers from one siloxane production facility in China and found levels much higher than those in general population. Considering that siloxanes are characterized by a high log K_{ow}, they are prone to transferring from plasma to fat [55, 56], which could be used as potential indicator for human exposure to MSs. Xu et al. [37] showed that cyclic (D4–D6) and linear (L5–L16) siloxanes in indoor air and dust samples from three general industries (building, textile industries and automobile) were 1–3 orders of magnitude higher than those found in residential areas, and both cyclic and lVMSs were detected in plasma of industrial workers, whereas only cyclic compounds were detected in plasma of the general population. Moreover, only lVMSs showed an apparent accumulation in abdominal fat. Recently, Tran et al. [57] estimated the human exposure to siloxanes through inhalation for various age groups and found mean inhalation exposure doses to total siloxanes of 352, 219, 188, 132 and 95.9 ng kg^{-1}-bw day^{-1} for infant, toddlers, children, teenagers and adults, respectively.

Regarding dust ingestion, based on an average dust intake rate and median exposure concentration, Lu et al. [41] calculated a daily intake of siloxanes of 15.9 ng day^{-1} for adults and of 32.8 ng day^{-1} for toddlers, respectively. Tran et al. [40] evaluated the exposure doses to total MSs through indoor dust ingestion or adults and toddlers/infants in different countries (China, Colombia, Greece, India, Japan, Kuwait, Pakistan, Romania, Saudi Arabia, South Korea, USA and Vietnam). Considering the median values for each country, exposure doses for adults ranged from 0.06 ng kg^{-1}-bw day^{-1} (Pakistan) to 2.48 ng kg^{-1}-bw day^{-1} (Greece), whereas for infant and toddlers were 11.1 and 11.9 ng kg^{-1}-bw day^{-1} and 0.26 and 0.27 ng kg^{-1}-bw day^{-1} in Greece and Pakistan, respectively. More recently, Liu et al. [50] reported exposure rates of 14.3 ng kg^{-1}-bw day^{-1} for hairdressers and 3.43, 2.00 and 222 ng kg^{-1}-bw day^{-1} for the general population, college students and toddlers (1–3 years), respectively. In a recent paper, Capela et al. [58] reported that organosiloxane concentrations were determined in all bestselling brands of PCPs in the Oporto region (Portugal) with total concentrations ranging between 0.003 and 1,203 µg g^{-1}, including adult and baby/children products. Authors estimated human dermal and inhalation exposures, and the latter (related to indoor air) were 1.56 µg kg-bw^{-1} day^{-1} for adults and 0.03 µg kg-bw^{-1} day^{-1} for children. Regarding the products, body moisturizer was the main contributor for adult inhalation exposure, whereas shower gel and shampoo were predominant for children exposure. Naturally the product formulations and the usage patterns can change from country to country or even within countries, which may result in different exposure rates and product sources.

4 Conclusions

The chemical-physical properties of siloxanes (low solubility, high Henry's law constant, high vapour pressure) suggest their strong preference to partition into the atmosphere (indoor and outdoor), which is the main environmental sink for this class of organic compounds.

Levels of MSs in dust samples were correlated with the quantities of the consumer products used daily, suggesting that the household PCPs are the major sources of MS contamination in the indoor microenvironment. The EU commission is currently evaluating D4 and D5 silicones for final PBT (persistent, bioaccumulative and toxic) classification. If both congeners were considered PBTs, they would be banned for use in personal care products under the REACH regulation within the EU.

The high levels of siloxanes in indoor environments (in particular cyclic VMSs) highlight the need to extend the research to new studies in order to identify sources, assess pathways, determine bioaccumulation and evaluate toxic effects of these organic compounds on human health. In particular, the high exposure levels of MSs through dust ingestion suggest that the potential health risks for occupational people, general population and toddlers should not be overlooked and thus both domestic and occupational environments should be taken into consideration in estimates of human exposure to these chemicals.

References

1. Sánchez-Brunete C, Miguel E, Albero B, Tadeo JL (2010) Determination of cyclic and linear siloxanes in soil samples by ultrasonic-assisted extraction and gas chromatography-mass spectrometry. J Chromatogr A 1217(45):7024–7030
2. Wang R, Moody RP, Koniecki D, Zhu J (2009) Low molecular weight cyclic volatile methylsiloxanes in cosmetic products sold in Canada: implication for dermal exposure. Environ Int 35:900–904
3. Companioni-Damas EY, Santos FJ, Galcerán MT (2012) Analysis of linear and cyclic methylsiloxanes in water by headspace-solid phase microextraction and gas chromatography-mass spectrometry. Talanta 89:63–69
4. Lassen C, Hansen CL, Mikkelson SJ, Maag J (2005) Siloxanes – consumption, toxicity and alternatives. Danish Ministry of the Environment, Environmental Protection Agency (Danish EPA). Environmental Project No. 1031
5. Will R, Löchner U, Masahiro Y (2007) CEH marketing research report siloxanes. SRI Consulting, Menlo Park
6. OECD (Organisation for Economic Co-operation and Development) (2007) Manual for investigation of HPV chemicals. OECD Secretariat. http://www.oecd.org/document/7/0,3343,en_2649_34379_1947463_1_1_1_1,00.html. 21 July 2019
7. USEPA (United States Environmental Protection Agency) (2007) High production volume (HPV) challenge program. Sponsored chemicals. http://www.epa.gov/hpv/pubs/update/spnchems.htm. Accessed Feb 2018
8. Danish Ministry of the Environment, EPA (2007) Survey of chemical substances in consumer products. Report 88
9. He B, Rhodes-Brower S, Miller MR, Munson AE, Germolec DR, Walker VR, Korach KS, Meade BJ (2003) Octamethylcyclotetrasiloxane exhibits estrogenic activity in mice via ERR. Toxicol Appl Pharmacol 192(3):254–261
10. Granchi D, Cavedagna D, Ciapetti G, Stea S, Schiavon P, Giuliani R, Pizzoferrato A (1995) Silicone breast implants: the role of immune system on capsular contracture formation. J Biomed Mater Res 29(2):197–202
11. Lieberman MW, Lykissa ED, Barrios R, Ou CN, Kala G, Kala SV (1999) Cyclosiloxanes produce fatal liver and lung damage in mice. Environ Health Perspect 107(2):161–165

12. Quinn AL, Dalu A, Meeker LS, Jean PA, Meeks RG, Crissman JW, Gallavan JRH, Plotzke KP (2007) Effects of octamethylcyclotetrasiloxane (D4) on the luteinizing hormone (LH) surge and levels of various reproductive hormones in female Sprague-Dawley rats. Reprod Toxicol 23 (4):532–540

13. Quinn AL, Regan JM, Tobin JM, Marinik BJ, McMahon JM, McNett DA, Sushynski CM, Crofoot SD, Jean PA, Plotzke KP (2007) In vitro and in vivo evaluation of the estrogenic, androgenic, and progestagenic potential of two cyclic siloxanes. Toxicol Sci 96(1):145–153

14. Danish EPA (Miljøstyrelsen) (2003) Liste over potentielle PBT og vPvB stoffer (List of potential PBT and vPvB substances)

15. McBean E (2008) Implications and trends of siloxanes in landfill and wastewater biogases. Can J Civil Eng 35:431–436

16. Genualdi S, Lee SC, Shoeib M, Gawor A, Ahrens L, Harner T (2010) Global pilot study of legacy and emerging persistent organic pollutants using sorbent-impregnated polyurethane foam disk passive air samplers. Environ Sci Technol 44:5534–5539

17. Lee SY, Lee S, Choi M, Kannan K, Moon HB (2018) An optimized method for the analysis of cyclic and linear siloxanes and their distribution in surface and core sediments from industrialized bays in Korea. Environ Pollut 236:111–118

18. Li B, Li WL, Sun SJ, Qi H, Ma WL, Liu LY, Zhang ZF, Zhu NZ, Li YF (2016) The occurrence and fate of siloxanes in wastewater treatment plant in Harbin. China Environ Sci Pollut Res Int 23(13):13200–13209

19. Pieri F, Katsoyiannis A, Martellini T, Hughes D, Jones KC, Cincinelli A (2013) Occurrence of linear and cyclic volatile methyl siloxanes in indoor air samples (UK and Italy) and their isotopic characterization. Environ Int 59:363–371

20. Tang X, Misztal KP, Nazaroff WW, Goldstein HA (2015) Siloxanes are the most abundant volatile organic compound emitted from engineering students in a classroom. Environ Sci Technol Lett 2:303–307

21. Tran TM, Abualnaja OK, Asimakopoulos GA, Covaci A, Gevao B, Johnson-Restrepo B, Kumosani AT, Malarvannan G, Minh BT, Moon BH, Nakata H, Sinha KR, Kannan K (2015) A survey of cyclic and linear siloxanes in indoor dust and their implications for human exposures in twelve countries. Environ Int 78:39–44

22. Horii Y, Kannan K (2008) Survey of organosilicone compounds, including cyclic and linear siloxanes, in personal-care and household products. Arch Environ Contam Toxicol 55 (4):701–710

23. Lu Y, Yuan T, Wang W, Kannan K (2011) Concentrations and assessment of exposure to siloxanes and synthetic musks in personal care products from China. Environ Pollut 159:3522–3528

24. Bruinen de Bruin Y, Koistinen K, Kephalopoulos S, Geiss O, Tirendi S, Kotzias D (2008) Characterisation of urban inhalation exposures to benzene, formaldehydes and acetaldehyde in the European Union: comparison of measured and modelled exposure data. Environ Sci Pollut Res 15:417–430

25. USEPA (2014) Introduction to indoor air quality [WWW Document]. US EPA. https://www.epa.gov/indoor-air-quality-iaq/introduction-indoor-airquality. Accessed 7 Dec 2017

26. Leva P, Katsoyiannis A, Barrero-Morero J, Kephalopoulos S, Kotzias D (2009) Evaluation of the fate of the active ingredients of insecticide sprays used indoors. J Environ Sci Health B 44 (1):51–57

27. Haug LS, Huber S, Schlabach M, Becher G, Thomsen C (2011) Investigation on per- and polyfluorinated compounds in paired samples of house dust and indoor air from Norwegian homes. Environ Sci Technol 45(19):7991–7998

28. Huber S, Haug LS, Schlabach M (2011) Per- and polyfluorinated compounds in house dust and indoor air from northern Norway –a pilot study. Chemosphere 84(11):1686–1693

29. Northcross AL, Katharine Hammond S, Canuz E, Smith KR (2012) Dioxin inhalation doses from wood combustion in indoor cook fires. Atmos Environ 49:415–418

30. Besis A, Katsoyiannis A, Botsaropoulou E, Samara C (2014) Concentrations of polybrominated diphenylethers (PBDEs) in central air-conditioner filter dust and relevance of non-dietary exposure in occupational indoor environments in Greece. Environ Pollut 188:64–70
31. Shields H, Weschler CJ (1992) Volatile organic compounds measured at a telephone switching center from 5/30/85–12/6/88: a detailed case study. J Air Waste Manage Assoc 42:792–804
32. Shields HC, Fleischer DM, Weschler CJ (1996) Comparisons among VOCs measured in three US commercial buildings with different occupant densities. Indoor Air 6:2–17
33. Yucuis RA, Stanier CO, Hornbuckle KC (2013) Cyclic siloxanes in air, including identification of high levels in Chicago and distinct diurnal variation. Chemosphere 92(8):905–910
34. Zhou SN, Chan CC, Zhy J (2012) Detection of volatile methylsiloxanes in indoor air using thermal desorption GC/MS method. 3rd workshop on organosilicones in the environment. Burlington, ON, Canada
35. Katsoyiannis A, Anda EE, Cincinelli A, Martellini T, Leva P, Goetsch A, Sandanger TM, Huber S (2014) Indoor air characterization of various microenvironments in the Arctic. The case of Tromsø, Norway. Environ Res 134:1–7
36. Meng F, Wu H (2015) Indoor air pollution by methylsiloxane in household and automobile settings. PLoS One 10(8):e0135509
37. Xu L, Shi Y, Liu N, Cai Y (2015) Methyl siloxanes in environmental matrices and human plasma/fat from both general industries and residential areas in China. Sci Total Environ 505:454–463
38. Kaj L, Schlabach M, Andersson J, Cousins AP, Schmidbauer N, Brorström-Lundén E (2005) Siloxanes in the Nordic environment. TemaNord. Nordic Council of Ministers, Copenhagen
39. Kaj L, Andersson J, Cousins AP, Revemberger M, Brorström-Lundén E, Cato I (2005) Results from the Swedish National Programme 2004. Subreport 4: siloxanes. IVL Report B1643. IVL Swedish Environmental Research Institute Ltd., Stockholm
40. Tran TM, Kannan K (2015) Occurrence of cyclic and linear siloxanes in indoor air from Albany, New York, USA and its implications for inhalation exposure. Sci Total Environ 511:138–144
41. Lu Y, Yuan T, Yun SH, Wang W, Wu Q, Kannan K (2010) Occurrence of cyclic and linear siloxanes in indoor dust from China, and implications for human exposures. Environ Sci Technol 44(16):6081–6087
42. Franzen A, Van Landingham C, Greene T, Plotzke K, Gentry R (2016) A global human health risk assessment for decamethylcyclopentasiloxane (D5). Regul Toxicol Pharmacol 74(Suppl): S25–S43
43. Cincinelli A, Martellini T, Amore A, Dei L, Marrazza G, Carretti E, Belosi F, Ravegnani F, Leva P (2016) Measurement of volatile organic compounds (VOCs) in libraries and archives in Florence (Italy). Sci Total Environ 572:333–339
44. Daisey JM, Angell WJ, Apte MG (2003) Indoor air quality, ventilation and health symptoms in schools: an analysis of existing information. Indoor Air 13:53–64
45. Yoon C, Lee K, Park D (2011) Indoor air quality differences between urban and rural pre-schools in Korea. Environ Sci Res 18:333–345
46. Laumbech R, Meng Q, Kipen H (2015) What can individuals do to reduce personal health risks from air pollution? J Thorac Dis 7(1):96–107
47. Villanueva F, Tapia A, Lara S, Amo-Salas M (2018) Indoor and outdoor air concentrations of volatile organic compounds and NO2 in schools of urban, industrial and rural areas in Central-Southern Spain. Sci Total Environ 622–623:222–235
48. Butte B, Heinzow B (2002) Pollutants in house dust as indicators of indoor contamination. Rev Environ Contam Toxicol 175:1–46
49. Lewis RG, Fortmann RC, Camann DE (1994) Evaluation of methods for monitoring the potential exposure of small children to pesticides in the residential environment. Arch Environ Contam Toxicol 26(1):37–46
50. Liu N, Xu L, Cai Y (2018) Methyl siloxanes in barbershops and residence indoor dust and the implication for human exposures. Sci Total Environ 618:1324–1330

51. Andersen M, Sarangapani R, Reitz RH, Gallavan RH, Dobrev ID, Plotzke KP (2001) Physiological modeling reveals novel pharmacokinetic behavior for inhaled octamethylcyclotetrasiloxane in rats. Toxicol Sci 60(2):214–231
52. Utell MJ, Gelein R, Yu CP, Kenaga C, Geigel E, Torres A, Chalupa D, Gibb FR, Speers DM, Mast RW, Morrow PE (1998) Quantitative exposure of humans to an octamethylcyclotetrasiloxane (D4) vapor. Toxicol Sci 44:206–213
53. Biesterbos JW, Beckmann G, van Wei L, Anzion RB, von Goets N, Dudzina T, Roelevend N, Ragas AM, Russel FG, Scheepers PT (2015) Aggregate dermal exposure to cyclic siloxanes in personal care products: implications for risk assessment. Environ Int 74:231–239
54. Xu L, Shi YL, Wang T, Dong ZR, Su WP, Cai YQ (2012) Methyl siloxanes in environmental matrices around a siloxanes production facility, and their distribution and elimination in plasma of exposed population. Environ Sci Technol 46(21):11718–11726
55. Kala SV, Lykissa ED, Neely MW, Lieberman MW (1998) Low molecular weight silicones are widely distributed after a single subcutaneous injection in mice. Am J Pathol 152(3):645–649
56. Flassbeck D, Pfleiderer B, Klemens P, Heumann KG, Eltze E, Hirner AV (2003) Determination of siloxanes, silicon, and platinum in tissues of women with silicone gel-filled implants. Anal Bioanal Chem 375:356–362
57. Tran TM, Le HT, Vu ND, Dang GHM, Minh TB, Kannan K (2017) Cyclic and linear siloxanes in indoor air from several northern cities in Vietnam: levels, spatial distribution and human exposure. Chemosphere 184:1117–1124
58. Capela D, Alves A, Homem V, Santos L (2016) From the shop to the drain-volatile methylsiloxane in cosmetics and personal care products. Environ Int 92–93:50–62

Levels of Volatile Methyl Siloxanes in Outdoor Air

Eva Gallego, Pilar Teixidor, Francisco Javier Roca, and José Francisco Perales

Contents

Abstract Field data showing volatile methyl siloxane (VMS) concentrations in the atmosphere is still limited. Outdoor air concentrations are highly conditioned by population, being VMS values much higher in urban locations than in remote regions, generally in the range of ng m^{-3}, one to three orders of magnitude lower than other volatile organic compounds commonly found in the atmosphere. Cyclic VMS (cVMS) are the most abundant compounds, with concentrations up to 2–3 orders of magnitude higher than those of the linear VMS (lVMS). This abundance is related to the large production and use of cVMS globally. In urban areas, lVMS are generally in the range of 1–20 ng m^{-3}. On the other hand, cVMS present much higher

E. Gallego (✉), F. J. Roca, and J. F. Perales
Laboratori del Centre de Medi Ambient, Escola Tècnica Superior d'Enginyeria Industrial de Barcelona (ETSEIB), Universitat Politècnica de Catalunya (LCMA-UPC), Barcelona, Spain
e-mail: eva.gallego@upc.edu; lcma.info@upc.edu; fco.javier.roca@upc.edu; lcma.info@upc. edu; jose.francisco.perales@upc.edu; lcma.info@upc.edu

P. Teixidor
Centres Científics i Tecnològics, Universitat de Barcelona (CCiTUB), Barcelona, Spain
e-mail: teixidor@ccit.ub.edu

V. Homem and N. Ratola (eds.), *Volatile Methylsiloxanes in the Environment*,
Hdb Env Chem (2020) 89: 201–226, DOI 10.1007/698_2018_343,
© Springer International Publishing AG, part of Springer Nature 2018,
Published online: 7 July 2018

concentrations, ranging from a few hundred to several thousand ng m^{-3}. A limited number of studies evaluating VMS in outdoor air include background and rural locations. Background regions generally present VMS levels one order of magnitude lower than those usually found in urban areas, with lVMS concentrations between 0.01 and 1 ng m^{-3}. In contrast, cVMS concentrations range from 1 to 100 of ng m^{-3}. In the Arctic, lVMS are seldom observed, but cVMS are usually found in the range of 0.1 and 4 ng m^{-3}. The regulation and establishment of air quality criterions for VMS are still very limited worldwide. Evaluations regarding human and environmental exposure to these compounds would be mandatory in the future, as well as the establishment of air quality standards.

Keywords Air quality, Cyclic volatile methyl siloxanes, Linear volatile methyl siloxanes, Outdoor air, Volatile methyl siloxanes

1   Introduction

Volatile methyl siloxanes (VMS), both linear (lVMS) and cyclic (cVMS), have been produced commercially for more than 70 years, being used in the manufacture of consumer and industrial products [1]. Taking into account the new potential uses for these compounds [2], such as halogen-free flame retardants and solvents, their production rate will likely increase in times to come [3]. Yearly, several millions of tonnes of VMS are used as intermediates in high-molecular-weight silicon polymer production and in several industrial applications, such as electrical insulators, dielectric and heat transfer fluids and defoaming and foam stabilizing agents [1, 3–10]. Due to their outstanding chemical characteristics (i.e. high vapour pressure, low surface tension and high degree of compatibility with numerous ingredients used in consumer products formulations) [9], they are also extensively employed in personal care products (antiperspirants, hair and body shampoos and lotions and cosmetics) [11, 12], as well as in household cleaning and coating agents (sealants, adhesives, coatings, paints) [3, 13–16]. VMS will eventually be released to the environment, in shorter or larger time spans, depending on their use and application. Additionally, VMS will be directly volatilized to the atmosphere from "leave-on" personal care products [12, 17]. These conditions lead to high and continuous emissions of these compounds to air, water and soil compartments [3]. Up to a 90–99.5% of the environmentally released siloxanes will be diffused directly into the atmosphere [10, 18–22], where they can be subject to long-range atmospheric transport [1, 10, 23, 24]. Because of their high volatility and low water solubility, their partitioning properties are different from regular persistent organic pollutants (POPs), being the atmosphere their eventual sink [19] and their atmospheric fate ruled mainly by gas-phase reactions [1, 7, 8, 21, 25–27].

VMS were not considered to have potentially adverse health and environmental effects when firstly introduced in industrial, medical, personal care and household uses and products [2, 3, 28]. However, even though nowadays they have been recognized as safe compounds in cosmetic formulations [29], the concern is growing in respect to their environmental distribution, due to their capacity of being transported long-range, persistence, possible toxic and estrogenic effects and tendency to bioaccumulate and/or biomagnify [1–3, 10, 13, 24, 30–39]. An exposure to high concentrations of VMS can lead to the incorporation of these compounds to human plasma [38], even though their half-lives in this matrix are much lower than the usually found for POPs – around 2–10 days [40]. They are considered "emerging contaminants" (EC) [18, 41–43], and further investigations to evaluate their potential human and environmental exposure risks have been claimed by several researchers [3, 18, 23, 44]. In this line, the US EPA included D4 as a chemical for its evaluation during 2013–2014 to determine whether or not this compound had to be regulated under the Toxic Substances Control Act. On the 26th of September 2016, the US EPA announced the receipt of the environmental testing information on D4, submitted following the Enforceable Consent Agreement (ECA) [45]. According to this agreement [46, 47], a final report on D4 was received on the 29th of September 2017. The US EPA is reviewing the gathered information, a task that will take several months due to the extension of the document (US EPA Office of Pollution Prevention and Toxics, personal communication, 15 November 2017). Additionally, the European Commission has adopted a decision where the European Union shall submit a proposal for the listing of D4 in Annex A, B and/or C to the Stockholm Convention on Persistent Organic Pollutants [31]. On the other hand, on the 11th of June 2016 the Committee for Socio-Economic Analysis (SEAC) of the European Chemicals Agency (ECHA) agreed on a restriction to the use of D4 and D5 in "wash-off" personal care products, such as shower gels, shaving foams and shampoos, due to the persistence, bioaccumulative potential and toxicity of these compounds [48].

Generally, the presence and concentrations of VMS in outdoor air have been barely studied [15, 18, 19, 49–51], being sampling methodologies relatively recent and limited [52, 53]. Additionally, VMS concentrations are generally found between one and three orders of magnitude lower than volatile organic compounds (VOCs) [53], and the potential for the contamination of samples during collection and analysis is another challenging condition in VMS monitoring [36]. Hence, these three aspects make the evaluation of VMS in outdoor air a complex task.

2 Handicaps of VMS Outdoor Air Sampling and Analytical Methodologies (Contamination Risk, Sample Handling and Storage, Analytical Quality Assurance/Quality Control)

The quantitative determination of siloxanes in ambient air is a difficult goal to achieve. As a result of the physicochemical nature of VMS, in addition to the first judgement as innocuous compounds, sound analytical methods have not been available until recent years [3]. Due to the fact that VMS are present in a great number of consumer and personal care products, extreme precautions are needed to prevent sample contamination, and meticulous protocols should be applied in order to minimize siloxane contamination during sampling and analysis of ambient air, as concentrations are generally in the range of ng m^{-3} in outdoor air [1, 49, 53–55]. As samples can be easily contaminated by indoor air, special attention has to be paid to the laboratory equipment and personal care products used by the employees [1, 50]. In order to avoid contamination, working in a clean air cabinet under a laminar flow of charcoal and particle filtered air is a preventive measure that can also be taken into account when preparing and processing VMS samples [27, 36, 51, 52].

Gloves, preferably nitrile, are mandatory to handle samples, and the use of fragrances and hand care products by laboratory workers must be avoided during sampling and analysis [15, 36, 49, 53]. Deodorants and hairstyle products used must be siloxane-free, and laboratory glassware used in standard preparation has to be thoroughly washed and/or baked (e.g. 250°C or 450°C overnight). Additionally, the materials used for sampling (e.g. inert glass thermal desorption tubes, unsilanized glass wool, solid-phase extraction cartridges) have to be meticulously cleaned to prevent the possible adsorption of the target VMSs onto silicon surfaces. Once sampled, tubes or cartridges are recommended to be sealed with PTFE caps and wrapped in aluminium foil or kept in clean glass jars and stored at 4°C or frozen (−18°C or −20°C), depending on the sampler employed [27, 36, 50, 52, 53, 56].

VMS can be lost from sampled tubes during storage [36], mainly due to the transformation of the target compounds into other siloxanes or to a decrease in extractability during storage [27], leading to an underestimation of VMS concentrations. cVMS can specially undergo ring opening, losing siloxane units [27, 36]. The more rings they have, the more probabilities to experience this phenomenon, being D4 the most stable cVMS [27]. This aspect can lead to the formation of other target cVMS, overestimating their concentrations in the samples. D3 and D4 are especially prone to be generated from D5 [36, 51]. On the other hand, D3, due to its high volatility, can be lost during storage by evaporation, as several studies have observed important D3 losses in experiments conducted immediately after spiking the samplers [36].

Furthermore, the analytical procedures used in siloxane determination also have to avoid sample contamination or loss of analytes. In this aspect, thermal desorption coupled to gas chromatography/mass spectrometry (TD-GC/MS) eludes sample manipulation, as the sorbent tubes are placed directly in the thermal desorber for analysis, not needing the common treatment of solvent-desorption protocols [53, 57], that can contribute to the existence of background levels of VMS and the contamination of samples with VMS coming from the injection ports (i.e. septa and glass liners) [1, 3, 54]. However, when the use of solvents is necessary for sample extraction, they have to be previously analysed in order to determine VMS background concentrations [1]. Additionally, analytical columns can also be a source of VMS, mainly due to the hydrolysis of silicone column material by humidity [3]. Nonetheless, low-bleed capillary columns release much less target compounds than certain septums [1]. Analyte losses can also occur during sample preparation and processing (e.g. during concentration of the extraction solutions using rotary evaporation or a gentle stream of nitrogen), decreasing the sensitivity of the methodology used [52]. Hence, it is advisable to simplify the analytical methodology as much as possible [56].

Given these aspects, field and procedural blanks are recommended to be taken with each batch of samples and to be analysed each day before every set of samples/ standards in order to evaluate the background levels of instruments and samplers, as well as the possible generation of artefacts (e.g. benzene and/or toluene) during analysis [36, 49, 50, 52, 53]. lVMS are generally not detected in blank samples, whereas trimethylsilanol (TMS) and cVMS are frequently found in blanks at trace levels [23, 27, 36, 49, 53, 58].

On the other hand, TMS is a highly volatile and unstable compound [59], and the use of organic solvents, activated charcoal, Tenax TA, XAD resins and polyurethane foam (PUF) for its determination lacked efficiency [60, 61]. However, its evaluation using methods based on adsorption in special cartridges followed by cryofocusing in cold traps and GC/MS [53, 60, 62, 63] or PTR-TOF-MS [61] proved to be effective.

Method detection limits (MDL) for several studies using different sampling and analytical methodologies are shown in Table 1.

3   VMS Concentrations in Outdoor Air

Field data showing VMS emissions and concentrations outdoors is still limited [10, 19, 69, 70], even though these compounds are nowadays being much more taken into account in air quality monitoring programmes [4–6, 30, 53]. As can be derived from worldwide publications, there is a real tendency emphasizing the need to monitor these compounds in environmental matrices, especially since 1996 [28].

Due to their high hydrophobicity and high Henry's Law constant, VMS will tend to be extremely distributed in the atmospheric compartment, with values of total mass as far as 99.5% in respect to other environmental niches [19]. Once released to the atmosphere, they will be distributed worldwide at ng m^{-3} concentrations, with higher levels near emission sources [20, 53, 71]. Their levels in the outdoor air fluctuate by a factor of 2–4 on a time scale of 1 day to 1 week approximately, as several studies have observed [24, 27, 50, 51, 53, 56]. Additionally, during a same day, VMS variations over time can be assigned to differences in wind speed, to an increase or decrease of the mixing of these compounds in ambient air [51], to oscillations in the height of the planetary boundary layer [44] and to changes in atmospheric circulation [24]. Finally, outdoor air concentrations are highly conditioned by population, being VMS values much higher in urban locations than in remote regions [71].

cVMS are the most abundant VMS in outdoor air, with concentrations up to 2–3 orders of magnitude higher than those of the lVMS [2, 15, 23, 50, 52, 53, 56, 72]. This abundance is related to the large production and use of cVMS globally. VMS patterns worldwide generally reflect the use of personal care products, as well as the production of these compounds in different industrial areas. Hence, a relative variability

Table 1 Limits of detection (LOD, pg m^{-3}) of VMS at trace levels from several studies regarding outdoor air

TMS	L2	L3	L4	L5	L6	L7	L8	L9	L10	L11	L12	L13	L14	L15	L16	D3	D4	D5	D6	Methodology	Reference
–	–	–	–	–	–	–	–	–	–	–	–	–	–	–	–	30–140	40–60	10–40	10–40	SPE cartridge+GC/MS	[36]a
–	–	3.8	7.0	8.9	16	–	–	–	–	–	–	–	–	–	–	270	210	150	130	SPE cartridge+GC/MS	[27]
–	–	–	–	–	–	–	–	–	–	–	–	–	–	–	–	–	–	120	–	SPE cartridge+GC/MS	[64]
–	–	–	–	–	–	–	–	–	–	–	–	–	–	–	–	–	400–1,200	800–2,900	800–3,800	SPE cartridge+GC/MS	[16]
500	900	900	600	800	–	–	–	–	–	–	–	–	–	–	–	700	800	900	700	Sorbent tubes+TD-GC/MS	[53]
–	180	20	20	10	–	–	–	–	–	–	–	–	–	–	–	100	150	100	80	SPE cartridges+GC/MS	[52]
–	–	11	0	0	–	–	–	–	–	–	–	–	–	–	–	1,100	2,500	7,100	2,300	SIP disk samplers+GC/MS	[23]b
–	–	<10	<10	<10	–	–	–	–	–	–	–	–	–	–	–	7,090	7,350	18,200	7,510	PUF/XAD-4 cartridges +GC/MS	[49]b
–	–	140	140	140	190	190	560	560	830	830	–	–	–	–	–	60	80	60	190	PUF cartridges+GC/MS	[58]
–	–	<171	<188	<293	–	–	–	–	–	–	–	–	–	–	–	12,200	6,050	4,370	5,730	XAD-2 cartridges+GC/MS	[56]b
–	–	<24	<35	<64	–	–	–	–	–	–	–	–	–	–	–	2,300	2,070	2,540	940	SPE cartridges+GC/MS	[56]
–	–	10	20	30	–	–	–	–	–	–	–	–	–	–	–	490	3,300	2,400	1,900	XAD-4 cartridges+GC/MS	[50]b
–	–	4	20	20	–	–	–	–	–	–	–	–	–	–	–	360	790	1,000	250	PUF/XAD-2 cartridges +GC/MS	[50]
–	–	–	–	–	–	–	–	–	–	–	–	–	–	–	–	–	50,000	50,000	–	Passive charcoal sampler +GC/MS	[65]b
–	–	–	–	–	–	–	–	–	–	–	–	–	–	–	–	–	6,000,000	4,000,000	–	Direct sampling/MS	[66]
2,800–9,500	–	–	–	–	–	–	–	–	–	–	–	–	–	–	–	–	–	–	–	Sorbent tubes+TD-GC/MS	[62]c

																			Method	Ref.
480	–	–	–	–	–	–	–	–	–	–	–	–	–	–	–	–	–	–	Canisters +preconcentration+GC/MS	[67]
–	70	60	20	180	160	180	140	180	190	180	330	290	360	350	360	–	720	280	100	SIP disk samplers+GC/MS	[68][b]
–	160	160	180	400	500	500	400	500	600	500	900	900	1,000	1,000	360	–	360	360	300	SPE cartridge+GC/MS	[40][d]
–	500	500	500	500	500	500	500	500	600	500	900	900	1,000	1,000	1,000	–	1,000	1,000	900	SPE cartridge+GC/MS	[40][e]

[a] LOD depending on the sampling volume (88 and 27 m^3, respectively)

[b] Passive sampling

[c] LOD depending on the sampling volume (20 and 6 L, respectively)

[d] Limit of quantification (24 h samples)

[e] Limit of quantification (8 h samples)

among concentrations and distribution profiles of VMS is noticed throughout the globe [18]. D5 generally dominates in urban areas influenced by industrial activities and personal care products [16, 23, 50, 72]. However, in the Pearl River Delta (China), D3 and D4 were found to be the most abundant compounds, being D5 only present in trace concentrations in some of the samples evaluated [73]. In the Nordic environment (i.e. Denmark, Faroe Islands, Finland, Iceland and Sweden), D4 was the dominating compound in outdoor air, whereas in Norway, D5 was the predominant one, failing to agree with the use patterns, in which D4, D5 and D6 presented quite comparable values in these territories [74]. On the other hand, a recent study observed higher D3 than D5 concentrations in several Catalan urban areas (Spain) [53], as opposed to most data published until now. Additionally, D3 was generally found at higher concentrations in Arctic sites [23] and in Chinese cities [1]. D4–D6 are the most abundant compounds in personal care product composition, and outdoor air VMS concentration patterns worldwide should reflect the use and production of these goods. Nevertheless, the presence of D3 as the most abundant VMS in several regions could be related to the degradation and/or transformation in the atmosphere of siloxane cyclic compounds with a higher number of rings. Concerning all this information, it has to be taken into account that D5 is one of the most studied VMS in the atmosphere [27] and that much more information regarding this compound is available in comparison with other VMS. As an example, TMS has not been evaluated in a great number of studies that determined VMS concentrations in outdoor air. These studies generally use sampling and analysis methodologies that are not suitable for the analysis of this compound [60, 61]; hence, these limitations can be an explanation of the little presence of TMS in outdoor air quality evaluations regarding VMS.

All in all, VMS concentrations in outdoor air, even in urban areas with industrial impacts, are generally in the range of ng m^{-3}, one to three orders of magnitude lower than other VOCs commonly found in the atmosphere, which are usually in the range of μg m^{-3}.

3.1   Urban and Industrial Areas

Worldwide VMS concentrations in urban and a small number of industrial areas are summarized in Table 2. As can be observed, cVMS have been much more investigated in outdoor air, being lVMS concentrations available only from a few research studies. Additionally, the results presented are quite recent, with a 50 and a 67% of them published within the last 5 and 7 years, respectively. Mojsiewicz-Pieńkowska and Krenczkowska [28] stated that until 1991 no scientific articles had been published regarding siloxanes in environmental samples, being an 83% of the articles concerning this topic published during the 2009–2017 period. In fact, the lack of reliable analytical methods for VMS determination in outdoor air in routine measurements could be one of the main reasons why VMS have not been generally included in air quality evaluations [3]. This absence of monitoring data from

Table 2 VMS outdoor air concentrations (ng m^{-3}) in several worldwide urban and industrial areas

TMS	L2	L3	L4	L5	D3	D4	D5	D6	Location	Reference
Outdoor air (ng m^{-3})										
Urban/industrial										
22–178	3–215	0.3–35	n.d.–12	n.d.–3	39–1,166	9–676	87–1,942	16–68	Urban areas, Spain	[53]
214	6	7	4	18	1,358	642	14,914	449	Urban area hotspot, Spain	[53][a]
–	12–22	14–16	16–17	6–8	2.2–5.0	73–79	375–439	45–60	Urban area, Barcelona, Spain	[53]
–	–	<lod–0.1	0.06–0.7	0.1–0.5	0.7–30	3.9–50	55–280	4.0–53	Urban areas (Europe and America)	[23][b]
–	–	–	–	–	–	18–190	100–1,100	n.d.–7.6	Urban area, Chicago, USA	[16]
–	–	0.1	0.1	0.6	0.1–0.3	1.1–1.2	1.9–2.6	1.1–1.5	Urban area, Braga, Portugal	[72][b]
–	–	0.1	0.1	0.1	0.7	0.6–1.0	1.4–2.9	1.7–3.4	Urban area, Porto, Portugal	[72][b]
–	–	–	–	–	–	–	100–650	10–79	Urban area, Zurich, Switzerland	[51]
–	–	–	–	–	–	100	500	–	Urban areas, USA	[65][b]
–	–	–	–	–	–	–	460–1,240	–	Urban area, San Francisco, USA	[75]
–	<19–79	<3–<5	<3–<6	<3–<6	–	18–230	<13–140	<12–42	Urban areas, Sweden	[76]
–	<4	<8	<6	<20	–	260–320	190–310	70–140	Urban areas, Denmark	[74]
–	<4	<8	<6	<20	–	2,100–4,000	930–2,400	390–2,100	Urban area, Tórshavn, Faroe Islands	[74]
–	<4	<8	<6	<20	–	320–2,100	130–1,600	80–420	Urban area, Reykjavik, Iceland	[74]
–	<4	<8	<6	<20	–	550–580	890–2,500	820–870	Urban areas, Norway	[74]
–	<4	<8	<6	<20	–	140–350	50–200	50–110	Urban areas, Sweden	[74]
–	–	–	–	–	0–11,300	0–3,300	n.d.–traces	–	Urban area, Guangzhou, China	[73]
–	–	–	–	–	2,100–5,800	800–4,300	n.d.–traces	–	Urban area, Macau, China	[73]

(continued)

Table 2 (continued)

TMS	L2	L3	L4	L5	D3	D4	D5	D6	Location	Reference
–	–	–	–	–	0–2,300	0–3,500	n.d.–traces	–	Urban area, Nanhai, China	[73]
–	–	–	–	–	–	5.1–37	22–65	3.3–9.3	Medium-sized urban, USA	[16]
–	–	–	–	–	–	–	45–160	7–16	Suburban site, Switzerland	[51]
–	–	1.8	1.2	0.5	<lod	24.2	93.5	65.9	Suburban site, Toronto, Canada	[56][b]
–	–	0.2–4.9	0.4–6.5	0.7–4.8	0.5–4.7	2.8–77	15–247	1.9–22	Semiurban site, Toronto, Canada	[50]
–	–	0.9 ± 0.3	1.4 ± 0.5	0.50 ± 0.04	19 ± 23	108 ± 50	268 ± 111	13 ± 4	Semiurban site, Ontario, Canada	[49][b]
–	–	–	–	–	100–1,000	0–1,600	n.d.–traces	–	Suburban area, Guangzhou, China	[73]
–	–	–	–	–	1,900–9,300	6,400–20,500	n.d.–traces	–	Industrial area, Guangzhou, China	[73]
–	–	0.03–0.04	0.03–0.04	0.2	0.6	1.3–1.5	0.9–1.3	0.4	Industrial area, Estarreja, Portugal	[72][b]
–	–	0.1	0.1	0.8–1.1	0.4	0.4–1.8	1.1–3.1	3.1	Industrial area, Z.I. Mota, Portugal	[72][b]

[a]Urban area located close to an important petrochemical industrial area
[b]VMS concentrations obtained by passive sampling

compounds used for decades is a drawback when the assessment of fate and environmental/human exposure to these compounds is needed [20]. In fact, values of both ambient concentrations and emission rates are essential input parameters for models predicting VMS environmental fate and risk assessment [70].

The lVMS evaluated are generally in the range of 1 to 20 ng m^{-3}, while cVMS present much higher concentrations, between a few hundred to several thousand ng m^{-3}, depending on the compound. Additionally, hotspot and Chinese industrial areas show concentrations as far as one order of magnitude higher than those found in urban and suburban sites, up to 20,000 ng m^{-3}. On the other hand, however, two Portuguese industrial areas presented much lower concentrations [72].

The evaluation of distribution profiles of VMS in the atmosphere suffers from an important flaw, as not all studies have focused on the same target compounds. The main compounds studied are L3–L5 and D4–D6 for lVMS and cVMS, respectively. VMS distributions in outdoor air in several worldwide urban and industrial areas are presented in Table 3. Due to the lack of data of several VMS from the studies in question, only L3–L5 and D4–D6 were evaluated. As it can be observed, the lVMS distributions are quite variable, with no trend being recognized. On the other hand, D5 is the most important cVMS generally found outdoors in urban areas (21–93%), followed by D4 (4–61%) and D6 (n.d.–53%). A predominance of D5 in outdoor air in an important number of urban areas in Europe and North America seems to be linked to the fact that D5 is the most abundant cVMS present in the formulations of personal care products, followed by D6 and D4 [3, 77–79]. The use of D5 has increased in recent years in detriment of D4, as lower toxicity potential is assumed for the first one [3]. D4–D6 are the major VMS found in domestic wastewater treatment plants (WWTP) water and sludge, derived mainly from the use of personal care products [80, 81]. Moreover, compared to the total burden of VMS in the atmosphere, on-site measurements of ambient air in WWTP also present profiles with D5 and D4 as major compounds [49].

On the other hand, D3 and D4 concentrations were observed in Chinese sites (Table 2) at much higher concentrations than D5, indicating a variation in siloxane sources between continents, mainly due to industrial emissions related to silicone production [23] and to different consumption patterns in each country [25, 82]. Additionally, a recent study conducted by Gallego et al. [53] in Catalonia (Spain) found a predominance of D3 in several of the studied urban areas. These findings were not in line with the data usually found in outdoor air from developed countries, clearly indicating that the studies on sources and behaviour of this kind of compounds in the atmosphere have to be considerably improved.

3.2 Background and Rural Sites

The limited number of studies evaluating worldwide VMS concentrations in background and rural sites is summarized in Table 4. Background regions generally present VMS levels one order of magnitude lower than the usually found in urban

Table 3 VMS outdoor air distributions (%) in several worldwide urban/industrial, background and polar areas

L3	L4	L5	D4	D5	D6	Location	Reference
Urban/industrial							
70–100	n.d.–24	n.d.–6	8–25	72–78	3–14	Urban areas, Spain	[53]
24	14	62	4	93	3	Urban area hotspot, Spain	[53][a]
39	41–44	17–20	14–15	76	9–10	Urban area, Barcelona, Spain	[52]
<lod-8	38–54	38–63	6–13	73–87	6–14	Urban areas (Europe and America)	[23][b]
–	–	–	15	85	n.d.–1	Urban area, Chicago, USA	[16]
13	13	74	23–27	46–49	27–28	Urban area, Braga, Portugal	[72][b]
33	33	33	14–16	38–40	46	Urban area, Porto, Portugal	[72][b]
29–33	33–35	33–35	42–56	30–34	10–28	Urban areas, Sweden	[76]
23	18	59	42–50	37–40	13–18	Urban areas, Denmark	[74]
23	18	59	47–61	27–28	11–25	Urban area, Tórshavn, Faroe Islands	[74]
23	18	59	51–60	25–39	10–15	Urban area, Reykjavik, Iceland	[74]
23	18	59	15–24	39–63	22–36	Urban areas, Norway	[74]
23	18	59	53–58	21–30	17–21	Urban areas, Sweden	[74]
–	–	–	17–33	58–72	8–11	Medium-sized urban, USA	[16]
52	34	14	13	51	36	Suburban site, Toronto, Canada	[56][b]
15–30	31–40	30–54	14–22	71–76	6–10	Semiurban site, Toronto, Canada	[50]
32	50	18	28	69	3	Semiurban site, Ontario, Canada	[49][b]
12–50	12–50	76	50–54	35–46	15	Industrial area, Estarreja, Portugal	[72][b]
7–10	7–10	80–86	9–22	24–39	39–53	Industrial area, Z.I. Mota, Portugal	[72][b]
<lod-100	n.d.–54	n.d.–86	4–61	21–93	n.d.–53	All urban/industrial areas	
Background							
<lod-3	<lod-92	<lod-5	70–100	<lod-23	<lod-7	Background (Europe, America and Australia)	[23][b]
86–88	9	3–5	21–23	71–72	6–7	Background, Sweden	[27]
–	–	–	31–33	59–64	5–8	Rural site, USA	[16]

(continued)

Table 3 (continued)

L3	L4	L5	D4	D5	D6	Location	Reference
33	33	33	55–64	16–31	14–20	Rural site, Sweden	[76]
23	18	59	63	25	12	Rural site, Denmark	[74]
<lod-88	<lod-92	<lod-59	21–100	<lod-72	<lod-20	All background sites	
Polar							
<lod	<lod-100	<lod	64–80	18–27	2–9	Arctic	[23][b]
–	–	–	n.d.–31	57–67	12–33	Arctic	[36]
<lod	<lod-100	<lod	n.d.–80	18–67	2–33	All Arctic locations	

Percentages calculated for ΣL3–L5 and ΣD4–D6
[a]Urban area located close to an important petrochemical industrial area
[b]VMS concentrations obtained by passive sampling

areas. lVMS exhibit concentrations between 0.01 and 1 ng m^{-3}, while cVMS range from 1 to 100 of ng m^{-3}, with a specific rural site located in Denmark presenting a D4 value up to 2,400 ng m^{-3}. Unlike urban areas, cVMS distributions in background sites are mainly dominated by D4 (21–100%) instead of D5 (<lod-72%) (Table 3), an aspect related to the VMS transport in the atmosphere and to the regional •OH concentrations in ambient air [71]. In fresh air masses, i.e. within a few days of emission, the emission rates would be the most important factor affecting VMS concentrations in outdoor air [87]. However, as VMS are emitted to the atmosphere, they move away from the sources together with the air masses, where they are exposed to OH radicals. As D5 is more reactive to •OH than D4 and has a lower half-life [36, 88–90], the final concentrations found in background regions are commonly dominated by D4.

3.3 Polar Sites

VMS concentrations in polar sites are summarized in Table 5. As can be seen from the available information from Arctic locations, lVMS are seldom observed. The lower concentrations of lVMS in these sites in respect to worldwide urban and background areas have been linked to lower long-range transport of these com-pounds to remote areas [23, 91] and to their lower half-lives comparing to cVMS [26, 88, 89]. On the other hand, cVMS are usually found in the range of 0.1 to 4 ng m^{-3}, concentrations one order of magnitude lower than those obtained in background locations and up to four orders of magnitude lower than the determined in urban and

Table 4 VMS outdoor air concentrations (ng m^{-3}) in several worldwide background and rural locations

TMS	L2	L3	L4	L5	D3	D4	D5	D6	Location	Reference
Outdoor air (ng m^{-3})										
Background										
–	–	<lod-0.02	<lod-0.7	<lod-0.04	0.5–117	0.9–45	<lod-15	<lod-4.5	Background (Europe, America and Australia)	[23]ᵃ
–	–	0.1–0.5	0.01–0.05	<0.003–0.03	0.4–2.4	1.8–8.0	5.6–28	0.5–2.7	Background, Sweden	[27]
–	–	–	–	–	–	–	0.7–8.3	–	Rural site, Sweden	[64]
–	–	–	–	–	–	–	0.3–9	–	Rural site, Sweden	[24]
–	–	–	–	–	–	5.6–14	10–29	<1.4–2.3	Rural site, USA	[16]
–	<20–73	<5–<6	<5–<6	<5–<6	–	35–300	9–170	11–77	Rural site, Sweden	[76]
–	<4	<8	<6	<20	–	2,400	950	440	Rural site, Denmark	[74]
–	–	–	–	–	–	n.d.–1.5	0.3–5.4	0.2–2.1	Rural sites, Norway	[83]
–	–	–	–	–	–	n. d.–5.7	1.6–10.0	0.1–1.0	Rural sites, Norway	[84]
–	–	–	–	–	–	n.d.–3.2	0.8–7.1	0.1–4.9	Rural sites, Norway	[85]
–	–	–	–	–	–	n.d.–2.1	0.5–4.7	0.2–2.6	Rural sites, Norway	[86]

ᵃVMS concentrations obtained by passive sampling

Table 5 VMS outdoor air concentrations (ng m^{-3}) in several polar worldwide locations

TMS	L2	L3	L4	L5	D3	D4	D5	D6	Location	Reference
Outdoor air (ng m^{-3})										
Polar										
–	–	<lod	<lod-0.01	<lod	0.5–21	0.7–18	0.3–4	0.1–0.5	Arctic	[23][a]
–	–	–	–	–	n.d.–2.8	n.d.–2.1	0.2–3.9	0.1–0.8	Arctic	[36]

[a]VMS concentrations obtained by passive sampling

industrial areas. However, three polar locations (i.e. Ny-Ålesund in Norway and Little Fox Lake and Alert in Canada) exhibited somewhat higher concentrations, with D3 in the range of 10–21 ng m^{-3} and D4 in the range of 12–18 ng m^{-3} [23], closer to background concentrations than to the ones found in other polar territories. D4 and D5 are the most abundant compounds in Polar sites, with contributions to ΣD4-D6 cVMS of n.d.–80% and 18–67%, respectively (Table 3). Higher D5 concentrations were observed in winter than in summer in Arctic locations [24, 36] due to the higher phototransformation of D5 in the atmosphere in the warmer months, a major mechanism of removal of this compound and one of the responsible for the spatial variability in VMS concentrations [24].

4 VMS Sources and Distribution Profiles in Outdoor Air

4.1 VMS Sources to Outdoor Air

VMS sources to outdoor air can be varied. One emission pathway of these compounds comes directly from VMS production/use facilities [38], where surrounding VMS concentrations can be up to three orders of magnitude higher than those typically found in not directly impacted urban areas [40, 53, 73]. However, even though emissions can be very high from these plants, their territorial distribution is limited. Regarding indoor environments, mainly due to the use of personal care and cleaning products [92], the levels are much higher (i.e. 5–500 times) than those observed in outdoor air [58]. However, the diffusive emissions from indoor atmospheres can be meaningful contributors to outdoor concentrations [52, 71]. Moreover, as a result of the extensive use of VMS in a great variety of products, their presence in wastewater, landfill waste, and the biogas generated in these facilities is considerable [18, 22, 69]. In this line, wastewater treatment plants and landfills, through water, sewage and landfill gas, as well as the sludge treatment into agricultural fertilizer, are important sources of VMS to ambient air [1, 3, 21, 36, 49, 80–82, 93, 94].

Several studies indicate that a higher population density could be a significant variable to explain higher VMS air concentrations outdoors [3, 16, 44, 50, 56, 71, 73,

87]. In this line, Gallego et al. [53] observed significant correlations between population density (inh. km^{-2}) and average concentrations of D5 ($r^2 = 0.622$, $p \leq 0.05$) and D6 ($r^2 = 0.697$, $p \leq 0.05$) in Spain, implying higher emissions of these cVMS in more populated areas. Additionally, significant correlations (Pearson correlation, $p < 0.05$) were observed between VMS concentrations near WWTP and the population served by each facility in Canada [68]. On the other hand, Ratola et al. [72] observed significant differences in VMS concentrations in two Portuguese beach sites between winter and summer seasons, the latter having the highest VMS incidence. This aspect indicates an anthropogenic fingerprint of the results, suggesting a substantial influence of the population fluctuations in touristic regions.

Due to the composition of "wash-off" personal care products, VMS end up in WWTP influent streams [95]. These facilities, however, have not been projected for the elimination of these kinds of compounds, and they would be distributed to the environment through direct emissions or retained in sludge [2, 68], as more than 90% of VMS remain unaltered during wastewater treatment due to their chemical and thermal stabilities [41]. VMS processing in these plants will depend on several factors such as the specific physicochemical characteristics of each compound, the water and sludge characteristics (pH, organic matter, cations concentration) and the specific operational parameters of each facility [19, 22, 96]. Due to their low aqueous solubility, high K_{AW} (air water partition coefficient) and high vapour pressures, VMS will tend to volatilize from the influent water treated in WWTPs to the atmosphere [20, 22, 97], and they can also be emitted through aerosol ejection [25, 68]. Depending on the compound, their emission pathway will preferably be one or the other. For example, it has been estimated by modelling that a 61% of D4 is emitted through aerosol ejection compared to a 13% that would be associated to volatilization. On the other hand, D6 would be emitted in a 17% through aerosol ejection and only in a 4% through volatilization [68]. Moreover, D4 and D5 can also be degraded through microbe catalysis hydrolysis, being transformed to hydrolysates of siloxanes such as dimethylsilanediol, in anaerobic compartments [22].

As VMS are predominantly partitioned into sludge in WWTP systems [3, 25, 69, 82], about 50% of VMS entering the influent will end up in biosolids [95], and the application of the sludge for agricultural purposes and/or sludge disposal will promote the distribution of these contaminants in extensive agricultural regions [20, 96], leading to their final diffusion to the atmosphere [3, 21, 81], as well as increase their potential to bioaccumulate [95]. Additionally, during the composting process of the sludge, VMS volatilization can also occur, liberating these compounds to ambient air [3, 21]. Regarding landfills, due to the fact that the use of VMS in consumer products has increased considerably in the last few years, the emission of these compounds to the atmosphere from newer landfills would likely be higher than that from older ones [2, 97].

In this type of facilities, cVMS are emitted between 20 and 600 times more than lVMS [49, 68, 94]. Among cVMS, D3, D4 and D5 are the predominantly emitted compounds [1, 73]. This aspect is related to the VMS composition of personal care products, where cVMS prevail over lVMS [12], having generally two times the concentrations of lVMS [77]. However, it has to be taken into account that VMS emissions from these types of facilities are quite variable depending on the raw materials buried in the landfills and the origin and water treatment processes adopted in each wastewater treatment station, as well as the process conditions at the biogas production sites [1, 2, 97, 98]. In WWTP, the temperature of wastewater will also be a crucial condition, as higher volatilization and re-volatilizations of VMS will be expected in summer due to an increase in water temperature [50, 51, 68].

The difference in consumption patterns of siloxanes among countries and seasons is another important aspect to consider [19, 22, 25, 81, 82], along with the possible differences in formulations between the same personal care products produced in different regions, despite being of the same brand name and apparently alike [9, 70]. Likewise, VMS influent concentrations can also be affected by seasonality [68, 82] and reveal particular activities of people during the different hours of the day [17] and week days [80] regarding the consumption of products containing VMS.

4.2   VMS Correlations in Outdoor Air

Weak or no correlations were observed between outdoor air lVMS and cVMS concentrations in research studies, an aspect that can be related to different temporal and/or spatial distributions of emissions [27, 50] and different volatilization dispositions of these compounds [40]. In conjunction with less knowledge concerning lVMS levels in outdoor air [53], information is also scarce regarding their emission sources [27]. Nevertheless, lVMS emissions were estimated to be much lower than cVMS [27], even though they are considerably more concentrated in urban areas comparing to background and Arctic sites [23] (Tables 2, 3 and 4). Apart from the different emission patterns, lVMS vapour pressures are lower than those of cVMS, which makes them less volatile [40]. Hence, their relative release to the atmosphere is expected to be minor in comparison to cVMS. Another important aspect that could influence the lower lVMS concentrations observed in outdoor air is their lower half-lives in the atmosphere, from 2.9 to 8.8 days for L5 and L2, respectively [26, 88, 89]. The rate constants for the reaction of VMS with OH radicals (considering 7.7×10^5 molecule cm^{-3} over a 24-h period) range from 0.52 to 2.66×10^{-12} cm^3 molecule^{-1} s^{-1}, with L4 > L3 > D6 > D5 > L2 > D4 > D3 [36, 88, 89]. These constants seem to be divided in two categories, the cVMS and the lVMS, and the rate constants increase with the number of methyl substituent groups in each group [88]. Hence, the atmospheric degradation half-lives, in days, decrease as: D3 = 20–30-> D4 = 10.3 > L2 = 8.8 > D5 = 6.7–6.9 > L3 = 6.3 > D6 = 4.4–5.8 > L4 = 3.2–4.3-> L5 = 2.9 [20, 36, 88–90].

On the contrary, significant correlations were observed between D5 and D6 and D3 and D4 [23, 27, 53, 73]. These correlations between congeners generally indicate common sources and transport mechanisms of these compounds [50], an aspect that can also be linked to their similar half-lives [53].

Apart from correlations between different VMS congeners, another interesting aspect to consider is the D5/D4 ratio. Cyclic siloxane ratios can give information about the location of the emission sources and •OH photochemical ageing [16, 71]. As D4 presents a higher half-life value, in common emission source points, i.e. urban areas, this ratio will be higher than in background/remote sites. Hence, in areas close to the emitting sources, the corresponding rates will determine the ratio in fresh air masses, as regular emissions in these areas reduce the effects of seasonality [56]. On the other hand, in more remote sites, where emissions are not the main source of VMS to outdoor air, the ageing of the air masses will decrease the D5/D4 ratio, as D5 will be degraded earlier than D4 in the atmosphere during its transport [16, 71]. D5/D4 ratios from worldwide locations are presented in Table 6. As can be observed, D5/D4 ratios range between 0.2 and 1.7 and 0.4 and 3.3 in polar sites and background areas, respectively. In contrast, urban areas present a range of values depending on the studied site. Genualdi et al. [23] found relatively high ratios in two US cities, Sidney (FL) and Groton (CT), with an average value of nearly 20. However, worldwide urban areas generally exhibit ratios between 3 and 5. It has to be noted, nonetheless, that urban areas located in Nordic countries (i.e. Sweden,

Table 6 D5/D4 ratios in several worldwide outdoor air environments

D5/D4 ratio	Location	Reference
Urban areas		
19.9 ± 6.7	USA	[23][a]
4.7 ± 8.0	Catalonia, Spain	[53]
5.3 ± 0.3	Barcelona, Spain	[52]
5.6	Paris, France	[23][a]
4.6 ± 1.5	Chicago, USA	[16]
5.0	USA	[65][a]
5.0	Ontario, Canada	[23][a]
4.3 ± 1.5	Toronto, Canada	[50]
3.9	Toronto, Canada	[56][a]
2.6 ± 0.6	Ontario, Canada	[49][a]
3.0 ± 2.1	Norway	[74]
0.9 ± 0.8	Sweden	[76]
0.57 ± 0.04	Denmark	[74]
0.5 ± 0.1	Faroe Islands	[74]
1.7 ± 2.2	Iceland	[74]
0.7 ± 0.4	Sweden	[74]
Background areas		
0.9 ± 0.9	Europe, USA and Australia	[23][a]
3.3 ± 0.3	Sweden	[27]
2.1 ± 0.7	USA	[16]
0.5 ± 0.4	Sweden	[76]
0.4	Denmark	[74]
Polar sites		
0.2 ± 0.2	Arctic	[23][a]
1.7 ± 0.5	Arctic	[36]

[a]VMS concentrations obtained by passive sampling

Denmark, Iceland) present lower D5/D4 ratios, more similar to those observed in background areas and polar sites, an aspect that could indicate a release of VMS relatively far from these Nordic cities.

5 Siloxane Quality Criteria for Outdoor Air

Even though both the US EPA, the ECHA and the European Commission have evaluated several VMS (mainly cVMS) in order to determine their potential toxicity to humans and their capacity for persistence in the environment [31, 45–48], they have mainly focused in the use of these compounds in personal care products. However, the negative potential effects derived from the direct or indirect release of VMS to the outdoor atmospheric environment is nowadays still barely regulated.

For several years, the establishment of appropriate legislation regarding VMS concentrations in the atmosphere and their inclusion in long-term monitoring programmes have been claimed by various scientists and research institutes [2, 30]. Any regulation will necessarily be linked to an increase in VMS monitoring in outdoor air, a fundamental aspect to assess more accurately the possible adverse effects that VMS concentrations can have both to human health and the environment [16].

Canada, for example, considers VMS not dangerous to human health or life [4–6]. Notwithstanding, this country has included D4–D6 in the environmental monitoring programme "Chemicals Management Plan Monitoring, Surveillance and Research Program", where their concentrations in air are monitored in the field to improve the comprehension of the distribution in the environment and the bioaccumulation potential of these compounds. As a final goal, this information is being used to inform on the development of risk management actions and to estimate their efficiency [99]. On the other hand, Norway has included D4, D5 and D6 in the "List of Priority Substances", which aims to reduce the emissions of these compounds until complete elimination by 2020 [100]. D4–D6 have been included in the campaigns "Monitoring of environmental contaminants in air and precipitation" since 2013, which determine environmental data from selected contaminants in air and precipitation at Norwegian background sites [86].

Two other countries have regulated emission concentrations of VMS in outdoor air. The Danish Environmental Protection Agency has classified L2 and D3–D6 as main group 2 compounds and set for them an air quality criterion of 10 $\mu g\ m^{-3}$, based on a tolerable concentration of 0.1 mg m^{-3} and assuming a 100% of absorption after inhalation [101]. Likewise, Ontario (Canada) has established standards for the maximum allowable concentrations of landfill gas emissions in order to evaluate the particular impact of a facility on its surrounding air quality. Guidelines of 16 and 15 $\mu g\ m^{-3}$ were defined for L2 and D4, respectively [102].

The regulation and establishment of air quality criteria for VMS are still very limited worldwide. Due to the fact that reliable sampling and analytical methodologies have been developed in the last few years, more information would be available regarding these compounds in outdoor air in the time to come. As VMS are highly volatile and a considerable part of their emissions diffuse directly to the atmosphere, evaluations with respect to human and environmental exposure to these compounds and their potential adverse effects would be mandatory in the future, as well as the establishment of air quality standards.

References

1. Wang DG, Norwood W, Alaee M, Byer JD, Brimble S (2013) Review of recent advances in research on the toxicity, detection, occurrence and fate of cyclic volatile methyl siloxanes in the environment. Chemosphere 93:711–725
2. Gaj K, Pakuluk A (2015) Volatile methyl siloxanes as potential hazardous air pollutants. Pol J Environ Stud 24:937–943
3. Rücker C, Kümmerer K (2015) Environmental chemistry of organosiloxanes. Chem Rev 115:466–524
4. Environment Canada (2008) Screening assessment for the challenge octamethylcyclotetrasiloxane (D4). Chemical abstracts service registry number 556-67-2, Health Canada
5. Environment Canada (2008) Screening assessment for the challenge decamethylcyclopentasiloxane (D5). Chemical abstracts service registry number 541-02-6, Health Canada
6. Environment Canada (2008) Screening assessment for the challenge dodecamethyl-cyclohexasiloxane (D6). Chemical abstracts service registry number 540-97-6, Health Canada
7. Brooke DN, Crookes MJ, Gray D, Robertson S (2009) Environmental risk assessment report: decamethylcyclopentasiloxane. Environmental Agency of England and Wales, Bristol
8. Brooke DN, Crookes MJ, Gray D, Robertson S (2009) Environmental risk assessment report: octamethylcyclotetrasiloxane. Environmental Agency of England and Wales, Bristol
9. Dudzina T, von Goetz N, Bogdal C, Biesterbos JWH, Hungerbühler K (2014) Concentrations of cyclic volatile methylsiloxanes in European cosmetics and personal care products: prerequisite for human and environmental exposure assessment. Environ Int 62:86–94
10. Xu S, Wania F (2013) Chemical fate, latitudinal distribution and long-range transport of cyclic volatile methylsiloxanes in the global environment: a modelling assessment. Chemosphere 93:835–843
11. Biesterbos JWH, Beckmann G, van Wel L, Anzion RBM, von Goetz N, Dudzina T, Roeleveld N, Ragas AMJ, Russel FGM, Scheepers PTJ (2015) Aggregate dermal exposure to cyclic siloxanes in personal care products: implications for risk assessment. Environ Int 74:231–239
12. Capela D, Alves A, Homem V, Santos L (2016) From the shop to the drain – volatile methylsiloxanes in cosmetics and personal care products. Environ Int 92-93:50–62
13. Dodson RE, Nishioka M, Standley LJ, Perovich LJ, Brody JG, Rudel RA (2012) Endocrine disruptors and asthma-associated chemicals in consumer products. Environ Health Persp 120:935–943
14. Nørgaard AW, Jensen KA, Janflet C, Lauritsen FR, Clausen PA, Wolkoff P (2009) Release of VOCs and particles during use of nanofilm spray products. Environ Sci Technol 43:7824–7830
15. Pieri F, Katsoyannis A, Martellini T, Hughes D, Jones KC, Cincinelli A (2013) Occurrence of linear and cyclic volatile methyl siloxanes in indoor air samples (UK and Italy) and their isotopic characterization. Environ Int 59:363–371
16. Yucuis RA, Stainer CO, Hornbuckle KC (2013) Cyclic siloxanes in air, including identification of high levels in Chicago and distinct diurnal variation. Chemosphere 92:905–910

17. van Egmond R, Sparham C, Hastie C, Gore D, Chowdhury N (2013) Monitoring and modelling of siloxanes in a sewage treatment plant in the UK. Chemosphere 93:757–765
18. Balducci C, Perilli M, Romagnoli P, Cecinato A (2012) New developments in emerging organic pollutants in the atmosphere. Environ Sci Pollut Res 19:1875–1884
19. Kim J, Mackay D, Whelan MJ (2018) Predicted persistence and response times of linear and cyclic volatile methylsiloxanes in global and local environments. Chemosphere 195:325–335
20. Mackay D, Cowan-Ellsberry CE, Powel DE, Woodburn KB, Xu S, Kozerski GE, Kim J (2015) Decamethylcyclopentasiloxane (D5) environmental sources, fate, transport, and routes of exposure. Environ Toxicol Chem 34:2689–2702
21. Panagopoulos D, MacLeod M (2018) A critical assessment of the environmental fate of linear and cyclic volatile methylsiloxanes using multimedia fugacity models. Environ Sci Process Impacts 20:183–194
22. Xu L, Shi Y, Cai Y (2013) Occurrence and fate of volatile siloxanes in a municipal wastewater treatment plant of Beijing, China. Water Res 47:715–724
23. Genualdi S, Harner T, Cheng Y, MacLeod M, Hansen KM, van Egmond R, Shoeib M, Lee SC (2011) Global distribution of linear and cyclic volatile methyl siloxanes in air. Environ Sci Technol 45:3349–3354
24. McLachlan MS, Kierkegaard A, Hansen KM, van Egmond R, Christensen JE, Skjøth CA (2010) Concentrations and fate of decamethylcyclopentasiloxane (D$_5$) in the atmosphere. Environ Sci Technol 44:5365–5370
25. Capela D, Ratola N, Alves A, Homem V (2017) Volatile methylsiloxanes through wastewater treatment plants – a review of levels and implications. Environ Int 102:9–29
26. Whelan MJ, Estrada E, van Egmond R (2004) A modelling assessment of the atmospheric fate of volatile methyl siloxanes and their reaction products. Chemosphere 57:1427–1437
27. Kierkegaard A, McLachlan MS (2013) Determination of linear and cyclic volatile methyl-siloxanes in air at a regional background site in Sweden. Atmos Environ 80:322–329
28. Mojsiewicz-Pieńkowska K, Krenczkowska D (2018) Evolution of consciousness of exposure to siloxanes – review of publications. Chemosphere 191:204–217
29. SCCS (Scientific Committee on Consumer Safety) (2010) Opinion on cyclomethicone octamethylcyclotetrasiloxane (cyclotetrasiloxane, D4) and decamethylcyclopentasiloxane (cyclopentasiloxane, D5). https://ec.europa.eu/health/scientific_committees/consumer_safety/docs/sccs_o_029.pdf. Accessed 22 Jan 2018
30. Cousins AP, Brorström-Lundén E, Hedlund B (2012) Prioritizing organic chemicals for long-term air monitoring by using empirical monitoring data – application to data from the Swedish screening program. Environ Monit Assess 184:4647–4654
31. European Commission (2016) Proposal for a Council Decision on the submission, on behalf of the European Union, of a proposal for the listing of additional chemicals in Annex A, B and/or C to the Stockholm Convention on Persistent Organic Pollutants, Brussels
32. ECHA (2015) Opinion on persistency and bioaccumulation of octamethylcyclotetrasiloxane (D4) and decamethylcyclopentasiloxane (D5) according to a MSC mandate. Helsinki, Member State Committee
33. He B, Rhodes-Brower S, Miller MR, Munson AE, Germolec DR, Walker VR, Korach KS, Meade BJ (2003) Octamethylcyclotetrasiloxane exhibits estrogenic activity in mice via ERα. Toxicol Appl Pharm 192:254–261
34. Jean PA, Plotzke KP (2017) Chronic toxicity and oncogenicity of octamethylcyclotetrasiloxane (D4) in the Fischer 344 rat. Toxicol Lett 279:75–97
35. Warner NA, Evenset A, Christensen G, Gabrielsen GW, Borgå K, Leknes H (2010) Volatile siloxanes in the European arctic: assessment of sources and spatial distribution. Environ Sci Technol 44:7705–7710
36. Krogseth IS, Kierkegaard A, McLachlan MS, Breivik K, Hansen KM, Schlabach M (2013) Occurrence and seasonality of cyclic volatile methyl siloxanes in Arctic air. Environ Sci Technol 47:502–509

37. McGoldrick DJ, Chan C, Drouillard KG, Keir MJ, Clark MG, Backus SM (2014) Concentrations and trophic magnification of cyclic siloxanes in aquatic biota from the Western Basin of Lake Erie, Canada. Environ Poll 186:141–148

38. Xu L, Shi Y, Liu N, Cai Y (2015) Methyl siloxanes in environmental matrices and human plasma/fat from both general industries and residential areas in China. Sci Total Environ 505:454–463

39. Schøyen M, Øxnevad S, Hjermann D, Mund C, Böhmer T, Beckmann K, Powell DE (2016) Levels of siloxanes (D4, D5, D6) in biota and sediments from the Inner Oslofjord, Norway, 2011–2014. In: The SETAC Europe 26th annual meeting, Nantes, 26 May 2016

40. Xu L, Shi Y, Wang T, Dong Z, Su W, Cai Y (2012) Methyl siloxanes in environmental matrices around a siloxane production facility, and their distribution and elimination in plasma of exposed population. Environ Sci Technol 46:11718–11726

41. Clarke BO, Smith SR (2011) Review of 'emerging' organic contaminants in biosolids and assessment of international research priorities for the agricultural use of biosolids. Environ Int 37:226–247

42. Farré M, Kantiani L, Petrovic M, Pérez S, Barceló D (2012) Achievements and future trends in the analysis of emerging organic contaminants in environmental samples by mass spectrometry and bioanalytical techniques. J Chromatogr A 1259:86–99

43. Genualdi S, Harner T (2012) Rapidly equilibrating micrometer film sampler for priority pollutants in air. Environ Sci Technol 46:7661–7668

44. Yucuis R (2013) Cyclic siloxanes in air including identification of high levels in Chicago and distinct diurnal variation. Master's thesis, University of Iowa

45. USEPA (2016) Receipt of information under the Toxic substances control act. https://www.regulations.gov/document?D=EPA-HQ-OPPT-2012-0209-0106. Accessed 22 Jan 2018

46. USEPA (2014) Enforceable consent agreement for environmental testing for Octamethylcyclotetrasiloxane (D4) (CASRN 556-67-2). Docket No. EPA-HQ-OPPT-2012-0209. https://www.epa.gov/sites/production/files/2015-01/documents/signed_siloxanes_eca_4-2-14.pdf. Accessed 22 Jan 2018

47. USEPA (2014) Consent agreements and testing consent orders: octamethylcyclotetrasiloxane (D4); export. https://www.regulations.gov/document?D=EPA-HQ-OPPT-2012-0209-0067. Accessed 22 Jan 2018

48. ECHA (2016) Committee for socio-economic analysis concludes on restricting D4 and D5. https://www.echa.europa.eu/-/committee-for-socio-economic-analysis-concludes-on-restricting-d4-and-d5. Accessed 19 Jan 2018

49. Cheng Y, Shoeib M, Ahrens L, Harner T, Ma J (2011) Wastewater treatment plants and landfills emit volatile methyl siloxanes (VMSs) to the atmosphere: investigations using a new passive air sampler. Environ Pollut 159:2380–2386

50. Ahrens L, Harner T, Shoeib M (2014) Temporal variations of cyclic and linear volatile methylsiloxanes in the atmosphere using passive samplers and high-volume air samplers. Environ Sci Technol 48:9374–9381

51. Buser AM, Kierkegaard A, Bogdal C, MacLeod M, Scheringer M, Hungerbühler K (2013) Concentrations in ambient air and emissions of cyclic volatile methylsiloxanes in Zurich, Switzerland. Environ Sci Technol 47:7045–7051

52. Companioni-Damas EY, Santos FJ, Galceran MT (2014) Linear and cyclic methylsiloxanes in air by concurrent solvent recondensation-large volume injection-gas chromatography-mass spectrometry. Talanta 118:245–252

53. Gallego E, Perales JF, Roca FJ, Guardino X, Gadea E (2017) Volatile methyl siloxanes (VMS) concentrations in outdoor air of several Catalan urban areas. Atmos Environ 155:108–118

54. Varaprath S, Stutts DH, Kozerski GE (2006) A primer on the analytical aspects of silicones at trace levels-challenges and artifacts – a review. Silicon Chem 3:79–102

55. Warner NA, Kozerski G, Durham J, Koerner M, Gerhards R, Campbell R, McNett DA (2013) Positive vs. false detection: a comparison of analytical methods and performance for analysis

of cyclic volatile methylsiloxanes (cVMS) in environmental samples from remote regions. Chemosphere 93:749–756

56. Krogseth IS, Zhang X, Lei YD, Wania F, Breivik K (2013) Calibration and application of a passive air sampler (XAD-PAS) for volatile methyl siloxanes. Environ Sci Technol 47:4463–4470

57. Lamaa L, Ferronato C, Fine L, Jaber F, Chovelon JM (2013) Evaluation of adsorbents for volatile methyl siloxanes sampling based on the determination of their breakthrough volume. Talanta 115:881–886

58. Tran TM, Kannan K (2015) Occurrence of cyclic and linear siloxanes in indoor air from Albany, New York, USA, and its implications for inhalation exposure. Sci Total Environ 511:138–144

59. Arnold M, Kajolinna T (2010) Development of on-line measurement techniques for siloxanes and other trace compounds in biogas. Waste Manag 30:1011–1017

60. Grümping R, Mikolajczak D, Hirner AV (1998) Determination of trimethylsilanol in the environment by LT-GC/ICP-OES and GC-MS. Fresenius J Anal Chem 361:133–139

61. Hayeck N, Temime-Roussel B, Gligorovski S, Mizzi A, Gemayel R, Tlili S, Maillot P, Pic N, Vitrani T, Poulet I, Wortham H (2015) Monitoring of organic contamination in the ambient air of microelectronic clean room by proton-transfer reaction/time-of flight/mass spectrometry (PTR-ToF-MS). Int J Mass Spectrom 392:102–110

62. Lee JH, Jia C, Kim YD, Kim HH, Pham TT, Choi YS, Seo YU, Lee IW (2012) An optimized adsorbent sampling combined to thermal desorption GC-MS method for trimethylsilanol in industrial environments. Int J Anal Chem 2012:1–10

63. Gallego E, Roca FJ, Perales JF, Guardino X, Gadea E (2015) Development of a method for determination of VOCs (including methylsiloxanes) in biogas by TD-GC/MS analysis using Supel™ inert film bags and multisorbent bed tubes. Int J Environ Anal Chem 95:291–311

64. Kierkegaard A, McLachlan MS (2010) Determination of decamethylcyclopentasiloxane in air using commercial solid phase extraction cartridges. J Chromatogr A 1217:3557–3560

65. Shields HC, Fleischer DM, Weschler CJ (1996) Comparisons among VOCs measured in three types of U.S. commercial buildings with different occupant densities. Indoor Air 6:2–17

66. Badjagbo K, Furtos A, Alaee M, Moore S, Sauvé S (2009) Direct analysis of volatile methylsiloxanes in gaseous matrixes using atmospheric pressure chemical ionization-tandem mass spectrometry. Anal Chem 81:7288–7293

67. Herrington JS (2013) Whole air canister sampling coupled with preconcentration GC/MS analysis of part-per-trillion levels of trimethylsilanol in semiconductor cleanroom air. Anal Chem 85:7882–7888

68. Shoeib M, Schuster J, Rauert C, Su K, Smyth S-A, Harner T (2016) Emission of poly and perfluoroalkyl substances, UV-filters and siloxanes to air from wastewater treatment plants. Environ Pollut 218:595–604

69. Kulkarni HV (2012) Occurrence of cyclo-siloxanes in wastewater treatment plants – quantification and monitoring. Thesis, Colorado State University

70. Buser A (2013) Siloxanes: emissions, properties and environmental fate. Thesis, ETH Zurich

71. Janecheck NJ, Hansen KM, Stainer CO (2017) Comprehensive atmospheric modeling of reactive cyclic siloxanes and their oxidation products. Atmos Chem Phys 17:8357–8370

72. Ratola N, Ramos S, Homem V, Silva JA, Jiménez-Guerrero P, Amigo JM, Santos L, Alves A (2016) Using air, soil and vegetation to assess the environmental behaviour of siloxanes. Environ Sci Pollut Res 23:3273–3284

73. Wang XM, Lee SC, Sheng GY, Chan LY, Fu JM, Li XD, Min YS, Chan CY (2001) Cyclic organosilicon compounds in ambient air in Guangzhou, Macau and Nanhai, Pearl River Delta. Appl Geochem 16:1447–1454

74. Kaj L, Schlabach M, Andersson J, Cousins AP, Schmidbauer N, Brörstrom-Lundén E (2005) Siloxanes in the Nordic environment. Nordic Council of Ministers, Copenhagen

75. Hodgson AT, Faulkner D, Sullivan DP, DiBartolomeo DL, Russell ML, Fisk WJ (2003) Effect of outside air ventilation rate on volatile organic compound concentrations in a call center. Atmos Environ 37:5517–5527

76. Kaj L, Andersson J, Cousins AP, Remberger M, Brörstrom-Lundén E (2005) Results from the Swedish national screening programme 2004. Subreport 4: Siloxanes. IVL Swedish Environmental Research Institute, Stockholm

77. Horii Y, Kannan K (2008) Survey of organosilicone compounds, including cyclic and linear siloxanes, in personal-care and household products. Arch Environ Contam Toxicol 55:701–710

78. Wang R, Moody RP, Koniecki D, Zhu J (2009) Low molecular weight cyclic volatile methylsiloxanes in cosmetic products sold in Canada: implication for dermal exposure. Environ Int 35:900–904

79. Lu Y, Yuan T, Wang W, Kannan K (2011) Concentrations and assessment of exposure to siloxanes and synthetic musks in personal care products from China. Environ Pollut 159:3522–3528

80. Bletsou AA, Asimakopoulos AG, Stasinakis AS, Thomaidis NS, Kannan K (2013) Mass loading and fate of linear and cyclic siloxanes in a wastewater treatment plant in Greece. Environ Sci Technol 47:1824–1832

81. Lee S, Moon H-B, Song G-J, Ra K, Lee W-C, Kannan K (2014) A nationwide survey and emission estimates of cyclic and linear siloxanes through sludge from wastewater treatment plants in Korea. Sci Total Environ 497–498:106–112

82. Li B, Li W-L, Sun S-J, Qui H, Ma W-L, Liu L-Y, Zhang Z-F, Zhu N-Z, Li Y-F (2016) The occurrence and fate of siloxanes in wastewater treatment plant in Harbin, China. Environ Sci Pollut Res 23:13200–13209

83. NILU (Norwegian Institute for Air Research) (2014) Monitoring of environmental contaminants in air and precipitation. Annual report 2013. Norwegian Environment Agency, Trondheim

84. NILU (Norwegian Institute for Air Research) (2015) Monitoring of environmental contaminants in air and precipitation. Annual report 2014. Norwegian Environment Agency, Trondheim

85. NILU (Norwegian Institute for Air Research) (2016) Monitoring of environmental contaminants in air and precipitation. Annual report 2015. Norwegian Environment Agency, Trondheim

86. NILU (Norwegian Institute for Air Research) (2017) Monitoring of environmental contaminants in air and precipitation. Annual report 2016. Norwegian Environment Agency, Trondheim

87. Navea JG, Young MA, Xu S, Grassian VH, Stainer CO (2011) The atmospheric lifetimes and concentrations of cyclic methylsiloxanes octamethylcyclotetrasiloxanes (D_4) and decamethylcyclopentasiloxane (D_5) and the influence of heterogeneous uptake. Atmos Environ 45:3181–3191

88. Atkinson R (1991) Kinetics of the gas-phase reactions of a series of organosilicon compounds with OH and NO_3 radicals and O_3 at 297 ± 2 K. Environ Sci Technol 25:863–866

89. Markgraf SJ, Wells JR (1997) The hydroxyl radical reaction rate constants and atmospheric reaction products of three siloxanes. In J Chem Kinet 29:445–451

90. Kim J, Xu S (2017) Quantitative structure-reactivity relationships of hydroxyl radical rate constants for linear and cyclic volatile methylsiloxanes. Environ Toxicol Chem 36:3240–3245

91. Lu Y, Yuan T, Yun SH, Wang W, Wu Q, Kannan K (2010) Occurrence of cyclic and linear siloxanes in indoor dust from China, and implications for human exposures. Environ Sci Technol 44:6081–6087

92. Lassen C, Hansen CL, Mikkelsen SH, Maag J (2005) Siloxanes – consumption, toxicity and alternatives. Environmental project no. 1031 2005. Environmental Protection Agency, Danish Ministry of the Environment, Denmark

93. Surita SC, Tansel B (2014) Emergence and fate of cyclic volatile polydimethylsiloxanes (D4, D5) in municipal waste streams: release mechanisms, partitioning and persistence in air, water, soil and sediments. Sci Total Environ 468-469:46–52

94. Tansel B, Surita SC (2014) Differences in volatile methyl siloxane (VMS) profiles in biogas from landfills and anaerobic digesters and energetics of VMS transformations. Waste Manag 34:2271–2277

95. Surita SC, Tansel B (2015) Contribution of siloxanes to COD loading at wastewater treatment plants: phase transfer, removal, and fate at different treatment units. Chemosphere 122:245–250

96. Fijalkowski K, Rorat A, Grobelak A, Kacprzak MJ (2017) The presence of contaminations in sewage sludge – the current situation. J Environ Manag 203:1126–1136

97. Rasi S (2009) Biogas composition and upgrading to biomethane. Thesis, University of Jyväskylä

98. Rasi S, Lehtinen J, Rintala J (2010) Determination of organic silicon compounds in biogas from wastewater treatment plants, landfills, and co-digestion plants. Renew Energ 35:2666–2673

99. Environment Canada (2010) Consultation document. Octamethylcyclotetrasiloxane (D4). Chemical abstracts service registry number 556-67-2

100. Norwegian Environment Agency (2018) List of priority substances. http://www.environment. no/List-of-Priority-Substances/. Accessed 26 Feb 2018

101. Danish Ministry of the Environment (2014) Siloxanes (D3, D4, D5, D6, HMDS). Evaluation of health hazards and proposal of health-based quality criterion for ambient air. Environmental project no. 1531. Copenhagen

102. Tetra Tech (2017) Air quality existing conditions report. Eastern Ontario waste handling facility landfill expansion. Lafleche Environmental, Boucherville

Atmospheric Fate of Volatile Methyl Siloxanes

Michael S. McLachlan

Contents

Abstract Volatile methyl siloxanes (VMS) are emitted primarily to air, and the bulk of VMS present in the environment resides in the atmosphere. Therefore, the atmospheric fate of VMS is a core component of the environmental chemistry of these chemicals. In this chapter the phase partitioning of VMS in the atmosphere is first examined, and then the different mechanisms by which they can be removed from the atmosphere are evaluated, both physical removal via deposition and chemical removal via reactions. We find that VMS are almost entirely present in gaseous form and that reaction with OH radicals is the dominant process for their removal. Consequently, for most purposes, the atmospheric fate of VMS can be simplified to three processes: the emission function, advection, and removal via reaction with OH radicals. However, each of these processes is complex, so we

M. S. McLachlan (✉)
Department of Environmental Science and Analytical Chemistry, Stockholm University, Stockholm, Sweden
e-mail: michael.mclachlan@aces.su.se

V. Homem and N. Ratola (eds.), *Volatile Methylsiloxanes in the Environment*,
Hdb Env Chem (2020) 89: 227–246, DOI 10.1007/698_2018_371,
© Springer Nature Switzerland AG 2018, Published online: 14 November 2018

explore how mathematical models have been used to capture this complexity, quantify the expected atmospheric fate, and describe the variability of VMS concentrations in time and space.

Keywords D4, D5, Deposition, Modeling, Phototransformation

1   Introduction

Volatile methyl siloxanes are a group of chemicals made up of $SiO(CH_3)_2$ units. They can be arranged cyclically (denoted D#, where # is the number of units) or linearly (denoted L#, whereby the end units are capped with methyl groups). D4 (octamethylcyclotetrasiloxane) and D5 (decamethylcyclopentasiloxane) have attracted the most attention from environmental chemists, and much of the information in this chapter will relate to those structures. However, more information is now emerging on the atmospheric fate of D3 and D6 as well as some of the linear structures (L2–L5).

D4, D5, and D6 are used primarily as intermediates in the manufacture of polydimethylsiloxane (PDMS), as carriers in personal care products, and in cleaning products [1–6]. Whereas the use as an intermediate occurs in closed systems, the use in personal care and cleaning products results in direct emission to the environment. It has been estimated that 90% of D4, D5, and D6 in personal care products is emitted to air. The remaining 10% initially enters wastewater, whereby 48%, 22%, and 9%, respectively, of the discharges to wastewater have been estimated to also volatilize [1–3]. Besides personal care products, the other large source of D4, D5, and D6 to the environment is volatilization of residues of these chemicals present in PDMS. This accounts for an estimated 66%, 5%, and 21% of direct emissions to air, respectively [1–3]. Combining this information, 95, 92, and 92% of the estimated total emissions of D4, D5, and D6 to the environment are directly or indirectly released to air. This indicates that the atmosphere plays a central role in the environmental fate of VMS.

The atmosphere is also an important compartment for the environmental distribution of VMS. When the environmental fate of D5 was assessed using the equilibrium criterion (EQC) model, a steady-state model of a hypothetical "unit world" environment, 64% of the chemical was present in the air compartment [7]. In a modeling study of the global fate of D4, D5, and D6, 99%, 92%, and 73%, respectively, of the total global residues were predicted to be in the atmosphere [8].

In the following, the atmospheric fate of VMS will be explored. We will begin by examining the phase distribution of these chemicals in the atmosphere. We will then study their physical removal from the atmosphere by various deposition processes. Transformation processes in the atmosphere, which are of central importance for VMS, will follow. We will close with a look at the consequences of these fate processes for the temporal and spatial variability of VMS concentrations in air.

2 Phase Distribution

The phase distribution of an organic chemical in the atmosphere has a decisive impact on its atmospheric fate [9]. In the atmosphere organic chemicals can be present as gases or associated with aerosols. Aerosols consist of inorganic material, organic material, and water. In principle, an organic chemical will associate with aerosols if it has a strong tendency to partition from the gas phase into one of these matrices. However, studies of gas/particle partitioning with ambient aerosols for a broad range of organic chemicals indicate that there are just two mechanisms of relevance: relative humidity (RH)-independent sorption to the water insoluble fraction of the aerosol and RH-dependent sorption to the aqueous fraction. Sorption to the water-insoluble fraction dominates for apolar compounds, while for polar and ionic compounds, both mechanisms can contribute, with the sorption to the aqueous fraction becoming more important at high RH. For sorption to the water insoluble fraction, organic material is the dominant sorbent; sorption to minerals plays a minimal role. Sorption to the aqueous fraction is dominated by partitioning into water [10]. Hence two partitioning processes are expected to govern the phase distribution of VMS in the atmosphere: gas/organic matter partitioning and gas/water partitioning.

Partitioning of many organic chemicals between aerosol organic matter and the gas phase can be predicted from the octanol/air partition coefficient K_{OA} [11], whereby such relationships are only applicable to apolar chemicals [12].

$$K'_P = 0.2 K_{OA} \qquad (1)$$

where K_P' is the volume normalized dimensionless particle/gas partition coefficient $(m^3\ m^{-3})$. Log K_{OA} at 25°C and the respective heats of phase change are given for L2, L3, L4, D4, D5, and D6 in Table 1 [13]. The distribution of the VMS between the aerosol and gas phases arising from partitioning to the organic fraction was estimated at different temperatures according to Mackay [15]

$$f_{\text{aerosol}} = \frac{v_Q K'_P}{1 + v_Q K'_P} \qquad (2)$$

where v_Q is the volume fraction of aerosol in the atmosphere $(m^3\ m^{-3})$. The fraction of chemical predicted to be present in the aerosol phase was very small, ranging from 3.5×10^{-9} (L2 at 40°C) to 2.3×10^{-3} (D6 at -40°C) (Table 1). The true values may be even lower, as the sorption of VMS to organic matter from the aqueous environment has been found to be two or more orders of magnitude lower than predicted from models based on sorption to octanol [16].

Gas/water partitioning can be predicted from the air/water partition coefficient K_{AW}. Log K_{AW} at 25°C and the respective heats of phase change are given in Table 2 [17]. The distribution of the VMS between the aerosol and gas phases arising from partitioning to the aqueous fraction was estimated in analogy to Eq. (2). The fraction

Table 1 Summary of the phase distribution of six VMS in the atmosphere arising from partitioning to the water-insoluble organic fraction of aerosols, expressed as the fraction associated with aerosol ($f_{Aerosol}$)

	Units	L2	L3	L4	D4	D5	D6
Log K_{OA} (25°C)[a]	m^3 m^{-3}	2.98	3.77	4.64	4.31	4.95	5.77
ΔU_{OA}[a]	kJ mol^{-1}	−25.5	−36.9	−45.5	−43.9	−47.0 [b]	−57.5
$f_{Aerosol}$ (−40°C)[c]		1.0×10^{-7}	2.2×10^{-6}	4.4×10^{-5}	1.7×10^{-5}	1.1×10^{-4}	2.3×10^{-3}
$f_{Aerosol}$ (25°C)[c]		5.7×10^{-9}	3.5×10^{-8}	2.6×10^{-7}	1.2×10^{-7}	5.3×10^{-7}	3.5×10^{-6}
$f_{Aerosol}$ (40°C)[c]		3.5×10^{-9}	1.7×10^{-8}	1.1×10^{-7}	5.2×10^{-8}	2.2×10^{-7}	1.2×10^{-6}

[a]From Xu and Kropscott [13]

[b]From Xu [14]

[c]Based on an average background volume fraction of aerosol in the atmosphere v_Q of 30×10^{-12} m^3 m^{-3} [11]

Table 2 Summary of the phase distribution of six VMS in the atmosphere arising from partitioning to the aqueous fraction of aerosols, expressed as the fraction associated with aerosol ($f_{Aerosol}$)

	Units	L2	L3	L4	D4	D5
Log K_{AW} (25°C)[a]	m^3 m^{-3}	2.49	3.06	3.45	2.74	3.16
ΔU_{AW}[a]	kJ mol^{-1}	53.0	39.5	65.5	73.9	123.9
$f_{Aerosol}$ (−40°C)[b]		7.5×10^{-12}	4.4×10^{-13}	3.4×10^{-12}	4.4×10^{-11}	4.7×10^{-9}
$f_{Aerosol}$ (25°C)[b]		1.9×10^{-14}	5.2×10^{-15}	2.1×10^{-15}	1.1×10^{-14}	4.2×10^{-15}
$f_{Aerosol}$ (40°C)[b]		7.0×10^{-15}	2.4×10^{-15}	6.0×10^{-16}	2.6×10^{-15}	3.8×10^{-16}

[a]Calculated from Xu and Kropscott [17]

[b]Based on an assumed water content of the aerosol of 20% and an average background volume fraction of aerosol in the atmosphere of 30×10^{-12} m^3 m^{-3} [11]

of chemical associated with the aqueous portion of the aerosol phase is even much smaller than the fraction associated with the insoluble organic portion, ranging from 3.8×10^{-16} (D5 at 40°C) to 4.7×10^{-9} (D5 at −40°C) (Table 2). This is a consequence of the highly hydrophobic properties of VMS.

VMS have been shown to sorb at 30% relative humidity to pure minerals and salts that are present in aerosols [18]. This is consistent with extensive literature showing strong sorption of other organic compounds to mineral surfaces at low relative humidity [19]. However, the same research group has found that this sorption phenomenon makes an insignificant contribution to gas/aerosol partitioning [10].

Combining the contributions from partitioning to the water-insoluble organic fraction and the organic fraction, the VMS are almost completely associated with the gas phase under all conditions in the troposphere. As we shall see below, this means that physical removal is dominated by gaseous deposition, while deposition associated with aerosols is negligible. Another consequence is that the VMS are completely available for gas phase reactions. However, despite there being limited partitioning into aerosols, VMS do interact with aerosol surfaces so mixed phase reactions are possible [20].

3 Deposition

The major mechanism for physical removal of chemicals from the atmosphere is deposition. Deposition can be classified into three categories: dry particle-bound deposition, wet deposition, and dry gaseous deposition.

3.1 Dry Aerosol-Bound Deposition

Chemicals are removed from the atmosphere when aerosols containing the chemicals are deposited on terrestrial and aquatic surfaces. Aerosols can be deposited by a number of mechanisms including diffusion, impaction, and sedimentation. The rate of deposition can be characterized by an aerosol deposition velocity U_d. Deposition velocities vary widely depending on particle properties (especially size), atmospheric turbulence, and surface roughness. Values of the order of 10 m h^{-1} have been measured for the deposition of semivolatile organic contaminants [21].

An illustrative way to assess the relevance of a removal process for the atmospheric fate of a chemical is to calculate a time constant for removal k

$$k = {}^V\!/_Q \tag{3}$$

where V is the rate or removal of the chemical (mol h^{-1}) and Q is the quantity of chemical in the air column Q (mol). For the dry deposition of aerosols, V can be estimated as

$$V_d = U_d A v_Q K'_P C \left(\frac{1}{1 + v_Q K'_P} \right) \qquad (4)$$

where A is the horizontal planar area of the surface being deposited to (m^2) and C is the chemical's concentration in air (mol m^{-3}). Q is equal to

$$Q = AhC \qquad (5)$$

where h is the height of the air column (m). The time constant for removal via dry aerosol-bound deposition k_d is then

$$k_d = \frac{U_d v_Q K'_P}{h} \left(\frac{1}{1 + v_Q K'_P} \right) \qquad (6)$$

k_d was estimated for different VMS by setting U_d to 10 m h^{-1}, v_Q to 30×10^{-12} m^3 m^{-3}, calculating K_P' according to Eq. (1), and setting h to 10,000 m (the approximate height of the troposphere) (Table 3). The time constant for removal via dry aerosol-bound deposition ranges from 3.5×10^{-12} to 2.3×10^{-6} h^{-1}. These values are very small, so this process will not contribute significantly to the atmospheric fate of VMS.

3.2   Wet Deposition

Chemicals are also removed from the atmosphere when they are deposited with rain or snow. Wet deposition can occur when gaseous chemical dissolves in water or when aerosols bearing chemical become entrained in the precipitation. For chemicals with high K_{AW} such as VMS, wet deposition due to gaseous dissolution V_{rg} can be estimated by

$$V_{rg} = \frac{U_r AC}{K_{AW}} \left(\frac{1}{1 + v_Q K'_P} \right) \qquad (7)$$

where U_r is the precipitation rate (m h^{-1}). Wet deposition due to the scavenging of aerosols V_{ra} can be estimated by

Table 3 Estimated time constant for removal of VMS from the troposphere via dry aerosol-bound deposition (h^{-1})

Temperature (°C)	L2	L3	L4	D4	D5	D6
-40	1.0×10^{-10}	2.2×10^{-9}	4.4×10^{-8}	1.7×10^{-8}	1.1×10^{-7}	2.3×10^{-6}
25	5.7×10^{-12}	3.5×10^{-11}	2.6×10^{-10}	1.2×10^{-10}	5.3×10^{-10}	3.5×10^{-9}
40	3.5×10^{-12}	1.7×10^{-11}	1.1×10^{-10}	5.2×10^{-11}	2.2×10^{-10}	1.2×10^{-9}

$$V_{ra} = U_r A Q v_Q K_P' C \left(\frac{1}{1 + v_Q K_P'} \right) \tag{8}$$

where Q is a scavenging coefficient ($m^3 \, m^{-3}$). These equations can be added and the time constant for removal by wet deposition calculated in analogy to Sect. 3.1. Assuming a scavenging ratio of 2×10^5 [15] and a generic global average precipitation rate of $9.7 \times 10^{-5} \, m \, h^{-1}$ (0.85 m year^{-1}) [15], the time constant for removal via wet deposition ranges from 4×10^{-11} to $7 \times 10^{-9} \, h^{-1}$. These values are also very small, so this process can be expected to play a negligible role in the atmospheric fate of VMS.

When the precipitation occurs as snow, a further mechanism can become relevant, namely, the sorption of the chemical to the surface of the snow crystals. Deposition due to sorption to snow surface can be estimated by

$$V_{rs} = U_{rS} A a_S \rho_W K_{SA} C \tag{9}$$

where U_{rS} is the snow precipitation rate (expressed as water precipitation, $m \, h^{-1}$), a_S is the specific surface area of the snow ($m^2 \, g^{-1}$), ρ_W is the density of water ($g \, m^{-3}$), and K_{SA} is the snow to air sorption coefficient ($m^3 \, m^{-2}$). K_{SA} values for VMS predicted from polyparameter free energy relationships range from 8×10^{-4} m for L3 to 0.56 m for D6 at $-7°C$ [22]. Using a typical specific surface area of 0.1 $m^2 \, g^{-1}$ [21] and the precipitation rate from above, the time constant for the removal of VMS from the atmosphere via snow ranges from 8×10^{-7} to $5 \times 10^{-4} \, h^{-1}$. Although faster than wet deposition, deposition via sorption to snow surface is still a slow removal process. However, the time constant for D6 is just one order of magnitude lower than the time constant for removal via reaction with OH radicals, the major atmospheric removal process for VMS (Sect. 4.2). It is thus possible that deposition with snow is the major removal process under conditions of heavy snowfall and low phototransformation, and lower concentrations of cyclic VMS in air have been observed during major snow events in Toronto (Canada) [23]. It is anticipated that most VMS in the snow pack will volatilize during snow melt, so the net deposition to surface media over a larger time scale will be small [22]. Snow scavenging of VMS is still a largely unexplored phenomenon.

3.3 Gaseous Deposition

Since VMS are primarily present in the gas phase, they diffuse as a gas between the atmosphere and solid or liquid surfaces in the terrestrial or aquatic environment. Diffuse transport results from the random movement of molecules in a medium. Net transport of a substance in a medium occurs when there is a spatial gradient in its concentration. Net diffusive exchange between the atmosphere and a surface medium will occur when the chemical potential of the substance is different in the atmosphere and the surface medium. If there is no difference in chemical potential, then the diffusion to the surface medium will equal the diffusion from the surface medium, and the net diffusive exchange will be zero. The net gaseous deposition from the atmosphere to a surface medium can be estimated by

$$V_g = U_g A(C - C_S/K_{SA}) \qquad (10)$$

where U_g is the gaseous deposition velocity to the surface medium, C_S is the concentration in the surface medium, and K_{SA} is the equilibrium partition coefficient of the VMS between the surface medium and the gas phase. The gaseous deposition velocity U_g arises from two resistances to transport, one in the gaseous phase and one in the surface medium. By applying the two resistance theory [15], U_g can be estimated by:

$$U_g = \left(1/U_{gA} - 1/K_{SA}U_{gS}\right)^{-1} \qquad (11)$$

where U_{gA} and U_{gS} are the mass transfer coefficients for transport from the free atmosphere to the phase interface and from phase interface to the bulk medium, respectively. U_g for deposition to water was estimated for different VMS (Table 4) using their air/water partition coefficients (Table 2) and typical values for U_{gA} and U_{gS} of 10 m h^{-1} and 0.1 m h^{-1} [15], respectively.

An upper estimate for gaseous deposition can be obtained by assuming that the concentration of the VMS in the water is negligible ($C_S = 0$). In this case, the time constant for removal of a chemical by gaseous deposition k_g can be simply calculated

$$k_g = \frac{U_g}{h} \qquad (12)$$

The estimated k_g values of the VMS range from 4×10^{-9} to 3×10^{-8} h^{-1} (Table 4). This is larger than for particle-bound deposition or wet deposition, but still very small.

For solid surfaces such as soil or vegetation, U_{gS} is difficult to estimate [24]. However, an upper limit of k_g can be estimated by setting U_{gS} to zero. This could represent a situation where the chemical of interest reacts when it encounters the surface. In this case, Eq. (10) reduces to

$$k_g = \frac{U_{gA}}{h} \qquad (13)$$

Table 4 Estimated gaseous deposition velocity to water and time constant for removal of VMS from the troposphere via gaseous deposition to water

	Units	L2	L3	L4	D4	D5
U_g	m h^{-1}	3×10^{-4}	9×10^{-5}	4×10^{-5}	2×10^{-4}	7×10^{-5}
k_g	h^{-1}	3×10^{-8}	9×10^{-9}	4×10^{-9}	2×10^{-8}	7×10^{-9}

U_{gA} can be much higher for foliage than it is for water, with values of the order of 100 m h^{-1} having been measured for forest canopies [24]. In this case k_g could conceivably be as high as 0.01 h^{-1}, which is markedly higher than the time constant for removal by any other deposition process.

However, gaseous deposition of VMS is generally expected to be lower. The comparatively low surface media/air partition coefficients for VMS mean that surface media equilibrate rapidly with the VMS in the atmosphere. Furthermore, degradation of VMS in surface media is slow compared to uptake from the atmosphere under many conditions. In such cases, C_S is non-negligible, and the removal of VMS via gaseous deposition is smaller than estimated in Table 4 and above.

3.4   Summary of the Impact of Deposition

Multimedia modeling has provided insight into the overall impact of deposition on the fate of VMS in the atmosphere. Xu and Wania employed a global multimedia fate and transport model that included parameterizations of the deposition processes listed above to evaluate the fate of D5 [8]. They assumed that D5 was emitted at a constant rate into the atmosphere for 30 years. After this period, they found that there was >1,000 times more chemical in the atmosphere than in surface media, which suggests that deposition from the atmosphere to surface media is small. This interpretation could be questioned if there was rapid degradation of the chemical in surface media, but modeled removal of the chemical via degradation was greater in the atmosphere than in surface media. The low overall rate of deposition of VMS to surface media is in agreement with the very slow rates of aerosol-bound deposition (3.1) and wet deposition (3.2) discussed above and corroborates the expectation of low net gaseous deposition (3.3).

4   Transformation

Chemical transformation is the second possibility to remove VMS from the atmosphere. Possible mechanisms include direct photolysis, indirect photolysis, and heterogeneous reactions.

4.1   Direct Photolysis

To be subject to direct photolysis, a substance must absorb high energy radiation in the spectrum present in the troposphere (290–800 nm). VMS do not absorb radiation at wavelengths >190 nm and are thus not subject to direct photolysis in the atmosphere [25]. This is corroborated by laboratory studies conducted under simulated atmospheric conditions, which have shown no direct photolysis [26].

4.2   Indirect Photolysis

VMS are subject to reaction with OH and nitrate radicals and ozone in the atmosphere. Laboratory studies have shown that reactions with nitrate radicals and ozone make negligible contributions to the removal of D3, D4, and D5 from the atmosphere, while reaction with OH radicals is an important removal mechanism [26]. The major products of the reaction of VMS with OH radicals are silanols. They are much more polar than VMS and can be readily scavenged from the atmosphere by wet deposition [27]. It has also been postulated that silanols can contribute to the formation of secondary aerosols, whereby reactions of the silanols with each other to form less volatile products are important [28–30].

Second order rate constants for reaction with the OH radical (k_2) have been measured for a range of VMS. Generally the measurements were done in laboratory chamber experiments, but there is one paper in which k_2 was estimated from field observations (Table 5). For the linear VMS, there is good agreement between the k_2 values reported in different studies, with the highest and lowest best estimates differing by <35%. The spread is considerably larger for the cyclic VMS, whereby the studies from Safron et al. [33] and Xiao et al. [35] display markedly higher values than the other studies. When these values are excluded, the difference between the highest and lowest best estimates is also <35% for D4 and D5.

Quantitative structure reactivity relationships have been developed to explore how the structure of the VMS molecule influences k_2. It was found that the number of methyl groups in the molecule and molecular descriptors of the size of the molecule were positively correlated with the reaction rate constant [36].

Chemical reactions generally depend on temperature. However, the temperature dependence of the reaction of OH radicals with VMS is low. Recent measurements indicate that k_2 will approximately double when the temperature increases by 100°C [37], but the small temperature effect is difficult to measure, and there is disagreement in the literature as to whether k_2 and temperature are positively or negatively correlated [33, 35].

A time constant for removal of VMS from the atmosphere due to reaction with OH radicals, k_{OH}, can be calculated in analogy to Eq. (3). The rate of removal of the chemical V_{OH} is given by

$$V_{OH} = 3{,}600 k_2 [OH] AhC \tag{14}$$

where [OH] is the OH radical concentration in the atmosphere (molecules cm^{-3}) and the constant 3,600 is to convert from seconds to hours. Combining Eqs. (3) and (14) yields

Table 5 Second order rate constants for the reaction of VMS with OH radicals at room temperature ($10^{12} k_2$, cm³ molecule⁻¹ s⁻¹). The best estimate and uncertainties as reported in each study are shown

Source	L2	L3	L4	L5	D3	D4	D5	D6
Atkinson [26][a]	1.38 ± 0.36				0.52 ± 0.17	1.01 ± 0.32	1.55 ± 0.49	
Sommerlade et al. [31][a]	1.19 ± 0.30					1.26 ± 0.40		
Markgraf and Wells [32][a]	1.32 ± 0.05	1.83 ± 0.09	2.66 ± 0.13					
Safron et al. [33]						1.9 (1.7–2.2)	2.6 (2.3–2.9)	2.8 (2.5–3.2)
MacLeod et al. [34][b]								5.1
Xiao et al. [35]					1.84 (1.76–1.93)	2.34 (1.93–2.85)	2.46 (2.20–2.74)	
Kim and Xu [36]	1.58 ± 0.16	2.15 ± 0.15	3.37 ± 0.40	4.03 ± 1.4	0.91 ± 0.23	0.95 ± 0.18	1.46 ± 0.12	2.44 ± 0.92
Bernard et al. [37]	1.24 ± 0.15	1.59 ± 0.20			0.70 ± 0.08	1.06 ± 0.15		

[a]Bernard et al. [37] provide revised values for these studies that are corrected for the updated value of the rate constant of the reference compounds employed. The revised values are several percent lower

[b]Derived from field observations

$$k_{OH} = 3{,}600k_2[OH] \qquad (15)$$

The OH radical concentration in the atmosphere is subject to strong diurnal and seasonal variations due to the role of solar irradiation in OH radical formation. For instance, the ratio of the monthly average [OH] between July and January is about 10:1 at 35° N and increases with increasing latitude [26]. As a result, the removal rate of VMS from the atmosphere will be highly variable in space and time. A global average [OH] of 1×10^6 molecules cm^{-3} is often assumed for calculating indicative atmospheric removal rates and half-lives. Using this value and the k_2 values from Kim and Xu [36] in Table 5, k_{OH} was estimated for several VMS (Table 6). The values range from 0.0033 to 0.015 h^{-1}, which corresponds to indicative atmospheric half-lives of 2–9 days. k_{OH} is much larger than the time constants for removal via deposition, confirming that reaction with OH radicals is the dominant sink for VMS in the atmosphere.

4.3   Heterogeneous Reactions

VMS can undergo heterogeneous reactions on mineral surfaces. Rapid degradation of VMS has been observed on soil at low relative humidity, and this has been attributed to the catalytic effect of mineral surfaces [38, 39]. Heterogeneous decay of D4 and D5 has also been observed on mineral dust aerosols under low humidity conditions [20]. Ozone was observed to accentuate the loss of D4 and D5 on the mineral aerosols, but the loss was reduced at relative humidities typical of conditions in the troposphere [40]. When these findings were incorporated into an atmospheric chemistry model with realistic aerosol loadings, heterogeneous reactions were found to reduce the atmospheric lifetime of the VMS by 3% [41]. This indicates that heterogeneous reactions are of minor relevance for the atmospheric fate of VMS.

5   Temporal and Spatial Variability of Concentrations

Summarizing the above, for most purposes the atmospheric fate of VMS can be simplified to three processes: the emission function, advection, and removal via reaction with OH radicals. However, despite this apparent simplicity, each of these processes bears complexity. VMS are emitted primarily from personal care products and PDMS, and thus emission fluxes are expected to be closely linked to population density. However, personal care product use can vary regionally, seasonally, and diurnally. In addition, there may be other major sources that we are currently unaware of. The second process, atmospheric advection, is highly complex, with strong variability on diurnal, day-to-week and seasonal time scales. With atmospheric half-lives of several days, these substances are subject to long-range

Table 6 Estimated time constant for removal of VMS from the troposphere via reaction with OH radicals

	Units	L2	L3	L4	L5	D3	D4	D5	D6
k_{OH}	h^{-1}	5.7×10^{-3}	7.7×10^{-3}	1.2×10^{-2}	1.5×10^{-2}	3.3×10^{-3}	3.4×10^{-3}	5.3×10^{-3}	8.8×10^{-3}

transport, and hence variability on all of these time scales will impact VMS concentrations at some distance from sources. As for reaction with OH radicals, the strong spatial and temporal variability in OH radical concentrations adds a further layer of complexity to the atmospheric fate of VMS.

To deal with this complexity, mathematical models have been used to simulate the interaction of these processes and to describe the temporal and spatial variability in VMS concentrations. The first model of the atmospheric fate of VMS employed the DEHM atmospheric chemistry and transport model of the Northern Hemisphere [42]. D5 was modeled using emissions estimates based on consumer product sales. Meteorological data were used to describe atmospheric circulation and predict OH radical concentrations over a 6-month period, and k_2 from Atkinson [26] was employed to quantify D5 degradation. The model simulation provided in the supporting information of the paper visualizes many features of the expected spatial and temporal variability of D5 concentrations in air at the Earth's surface including:

- Strong diurnal variation in major population centers at low latitudes
- Comparatively stable concentrations in major population centers at high latitudes
- Variation on the scale of days outside highly populated regions at low and mid latitudes caused by transport events (weather systems)
- A pronounced decrease in concentration outside major population regions from winter to summer at all latitudes
- Increasing concentrations with increasing latitude
- Steepening of the spatial gradients in concentration going from winter to summer in mid and high latitudes

Empirical evidence to support the DEHM model predictions was provided by daily measurements of D5 over a 5-month period at a rural site in Sweden (average difference between model and measurement of 36% during the last 6 weeks of the study when atmospheric reaction had the strongest influence on concentrations) and measurements of D5 during 46 days in late summer and 18 days in early winter in Svalbard (in the Arctic, average difference between model and measurement of 30%) [42, 43]. These studies support the model's ability to predict the temporal variability in VMS concentrations in rural/remote areas at mid to high latitudes as well as the latitudinal gradient. Less satisfactory agreement was obtained in a study using passive samplers deployed for 3 months during late spring/early summer at various locations around the Northern Hemisphere (average difference between model and measurement of 470%), whereby it is not clear what portion of the disagreement is attributable to the model [44].

Another step forward in modeling the atmospheric fate of VMS was provided by Janechek et al. [28]. They showed that a higher spatial resolution is required if the model is to predict VMS concentrations close to large local sources. Assessing VMS concentrations in source areas has been a major focus of empirical work. Strong urban to rural gradients in concentrations and comparatively low seasonal variability in large cities have been measured [23, 45–49]. This is in agreement with models which predict that concentrations in strong source regions will be largely controlled by source strength and transport events because the residence time of the air is too short for significant degradation to occur [28, 41].

The investigation of temporal variability has also progressed to the diurnal time scale. By combining measurements conducted with a high temporal resolution and atmospheric modeling, it was shown that there is a pronounced diurnal pattern to D5 emissions in Boulder, Colorado. Emissions peaked in early morning and then showed an exponential decay with a time constant of 9.2 h [50]. This diurnal variation of emissions was reflected in a diurnal variation in VMS concentrations in urban air, and it is likely that some variation will also be propagated to regions around cities.

A less explored aspect of the atmospheric fate of VMS is the spatial and temporal variability of the relative concentrations of different compounds. It has been postulated that during the winter in higher latitudes, the concentration ratios should reflect the relative strength in emissions of the respective VMS. Presuming that emissions are correlated with population, the concentration ratios should thus be relatively constant over larger spatial scales [51]. On the other hand, passive samplers exposed for 3 months during late spring and early summer at various mid and high latitude locations show D4:D5 ratios that vary over almost three orders of magnitude [44]. More work to further illuminate these questions is warranted, as it can provide insight into the relative magnitude, spatial distribution, and temporal variability of VMS emissions.

References

1. Brooke DN, Crookes MJ, Gray D, Robertson S (2009) Environmental risk assessment report: octamethylcyclotetrasiloxane. UK Environment Agency, Bristol
2. Brooke DN, Crookes MJ, Gray D, Robertson S (2009) Environmental risk assessment report: decaamethylcyclopentasiloxane. UK Environment Agency, Bristol
3. Brooke DN, Crookes MJ, Gray D, Robertson S (2009) Environmental risk assessment report: dodecamethylcyclohexasiloxane. UK Environment Agency, Bristol
4. Environment Canada, Health Canada (2008) Screening assessment for the challenge: octamethylcyclotetrasiloxane (D4). CAS RN 556-67-2, Environment Canada
5. Environment Canada, Health Canada (2008) Screening assessment for the challenge: decamethylcyclopentasiloxane (D5). CAS RN 541-02-6, Environment Canada
6. Environment Canada, Health Canada (2008) Screening assessment for the challenge: dodecamethylcyclohexasiloxane (D6). CAS RN 540-97-6, Environment Canada
7. Hughes L, Mackay D, Powell DE, Kim I (2012) An updated state of the science EQC model for evaluating chemical fate in the environment: application to D5 (decamethylcyclopentasiloxane). Chemosphere 87:118–124
8. Xu S, Wania F (2013) Chemical fate, latitudinal distribution and long-range transport of cyclic volatile methylsiloxanes in the global environment: a modeling assessment. Chemosphere 93:835–843
9. Mackay D, Paterson S, Schroeder WH (1986) Model describing the rates of transfer processes of organic chemicals between atmosphere and water. Environ Sci Technol 20:810–816
10. Arp HPH, Schwarzenbach RP, Goss K-U (2008) Ambient gas/particle partitioning. 1. Sorption mechanisms of apolar, polar, and ionisable organic compounds. Environ Sci Technol 42:5541–5547
11. Bidleman TF, Harner T (2000) Sorption to aerosols. In: Mackay D, Boethling RS (eds) Property estimation methods for chemicals – environmental and health sciences. CRC Press, Boca Raton, pp 233–260

12. Arp HPH, Schwarzenbach RP, Goss K-U (2008) Ambient gas/particle partitioning. 2. The influence of particle source and temperature on sorption to dry terrestrial aerosols. Environ Sci Technol 42:5951–5957

13. Xu S, Kropscott B (2013) Octanol/air partition coefficients of volatile methylsiloxanes and their temperature dependence. J Chem Eng Data 58:136–142

14. Xu S (2013) Correction to "Octanol/air partition coefficients of volatile methylsiloxanes and their temperature dependence". J Chem Eng Data 58:2136

15. Mackay D (2001) Multimedia environmental models: the fugacity approach. CRC Press, Boca Raton

16. Xu S, Kozerski G, Mackay D (2014) Critical review and interpretation of environmental data for volatile methylsiloxanes: partition properties. Environ Sci Technol 48:11748–11759

17. Xu S, Kropscott B (2014) Evaluation of the three-phase equilibrium method for measuring temperature dependence of internally consistent partition coefficients (K_{OW}, K_{OA}, and K_{AW}) for volatile methylsiloxanes and trimethylsilanol. Environ Toxicol Chem 33:2702–2710

18. Kim J, Xu S (2016) Sorption and desorption kinetics and isotherms of volatile methylsiloxanes with atmospheric aerosols. Chemosphere 144:555–563

19. Goss K-U, Buschmann J, Schwarzenbach RP (2003) Determination of the surface sorption properties of talc, different salts, and clay minerals at various relative humidities using adsorption data of a diverse set of organic vapors. Environ Toxicol Chem 22:2667–2672

20. Navea JG, Xu S, Stanier CO, Young MA, Grassian VH (2009) Heterogeneous uptake of octamethylcyclotetrasiloxane (D4) and decamethylcyclopentasiloxane (D5) onto mineral dust aerosol under variable RH conditions. Atmos Environ 43:4060–4069

21. MacLeod M, Scheringer M, Götz C, Hungerbühler K, Davidson CI, Holsen TM (2011) Deposition form the atmosphere to water and soils with aerosol particles and precipitation. In: Thibodeaux LJ, Mackay D (eds) Handbook of chemical mass transport in the environment. CRC Press, Boca Raton, pp 103–136

22. Mackay D, Gobas F, Solomon K, Macleod M, McLachlan M, Powell DE, Xu S (2015) Comment on "Unexpected occurrence of volatile dimethylsiloxanes in Antarctic soils, vegetation, phytoplankton, and krill". Environ Sci Technol 49:7507–7509

23. Ahrens L, Harner T, Shoeib M (2014) Temporal variations of cyclic and linear volatile methylsiloxanes in the atmosphere using passive samplers and high-volume air samplers. Environ Sci Technol 48:9374–9381

24. McLachlan MS (2011) Mass transfer between the atmosphere and plant canopy systems. In: Thibodeaux LJ, Mackay D (eds) Handbook of chemical mass transport in the environment. CRC Press, Boca Raton, pp 103–136

25. Gaj K, Pakuluk A (2015) Volatile methyl siloxanes as potential hazardous air pollutants. Pol J Environ Stud 24:937–943

26. Atkinson R (1991) Kinetics of the gas-phase reactions of a series of organosilicon compounds with OH and NO_3 radicals and O_3 at 297 ± 2 K. Environ Sci Technol 25:863–866

27. Whelan MJ, Estrada E, van Egmond R (2004) A modelling assessment of the atmospheric fate of volatile methyl siloxanes and their reaction products. Chemosphere 57:1427–1437

28. Janechek NJ, Hansen KM, Stanier CO (2017) Comprehensive atmospheric modeling of reactive cyclic siloxanes and their oxidation products. Atmos Chem Phys 17:8357–8370

29. Wu Y, Johnston MV (2016) Molecular characterization of secondary aerosol from oxidation of cylic methylsiloxanes. J Am Soc Mass Spectrom 27:402–409

30. Wu Y, Johnston MV (2017) Aerosol formation from OH oxidation of the volatile cyclic methyl siloxane (cVMS) decamethylcyclopentasiloxane. Environ Sci Technol 51:4445–4451

31. Sommerlade R, Parlar H, Wrobel D, Kochs P (1993) Product analysis and kinetics of the gas-phase reactions of selected organosilicon compounds with OH radicals using a smog chamber-mass spectrometer system. Environ Sci Technol 27:2435–2440

32. Markgraf SI, Wells JR (1997) The hydroxyl radical reaction rate constants and atmospheric reaction products of three siloxanes. In J Chem Kinet 29:445–451

33. Safron A, Strandell M, Kierkegaard A, MacLeod M (2015) Activation energies for gas-phase reactions of three cyclic volatile methyl siloxanes with the hydroxyl radical. In J Chem Kinet 47:420–428

34. MacLeod M, Kierkegaard A, Genualdi S, Harner T, Scheringer M (2013) Junge relationships in measurement data for cyclic siloxanes in air. Chemosphere 93:830–834

35. Xiao R, Zammit I, Wei Z, Hu W-P, MacLeod M, Spinney R (2015) Kinetics and mechanism of the oxidation of cyclic methylsiloxanes by hydroxyl radical in the gas phase: an experimental and theoretical study. Environ Sci Technol 49:13322–13330

36. Kim J, Xu S (2017) Quantitative structure-reactivity relationships of hydroxyl radical rate constants for linear and cyclic volatile methylsiloxanes. Environ Toxicol Chem 36:3240–3245

37. Bernard F, Papanastasiou DK, Papadimitriou VC, Burkholder JB (2018) Temperature dependent rate coefficients for the gas-phase reaction of the OH radical with linear (L2, L3) and cyclic (D3, D4) permethylsiloxanes. J Phys Chem 122:4252–4264

38. Xu S (1998) Hydrolysis of poly(dimethylsiloxanes) on clay minerals as influenced by exchangeable cations and moisture. Environ Sci Technol 32:3162–3168

39. Xu S (1999) Fate of cyclic methylsiloxanes in soils. 1. The degradation pattern. Environ Sci Technol 33:603–608

40. Navea JG, Xu S, Stanier CO, Young MA, Grassian VH (2009) Effect of ozone and relative humidity on the heterogeneous uptake of octamethylcyclotetrasiloxane and decamethylcyclopentasiloxane on model mineral dust aerosol components. J Phys Chem A 113:7030–7038

41. Navea JG, Young MA, Xu S, Grassian VH, Stanier CO (2011) The atmospheric lifetimes and concentrations of cyclic methylsiloxanes octamethylcyclotetrasiloxane (D4) and decamethylcyclopentasiloxane (D5) and the influence of heterogeneous uptake. Atmos Environ 45:3181–3191

42. McLachlan MS, Kierkegaard A, Hansen KM, van Egmond R, Christensen JH, Skjøth CA (2010) Concentrations and fate of decamethylcyclopentasiloxane (D5) in the atmosphere. Environ Sci Technol 44:5365–5370

43. Krogseth IS, Kierkegaard A, McLachlan MS, Breivik K, Hansen KM, Schlabach M (2013) Occurrence and seasonality of cyclic volatile methyl siloxanes in Arctic air. Environ Sci Technol 47:502–509

44. Genualdi S, Harner T, Cheng Y, MacLeod M, Hansen KM, van Egmond R, Shoeib M, Lee SC (2011) Global distribution of linear and cyclic volatile methyl siloxanes in air. Environ Sci Technol 45Ö:3349–3354

45. Buser AM, Kierkegaard A, Bogdal C, MacLeod M, Scheringer M, Hungerbühler K (2013) Concentrations in ambient air and emissions of cyclic volatile methylsiloxanes in Zurich, Switzerland. Environ Sci Technol 47:7045–7051

46. Companioni-Damas EY, Santos FJ, Galceran MT (2014) Linear and cyclic methylsiloxanes in air by concurrent solvent recondensation – large volume injection – gas chromatography – mass spectrometry. Talanta 118:245–252

47. Gallego E, Perales JF, Roca FJ, Guardino X, Gadea E (2017) Volatile methyl siloxanes (VMS) concentrations in outdoor air of several Catalan urban areas. Atmos Environ 155:108–118

48. Krogseth IS, Zhang X, Lei YD, Wania F, Breivik K (2013) Calibration and application of a passive air sampler (XADE-PAS) for volatile methyl siloxanes. Environ Sci Technol 47:4463–4470

49. Yucuis RA, Stanier CO, Hornbucke KC (2013) Cyclic siloxanes in air, including identification of high levels in Chicago and distinct diurnal variation. Chemosphere 92:905–910

50. Coggon MM, McDonald BC, Vlasenko A, Veres PR, Bernard F, Koss AR, Yuan B, Gilman JB, Peischl J, Aikin KC, DuRant J, Warneke C, Li S-M, de Gouw JA (2018) Diurnal variability and emission pattern of decamethylcyclopentasiloxane (D5) from the application of personal care products in two North American cities. Environ Sci Technol 52:5610–5618

51. Kierkegaard A, McLachlan MS (2013) Determination of linear and cyclic volatile methylsiloxanes in air at a regional background site in Sweden. Atmos Environ 80:322–329

Bioconcentration, Bioaccumulation, and Biomagnification of Volatile Methylsiloxanes in Biota

Sofia Augusto

Contents

Abstract Volatile methylsiloxanes (VMS) are synthetic chemicals that have been extensively used in the manufacture of many industrial and consumer products and in the formulation of personal and health-care products. Due to their extensive use, VMS have been found in a diversity of abiotic media (air, soil, water, sediments) and in a wide range of aquatic and terrestrial organisms. The ubiquitous presence of VMS has raised concerns regarding whether these chemicals are prone to accumulate in aquatic and terrestrial life to levels higher than those found in the environment and ultimately to affect human and ecosystem health. The purpose of this chapter is to provide an overview of the studies that have been developed to understand if VMS

S. Augusto (✉)
EPIUnit, Instituto de Saúde Pública, Universidade do Porto (ISPUP/UP), Porto, Portugal

Centre for Ecology, Evolution and Environmental Changes (cE3c), Faculdade de Ciências, Universidade de Lisboa, Lisboa, Portugal
e-mail: s.augusto@fc.ul.pt

V. Homem and N. Ratola (eds.), *Volatile Methylsiloxanes in the Environment*,
Hdb Env Chem (2020) 89: 247–278, DOI 10.1007/698_2019_387,
© Springer Nature Switzerland AG 2019, Published online: 25 July 2019

have the potential to bioconcentrate, bioaccumulate, and biomagnify. Key factors affecting bioaccumulation of VMS by different organisms will be described, including physicochemical properties, environmental conditions, characteristics of the exposed organism, and the respective food chains. A review of the studies reporting VMS in different biota samples will be provided.

Keywords Bioconcentration, Biota-sediment accumulation, Multimedia bioaccumulation, Trophic magnification, Volatile methylsiloxanes (VMS)

1   Introduction

Volatile methylsiloxanes (VMS) are synthetic organic silicon compounds that have been produced commercially since 1940 [1]. The basic chemical structure of VMS is a backbone of silicon-oxygen (Si-O) units with methyl groups ($-CH_3$) attached to each silicon atom. Linear methylsiloxanes (lVMS) are usually expressed as Ln, where L means linear structure and n is the number of silicon atoms. Cyclic methylsiloxanes (cVMS) are usually expressed as Dn, where D means cyclic structure and n is the number of silicon atoms [2, 3].

The chemical structure of VMS provides these compounds with important properties of dynamic flexibility and adhesion to surfaces (including human skin), which are the reason for their use in many industrial and consumer products and in the formulation of personal and health-care products [1, 4–6]. VMS have been extensively used in the manufacture of silicones, cooking utensils, plastics, paper, electronic products, building and decoration materials, cosmetics, skin- and hair-care products, and pharmaceutics [1, 5].

Since VMS were first produced, their use has increased exponentially [7]. As a result, VMS have been detected in almost all kinds of environmental media, including air [6, 8, 9], freshwater [10–12], seawater [13], sediments and soil [10, 13–16], and biota (e.g., algae, molluscs, crustaceans, fish, birds, mammals, vegetation, moss, and lichens) [13, 17–21].

Among the VMS, D4 (octamethylcyclotetrasiloxane), D5 (decamethylcyclopentasiloxane), and D6 (dodecamethylcyclohexasiloxane) have been classified as high production volume chemicals by the US Environmental Protection Agency [7] and the Organisation for Economic Co-operation and Development [22]. These chemicals have been subjected to several regulatory reviews in the United Kingdom [23–25], Canada and Health Canada [26–28], and Nordic countries [11, 29], as well as a judicial review [30, 31] and an evaluation for the European Chemicals Agency (ECHA) [32, 33].

Several toxicological studies suggested that, depending on the concentration, VMS can have estrogenic effects on biota [34, 35], cause immune system disorders and connective tissue diseases [36, 37], and lead to liver and lung injury [38].

The main emission route of VMS into the environment is the direct release into the atmosphere during the manufacturing process and product formulation, as well as through the common use of VMS-containing consumer products [23–25, 39, 40]. However, other important sources of emission are disposal in landfills, solid waste incineration plants [41], and sewage (or wastewater) treatment plants [16, 40, 42], being the latter the most important source to the aquatic environment. Environmental monitoring studies have found highest concentrations of VMS in the air of densely populated areas [6] and in the water from sewage treatment plants, where VMS are released sorbed to the suspended organic matter [43].

VMS possess an unusual combination of physicochemical properties that makes the prediction of their environmental fate after being emitted a difficult task [44]. Properties such as low water solubility and high volatility result in a high tendency for these compounds to escape to the atmosphere from either water or wet soil, while their lipophilicity points toward the potential to sorb to organic matter and accumulate in living organisms.

The ubiquitous presence of VMS has raised concerns regarding whether these chemicals have a propensity to concentrate in aquatic and terrestrial life to levels higher than those found in the environment (i.e., to bioaccumulate) and to ultimately affect human and ecosystem health.

Bioaccumulation is a function of bioconcentration and biomagnification, where bioconcentration is the non-trophic uptake and accumulation of a chemical in a specific organism from abiotic media (primarily water, but also sediment and air) and biomagnification is the change in concentration across organisms in a food web related to the change in trophic level of the organisms [45]. The ability of a chemical to bioaccumulate in an organism depends on several factors, including the physicochemical properties, the environmental conditions which will affect the bioavailability of the chemical to the organism, the characteristics of the exposed organism (e.g., warm-blooded air breathing versus non-air breathing), and the organism's food chain. The role of each of these factors in the bioaccumulation potential of VMS in both aquatic and terrestrial organisms will be explored in Sect. 2.

Several studies have been performed to evaluate the potential of VMS to bioconcentrate and biomagnify in biota. A number of bioaccumulation factors (notably, bioconcentration, biota-sediment accumulation, multimedia bioaccumulation, biomagnification, and trophic biomagnification) were derived for different organisms to understand the equilibrium partitioning between the VMS in the environment and the organism. Though these studies are complex to perform and many of them lack environmental relevance [46], the most pertinent ones will be reviewed and discussed in Sect. 3. Finally, in Sect. 4, an overview of the studies reporting VMS in different biota samples will be provided.

2 Factors Affecting VMS Bioaccumulation

The potential of a chemical for bioaccumulation is affected by four types of variables: physicochemical properties of the chemical, environmental conditions, characteristics of the exposed organism, and its respective food chain. These factors may act in concomitance or in opposition, resulting in a range of bioaccumulation potentials.

2.1 Physicochemical Properties

The physicochemical properties of a chemical play an important role in the bioaccumulation process. Features of chemicals that confer the tendency to bioaccumulate include (1) lipophilicity, which is directly related to the magnitude of a chemical's solubility in octanol and characterized by the magnitude of the octanol-water partition coefficient (K_{OW}); (2) low water solubility or hydrophobicity due to the lack of polar functional groups; and (3) structural stability resulting in environmental persistence (years instead of days). The role of these features in the bioaccumulation process depends on the type of organism and its structure and metabolism. Warm-blooded air-breathing animals (e.g., humans, birds), for example, are metabolically different from aquatic non-air-breathing animals (e.g., fish), and therefore, their ability to bioaccumulate VMS is also different.

VMS possess an unusual combination of physicochemical properties that confers a distinct profile from other chemicals and which affects their environmental distribution and bioaccumulation potential. One of these properties is related to their hydrophobic and volatile character. The solubility of VMS in distilled water is very low, and the vapor pressures are relatively high [1] (Table 1). This combination results in large air-water partition coefficients (K_{AW}) and large octanol-water partition coefficients (K_{OW}). Therefore, VMS have the propensity to escape from water and wet soil into air while seemingly having an affinity toward lipids (as confirmed

Table 1 Key physicochemical properties for cyclic VMS (D3, D4, D5, and D6) and linear VMS (L2, L3, L4, and L5) [23–25, 44, 46, 47]

Acronym	MW (g mol^{-1})	Log K_{OW}	Log K_{OC} (L kg^{-1})	Log K_{OA}	Water solubility (μg L^{-1})	H_C (atm m^3 mol^{-1})
L2	162	4.2	2.6	2.9	930	2.90
L3	236	6.6	4.3	3.8	34	7.27
L4	311	8.2	5.1	4.6	7	28.4
L5	385	6.0	5.3	4.5	0.30	116
D3	222	4.5	3.3	4.1	1.56	0.61
D4	297	6.5	4.2	4.3 4.1 (37.5°C)	56	11.8
D5	371	8.0	5.2	4.9 4.7 (37.5°C)	17	33.0
D6	445	9.0	6.0	5.8 5.3 (37.5°C)	5	48.8

H_C Henry's constant law, Log K_{OA}, and water solubility measured at 25°C, unless otherwise stated. Log K_{OW} and Log K_{OA} are unitless by the large K_{OW}). However, VMS have small octanol-air partition coefficients (K_{OA}) and large Henry's law constants (H_C), and this has a direct effect on their bioaccumulation potential in air-breathing animals. Chemicals with small K_{OA} (log $K_{OA} < 6$) [48] are not expected to bioaccumulate in air-breathing animals (independently of the exposure route being ingestion, contact with skin, or inhalation), as they are easily eliminated by exhalation [49]. The volatilization of VMS in this case depends on the capacity of lipids to retain the chemical relative to the tendency of volatilization from water. This volatilization can be predicted from the ratio K_{LW} (lipid-water partition coefficient) to K_{AW}, designated by K_{LA} (lipid-air partition coefficient). If all types of lipids are considered together, K_{LW} will be similar to K_{OA} [50]. In this case, K_{LA} will be equal to K_{OA}. At 37.5°C (close to the body temperature of mammals), log K_{OA} values for VMS are relatively small [44]. Based on an average content of lipid in plasma of 0.3%, plasma-air partition coefficients range from tens to hundreds, resulting in rapid depuration via respiration [46], making the potential for bioaccumulation in warm-blooded air-breathing animals including humans very limited.

The depuration via respiration increases with the body temperature. Birds, with a higher body temperature (around 42°C), can eliminate VMS more efficiently than aquatic non-air-breathing organisms, with lower body temperatures, which is confirmed by the greater VMS concentrations measured in the latter organisms in comparison with others (see Sect. 4) [46].

But despite the limited potential for bioaccumulation, VMS have been measured in warm-blooded air-breathing animals (birds, rats, humans, seals), as reported in several publications [1]. Even for non-bioaccumulative compounds, when the rate of exposure is greater than the rate of excretion, the chemical remains in the organism's tissues for some time. During this period, VMS may trigger processes leading to toxicity [1]. In mammals, liver and lungs have been reported to be target organs for D4 and D5 inhalation exposures, due to the role of these organs in the detoxification mechanisms. Increases in liver weight, as well as increases in focal macrophage accumulation and interstitial inflammation in lungs, were reported for rats after inhalation exposures [1]. Moreover, D5 showed a potential carcinogenic effect in a 2-year chronic toxicity and carcinogenicity study in rats exposed to vapor concentrations of 450 mg m^{-1} for 6 h/day, 5 days/week, for 24 months [1].

Organisms exposed to VMS via sediments or soils tend to present higher concentrations of these compounds than organisms occupying upper trophic levels. Unlike other neutral organic chemicals, the water-soil partition coefficient (corrected for organic carbon content, K_{OC}) of VMS is more than two orders of magnitude (about 200-fold) lower than would be predicted from the K_{OW}. This results in a shift of the partitioning equilibrium from soil and sediments into air. However, the K_{OC} still remain large enough to have a potential effect on the bioavailability to organisms exposed via sediments or soil [46].

Several studies have demonstrated that VMS are degraded in air, water, soil, and sediment, resulting in products that are less volatile and more water-soluble and with less potential for bioaccumulation and toxicity [1]. Although the degradation potential depends on the physicochemical properties of the chemical, the degradation rate depends on the environmental conditions.

2.2 Environmental Conditions

The environmental presence of chemicals that meet the criteria of lipophilicity, low water solubility and persistence, does not always lead to high degrees of bioaccumulation, for which chemicals need to contact a biomembrane and move through it onto lipid-rich storage sites. The amount of chemical contacting an organism's absorbing membranes is dependent not only on its environmental concentration in the abiotic media (e.g., water, including particulates) but also on the fraction available for uptake (i.e., the bioavailable fraction) [45].

In aquatic environments, this usually corresponds to the fraction of chemical that is truly dissolved in water. In the case of VMS, the maximum concentrations of bioavailable VMS in water are constrained by the poor solubility of these compounds (Table 1). Most of the VMS in the aquatic media are usually not dissolved in water, but rather sorbed on particulate organic carbon (organic matter). The maximum concentrations of VMS in dissolved or suspended organic matter in water, sediments, and soils are constrained by the sorption capacity of the matrix, which is governed by the K_{OC}, solubility in water, and the amount of organic carbon (OC) in wet soils or sediments. The fraction of OC varies from one soil or sediment to another, affecting the specific maximum sorption capacity [46]. Consequently, the extent of the organism's exposure to bioavailable VMS in soil and sediments depends on the OC of these matrices (see Sect. 3.2).

In lakes that are ice-covered throughout the winter, the exchange between water and air is limited, and volatilization of VMS may be low. This could result in higher dissolved concentrations of VMS in water and hence increased exposure to the entire aquatic ecosystem [51].

Another factor that greatly affects the potential of a chemical to bioaccumulate is its environmental stability or persistence. The effects of environmental degradation processes (e.g., hydrolysis, photolysis, and microbial degradation) on contaminants typically result in more hydrophilic or polar products, which have lower bioaccumulation potential than the parent compounds [45]. VMS have been found to degrade in the environment within days to months and years, depending on the compound (degradation decreases with the increase in the number of Si atoms) and on the matrix. Usually, for the same compound, the degradation rates follow the trend Air (days) → Soil → Water → Sediment (months–years).

In the atmosphere, gas-phase VMS are degraded due to reaction with hydroxyl radicals to form hydroxy-substituted silanols that are less volatile and more water-soluble and that have less potential to bioaccumulate and be toxic [52, 53]. The atmospheric half-lives for D4, D5, and D6 were estimated to be 10.3, 6.7, and 5 days, respectively [54]. These half-lives are above the criterion of 2 days used to identify long-range atmospheric transport (LRAT) substances, and thus, D4, D5, and D6 can undergo LRAT [54]. An increasing number of publications have been reporting the presence of VMS in remote regions [21, 51]. However, some authors argue that despite the potential for LRAT, the inherent physicochemical properties of VMS do not favor their deposition in surface matrices in remote regions [46, 55], where the presence of VMS would be due to local anthropogenic sources and sample contamination rather than to LRAT [46].

VMS that are bound to particles are additionally susceptible of undergoing heterogeneous reactions with ozone (O_3) at the mineral dust aerosol surface [1], being half-lives of cVMS estimated at 27 days [56].

In water, VMS can undergo hydrolysis, though the process is relatively slow under environmentally realistic conditions. Half-lives of 16.7 and 315 days were reported for D4 and D5 in freshwater (pH 7 and 12°C). In marine water, shorter half-lives for D4 and D5 were estimated (2.9 and 64 days, respectively, at pH 8 and 9°C). Factors such as the pH and temperature have been found to influence hydrolysis rate [1, 57, 58].

Degradation of VMS has also been demonstrated in dry soils, most probably by abiotic processes: hydrolysis and volatilization [59, 60]. Half-lives of 4.1–5.3, 9.7–12.5, and 158–202 days were estimated for D4, D5, and D6 for dry temperate soils in equilibrium with air (relative humidity of 50–90%) [61].

A degradation study for D4 in a water/sediment system showed that the calculated half-life for D4 degradation in sediment was 49–588 days at 5–25°C, being hydrolysis the major degradation process [57, 62].

Recently, the degradation of ^{14}C-labeled D5 in aquatic sediment was tested under both aerobic and anaerobic conditions [63] and was evident under both conditions, but under a very slow rate. The half-lives at 24°C were estimated to be about 1,200 days under biotic, aerobic conditions; 2,700 days under sterile aerobic conditions; 3,100 days under biotic, anaerobic conditions; and 800 days under sterile anaerobic conditions.

D4 was biodegraded under anaerobic conditions in composted sewage biosolids; after 100 days of incubation, only 3% of the spiked D4 was converted into dimethylsilanediol [64].

2.3 Characteristics of the Exposed Organism

As previously discussed, the metabolism depends on the type of organism, leading the exposure to VMS to occur through different routes.

Lipophilic chemicals are usually accumulated by aquatic organisms from water via respiration and from ingested food or sediments [45]. The main source of VMS in the aquatic environment is particles bound to effluent material emitted by sewage treatment plants. The VMS tend to remain attached to the particulate matter as they distribute in the water body. Consequently, the concentration in water remains below the solubility limits because of the unfavorable partitioning from sediment and loss of VMS from water to air [46]. As the low solubility of VMS will limit any significant uptake from water through respiratory and dermal surfaces of aquatic organisms, the uptake of these chemicals in aquatic species at the base of the food chain will most likely be from ingestion of sediment and/or ingestion of sediment-dwelling organisms. Nevertheless, the potential of VMS for bioconcentration (uptake from water) in different aquatic species was reported in several laboratory-based experiments, although the environmental relevance of these studies has been debated (see Sect. 3.1).

Diet is more likely to be the major route of uptake when chemicals have high K_{OW}, such as VMS. This is especially true for top predators. The assimilation efficiency of lipophilic chemicals across the gut is dependent on the ingested materials. If these are largely nondigestible, such as most natural sediment organic carbon, the likelihood of gastrointestinal uptake is diminished. Gut assimilation efficiencies for a series of lipophilic chemicals, from fish food (e.g., animal or plant tissues), were shown to range from about 50 to 85%. Note that lipid content of the consumer organism has little or no effect on dietary and respiratory rates of chemicals, but it does affect the ultimate capacity of an organism to accumulate a given chemical [45].

Bioaccumulation occurs only if the rate of a chemical's uptake exceeds its elimination rate. In aquatic organisms, depuration of many lipophilic chemicals occurs passively across the gills. This route of elimination appears to be the most important for nonpolar compounds that are not biotransformed. The rates of elimination for these compounds are generally inversely related to their K_{OW}.

The ability to eliminate VMS by all processes (including elimination and biotransformation) varies among species. Springer [64] conducted a study on the elimination and biotransformation of orally gavaged radiolabeled D5 in mature rainbow trout. Samples of blood from fish were collected via an aortic cannula at selected points, after they were fed with an oral dose of ^{14}C radiolabeled D5 in corn oil. The highest concentrations of ^{14}C were found in the bile of the fish, with only 4% of the total ^{14}C being parent D5. In the liver, 46% of the measured radioactivity was identified as parent D5 (50% in the intestinal tract). All radioactivity detected in urine was attributable to biotransformation products of D5, including $Me_2Si(OH)_3$ (where Me represents a methyl group), which indicates demethylation of the silicon-methyl bounds.

As previously explained, in air-breathing animals, VMS are rapidly depurated by exhalation, independently of the exposure route [49]. The inhalation pharmacokinetics of D4 and D5 were studied in male and female Fischer 344 rats following 6-h exposure, daily 6-h exposure for 15 consecutive days, and 6-month exposure (6 h/day, 5 days/week) [49]. Although D4 and D5 are lipophilic and can be stored in fat, tissue concentrations (plasma, liver, and fat) upon repeated exposures to these compounds did not continue to increase upon repeated exposures. Jovanovic et al. [65] reported that in the 20% radiolabeled D5 absorbed in rats after administrating an oral (gavage) dose of ^{14}C-D5 in corn oil, 50–60% was eliminated as parent D5 in exhaled air.

In air-breathing animals, metabolism also plays a role in the depuration of VMS in situations of repeated exposure. Metabolites of D5 were found in the urine of

Fischer 344 rats that were intravenously and orally exposed to D5 [66]. D4 and D5 are cleared by metabolism and exhalation, with the latter being the major pathway of elimination of free D4 and D5 in blood due to their low values of blood-air partition coefficients. During an exposure, metabolism is a significant mechanism in the elimination of D4 and D5. However, after cessation of the exposure, the primary mechanism of elimination is exhalation. Because D4 and D5 have unusually low levels of blood-air partition coefficients, they also have considerably higher pulmonary clearance than other volatile compounds (e.g., benzene) [49].

2.4 Food Chains

Trophic magnification (or simply biomagnification) is the increase in the bioaccumulation of certain chemicals in organisms occupying sequentially higher trophic positions in a food chain. This phenomenon occurs because of the following sequence of events: as lipids of contaminated prey are digested in the gut of predators, the capacity of the digestate (due to its increasing polarity) to retain nonmetabolized lipophilic contaminants is reduced, resulting in the net transfer of these chemicals to the predator's lipid-rich tissues. Then, if the predator continues to consume numerous prey, the uptake rates by diet can exceed the elimination rate, resulting in contaminant concentrations higher than those that would be found in the predator's fatty tissues at equilibrium. If this animal is, in turn, consumed by a predator in higher trophic level, further magnification in residue concentrations can occur [45].

Biomagnification is more likely to occur with chemicals that are lipid-soluble, well absorbed, and poorly metabolized, thereby limiting excretion. In the case of VMS, for biomagnification to occur, the rate of ingestion of these chemicals needs to surpass the rate of clearance. In cases where predators are fish-eating birds and mammals having high consumption rates of contaminated fatty prey, biomagnification can result in relatively high residue concentrations. Similarly, in the aquatic environment (where VMS bind to organic matter in sediment), assuming that the bound-VMS remains bioavailable, organisms feeding in contaminated sediment might bioaccumulate the chemicals. Other biota that feeds on sediment-dwelling organisms might bioaccumulate VMS if the rate of ingestion is greater than the rate of clearance, and in this case, biomagnification occurs.

3 Bioaccumulation

Bioaccumulation is a function of bioconcentration and biomagnification [67], where bioconcentration is the non-trophic uptake and accumulation of a chemical in a specific organism from abiotic media (primarily water, but also sediment and air) and biomagnification is the change in concentration across organisms in a food web related to the change in trophic level.

The research on bioaccumulation of VMS has yielded apparently contradictory results, with high laboratory bioconcentration factors on one hand and low field trophic magnification factors on the other. These studies are discussed below.

3.1 Bioconcentration

Most studies on bioconcentration of VMS have been conducted in aquatic environments, including the uptake and accumulation from water (bioconcentration) and from sediment (biota-sediment accumulation). Studies on water bioconcentration were restricted to laboratory-based experiments, where organisms are exposed to test chemicals in water, but not in the diet. In these studies, water is spiked to a known concentration of VMS, and VMS accumulated in the organisms are quantified thereafter. The ratio between the concentration accumulated in the organism ($C_{organism}$) and the concentration in water (or in other abiotic media) (C_{water}) is termed bioconcentration factor (BCF) and is usually expressed by L water kg^{-1} organism (1).

$$\text{Bioconcentration factor (BCF)} = \frac{C_{organism}}{C_{water}} \tag{1}$$

The BCF is considered a key parameter that reflects the equilibrium partitioning of a given compound between the environment and organisms.

In the regulatory context, the BCF has been used, together with other variables (e.g., K_{OW}), as a surrogate of the bioaccumulation potential of a chemical [68]. The United Nations Stockholm Convention on Persistent Organic Pollutants (POPs) considers a BCF in aquatic species above 5,000 L kg^{-1} as an indicator of bioaccumulation (or in the absence of such data, a log K_{OW} greater than 5) [68]. National regulations adopted this value (e.g., Canada) [69] and some added criteria to distinguish between bioaccumulative and very bioaccumulative substances [68]. A BCF $\geq$ 5,000 L kg^{-1} was adopted by both the European Union and the United States as indicative of very bioaccumulative substances [70, 71]. However, while the European Union assumed a BCF $\geq$ 2,000 L kg^{-1} as indicative of a bioaccumulative substance, the United States adopted a BCF $>$ 1,000 L kg^{-1} [70, 71].

The few studies that have calculated BCF for VMS focused on cVMS (mainly D3, D4, D5, and D6) in freshwater fish (goldfish, rainbow trout, and fathead minnow) and in the aquatic invertebrate *Daphnia magna* (water flea) (Table 2). These species have commonly been used as models in toxicological studies, as they are fast-growing organisms that allow short-time experiments.

A study on the bioaccumulation potential of D3 in aquatic biota demonstrated some accumulation in the tissues of rainbow trout (*Oncorhynchus mykiss*) after

Table 2 Bioconcentration factors (BCF) (L kg^{-1}), biota-sediment accumulation factors (BSAF) (kg organic carbon/kg lipid, unless otherwise stated), and multimedia bioaccumulation factors (mmBAF) (unitless) for cyclic VMS: D3, D4, D5, and D6

Factor	Species	Type of organism	Type of ecosystem	Type of study	D3	D4	D5	D6	Comments	References
BCF	Goldfish	Fish	Freshwater	Lab-based		1,090	1,010	1,200		[72]
BCF	Rainbow trout (Oncorhynchus mykiss)	Fish	Freshwater	Lab-based	100 ± 49	1,875–10,000	3,362	>1,000		[73, 74]
BCF	Fathead minnow (Pimephales promelas)	Fish	Freshwater	Lab-based		12,400	4,450–13,300	240–1,660		[75–79]
BCF	Water flea (Daphnia magna)	Aquatic invertebrate	Freshwater	Lab-based				2,400		[79]
BSAF	Midge (larvae) (Chironomus tentans)	Insect	Freshwater	Lab-based		2.2[a] 1.3[b] 0.7[c]			Sediment with low[a], medium[b], and high organic carbon content[c]	[80]
BSAF	Midge (Chironomus riparius)	Insect	Freshwater (eggs)	Lab-based			1.20[d] 1.10[e] 0.83[f] 0.46[g]		At treatment concentrations of 13[d], 30[e], 73[f], and 180[g] mg kg^{-1}	[81]
BSAF	Hyalella azteca	Amphipod crustacean	Freshwater	Lab-based			0.05 0.87			[82]
BSAF	Atlantic cod (Gadus morhua)	Fish	Marine	Field (European Arctic)			2.1			[83]
BSAF	Shorthorn sculpin (Myoxocephalus scorpius)	Fish	Marine	Field (European Arctic)			1.5			[83]

(continued)

Table 2 (continued)

Factor	Species	Type of organism	Type of ecosystem	Type of study	D3	D4	D5	D6	Comments	References
BSAF	Blackworm (*Lumbriculus variegatus*)	Worm	Freshwater	Lab-based			4.4[h] 7.1		Expressed as kg dry weight sediment/kg wet weight[h]	[84]
BSAF	Benthic invertebrates	Mix	Freshwater	Field (Lake Pepin, USA)		>1	>1	>1		[85]
BSAF	Benthic invertebrates	Mix	Freshwater	Estimated based on published lab-based data		0.7	1.0	0.5		[86]
mmBAF	Flounder (*Pleuronectes flesus*)	Fish	Marine	Field (Humber Estuary, England)		6	2		PCB180 used as benchmark	[15]
mmBAF	Ragworm (*Hediste diversicolor*)	Worm	Marine	Field (Humber Estuary, England)		14	2	−5 to −10	PCB180 used as benchmark	[15]

14 days of continuous exposure to both D3 and its hydrolysis products. A BCF for D3 of 100 ± 49 L kg^{-1} was derived based on the tissue content of the chemical measured at the end of the exposure period [73].

D4 has revealed potential to bioconcentrate in fish under optimized exposure conditions. The steady-state BCF for D4 in fathead minnows (*Pimephales promelas*), a freshwater fish, was determined to be 12,400 L kg^{-1} after 28 days continuous exposure to radiolabeled test material in soft water at 21–22°C in an enclosed flow-through system [75]. In other study, the BCF for D4 was estimated to be in the range 1,875–10,000 L kg^{-1} for the same species [74]. Using goldfish, the BCF for D4 was estimated to be 1,090 L kg^{-1} [72]. These studies indicate that D4 has potential for bioaccumulation, as most of the determined BCF values significantly exceed the criterion for bioaccumulation (BCF $\geq$ 2,000 L kg^{-1}) [87, 88].

Regarding D5, BCF values were reported to be 4,450 L kg^{-1} [76] and 13,300 L kg^{-1} [77] in different exposure toxicity studies using fathead minnows, whereas in rainbow trout and goldfish, the BCF values for D5 were estimated to be 3,362 L kg^{-1} [74] and 1,010 L kg^{-1} [72], respectively. A recent review of the available studies on the bioconcentration of D5 in fish recalculated BCF values for this chemical and reported values in the range 1,040–4,920 L kg^{-1} wet weight [68]. The review highlighted the fact that depuration of D5 in fish is more rapid than would be expected for a very hydrophobic organic chemical. The expected faster elimination of D5 in fish is attributed to its biotransformation into polar metabolites.

For D6, the BCF was estimated to be 1,660 L kg^{-1} [78] and $>$1,000 L kg^{-1} [74] using fathead minnows. Bioaccumulation studies using the water flea *Daphnia magna* (an aquatic invertebrate), resulted in a BCF value of 2,400 L kg^{-1} based solely on the uptake of the chemical [79]. Using goldfish, the BCF for D6 was estimated to be 1,200 L kg^{-1} [72].

Most studies reported BCF values for D4 and D5 that meet the bioaccumulative or very bioaccumulative criteria [87, 88]. The BCF values for D6 between 1,000 and 2,400 L kg^{-1} indicate a moderate bioaccumulation potential in aquatic life [79].

The environmental relevance of these laboratory-based studies was discussed [46]. A key consideration is whether the concentrations used exceeded the water solubility due to the use of solvents for the addition of the cVMS to the test medium (water). The main source of cVMS in the environment is from particles bound to effluent material emitted by sewage treatment plants. The cVMS tend to be attached to particulate matter as they distribute in the water body. Consequently, the concentration in water remains below the solubility limit because of the unfavorable partitioning from sediment and the loss of cVMS from water to air [46]. This low solubility will limit the uptake from water through respiratory and dermal surfaces of aquatic organisms. It is most likely that the uptake of VMS in aquatic species at the base of the food chain will be primarily from ingestion of sediment and/or ingestion of sediment-dwelling biota.

3.2   Biota-Sediment Accumulation

The magnitude of chemical accumulation from sediments is often expressed by the biota-sediment accumulation factor (BSAF), defined by the ratio between the concentration of a given chemical accumulated in an organism ($C_{organism}$) and the concentration of the chemical in sediment ($C_{sediment}$) (2). The BSAF has units of kg sediment dry weight kg organism^{-1} wet weight or kg organic carbon kg lipid^{-1}.

$$\text{Biota-sediment accumulation factor (BSAF)} = \frac{C_{organism}}{C_{sediment}} \qquad (2)$$

Studies reporting BSAF for VMS under laboratorial and field conditions are summarized in Table 2.

Kent et al. [80] calculated the BSAF for D4 by exposing larvae of midge *Chironomus tentans* for 14 days to increasing concentrations of radiolabeled ^{14}C D4 in each of three sediments (low, medium, and high organic carbon content), finding values of 2.2, 1.3, and 0.7, respectively.

In a similar study using a different midge species (*Chironomus riparius*), the average BSAF for D5 was calculated to be 1.2, 1.1, 0.83, and 0.46 at exposure concentrations of 13, 13, 73, and 180 mg kg^{-1} [81], respectively.

Norwood et al. [82] studied the bioaccumulation of D5 in the amphipod crustacean *Hyalella azteca* in a series of spiked sediment exposures. Two different types of natural sediments were spiked: one from Lake Erie (with low organic carbon content ≈0.5%) and the second from Lake Restoule (with higher organic carbon content ≈11%), Ontario. Juvenile *H. azteca* were exposed for 28 days, and the derived BSAF was 0.05 and 0.87 kg organic carbon kg lipid^{-1} for Lake Erie and Lake Restoule, respectively.

Warner et al. [83] reported BSAF for D5 using field sample data from fish (Atlantic cod and shorthorn sculpin) and sediment in European Arctic environment. Based on the concentrations of the field samples, median BSAF values (kg organic carbon kg lipid^{-1}) for D5 were reported to be 2.1 and 1.5 for cod and sculpin, respectively.

Powell et al. [85] reported that BSAF values (kg organic carbon kg lipid^{-1}) for D4, D5, and D6 were greater than 1.0 in many species (especially for some benthic invertebrate species) in a freshwater lake. The values were higher in organisms having a close association with the sediment compartment, indicating that the source of cVMS in this lake food web was sediment.

Krueger et al. [84], in a laboratory-based bioaccumulation test, exposed blackworms (*Lumbriculus variegatus*) to sediments spiked with D5 for 28 days, corresponding to the sediment uptake phase. The BSAF determined in the experiment was 4.4 kg dry weight sediment/kg wet weight, which corresponded to 7.1 kg organic carbon kg lipid^{-1}.

Woodburn et al. [86] estimated BSAF for D4, D5, and D6 for invertebrate biota, based on published records. For D4, six benthic invertebrate toxicological studies were considered, including *Chironomus dilutus*, *Chironomus riparius*, and *Lumbriculus variegatus*. For D5, six benthic invertebrate studies and two soil organism toxicity studies were considered. The species included in these studies were *Hyalella azteca*, *C. riparius*, *L. variegatus*, *Folsomia candida*, *Eisenia andrei*, and *Eisenia fetida*. For D6, three benthic invertebrate studies were considered and included the species *C. riparius* and *L. variegatus*. The estimated BSAF for D4, D5, and D6 were all equal to or lower than 1.0.

3.3 Multimedia Bioaccumulation

Bioaccumulation may be evaluated using multimedia bioaccumulation factors (mmBAF), which quantify the fraction of the contaminant in the environment that is transferred to the biota [89]. In aquatic environments, the calculation of mmBAF commonly considers the concentration of the contaminants in the water column and the underlying surface sediments:

$$\text{Multimedia bioaccumulation factor (mmBAF)} = \frac{C_{\text{organism}} \times A_{\text{env}}}{C_{\text{water}} + C_{\text{sediment}}} \qquad (3)$$

where C_{organism} corresponds to the concentration of the chemical in the organism under study, A_{env} is the area or spatial extent of the environment, and C_{water} and C_{sediment} are the concentration of the chemical in water and sediment, respectively. The mmBAF units are expressed by m^2 organism^{-1}.

Kierkagaard et al. [15] applied the concept of mmBAF to study the bioaccumulation of cVMS in a food chain in Humber Estuary, in the east coast of England. Samples of sediment, common ragworm (*Hediste diversicolor*), and flounder (*Pleuronectes flesus*) were collected from six different locations and analyzed for cVMS as well as for polychlorinated biphenyls (PCB). The mmBAF concept requires a previous decision on whether a mmBAF value indicates that the chemical is highly bioaccumulative or not. The authors opted for a benchmarking approach, where the mmBAF of each VMS was compared with the mmBAF of PCB 180 (a chemical that is known to be strongly bioaccumulative). The mean mmBAF of D5 was about twice that of PCB 180 in both ragworm and flounder, while for D4, it was 6 and 14 times higher, respectively (Table 2). The mmBAF of D6 was a factor 5–10 lower than that of PCB 180. The comparatively strong multimedia bioaccumulation of D4 and D5, even in the absence of biomagnification, was explained by both compounds having a >100 times stronger tendency (i.e., $K_{OW}/K_{OC} > 100$) to partition into lipid rather than into organic carbon, while PCB 180 partitions to a similar extent (i.e., $K_{OW}/K_{OC} \approx 1$) into both matrices [15].

3.4   Biomagnification and Trophic Magnification

The trophic magnification factor (TMF) describes the change in concentration of a chemical in organisms that occupy successively higher trophic levels, to evaluate food web magnification. These studies are complex due to the numerous variables that may influence the results.

In some studies, authors calculated dietary biomagnification factors (BMF), which are estimated by exposing organisms (usually aquatic) in controlled conditions, to food with known concentration of VMS [90]. These are laboratory-based experiments that shed light on the metabolic processes behind the excretion of VMS from aquatic organisms.

The TMF, on the other hand, is determined from field-derived chemical concentration data in different species across a defined food web and thus has a more significant environmental relevance. The TMF is calculated from the slope (m) of a linear regression between the logarithm of the concentration of the chemical in the organisms of the food web and the estimated trophic position of the organism (i.e., $TMF = 10^m$) [68]. In the case of VMS, due to their lipophilicity, the concentrations are normalized for lipid content. In fact, organisms with higher lipid content are expected to contain higher concentrations than organisms with lower lipid content, when subjected to the same exposure concentration and environmental conditions [68].

The trophic position can be determined by stable $^{15}N/^{14}N$-isotope ratios in animal tissues, which tend to increase with the trophic position in food webs. It is usual to confirm the results with analyses of the intestinal content of the organisms. Bioaccumulative substances able to biomagnify are characterized by a TMF above 1.0.

A few studies in both freshwater and marine aquatic food webs were conducted to determine whether VMS biomagnify or not [68]. Trophic magnification was demonstrated for some VMS in some food webs, whereas in others, trophic dilution occurs (Table 3).

Jia et al. [93] investigated the trophic transfer of cVMS in a marine food web from the coastal area of Northern China – the Dalian Bay. The food web was composed of five fish species (including pacific herring, mackerel, greenling, Schlegel's black rockfish, sea catfish), one crustacean species (mud crab), five molluscs (*Mactra quadrangularis*, short-necked clam, mussel, arthritic Neptune, black fovea snail), and sea lettuce. TMF for D4, D5, and D6 were estimated to be 1.2, 1.8, and 1.0, respectively (Table 3), revealing the existence of trophic magnification for D5.

Borgå et al. [17] also reported trophic magnification for D5 (TMF = 2.3) while studying the Lake Mjøsa pelagic food web in Norway in 2010 (Table 3). Assays included zooplankton (predominantly water fleas *Daphnia galeata* and copepods *Limnicalanus macrurus*), *Mysis relicta* (a shrimp-like crustacean), and fish vendace (*Coregonus albula*), smelt (*Osmerus eperlanus*), and brown trout (*Salmo trutta*). The TMF derived for D5 was found sensitive to the species on the higher trophic levels, namely, smelt and brown trout. This sensitivity was not seen for other

Table 3 Trophic magnification factors (TMF) for cycle VMS: D4, D5, and D6

Type of food web	Species	Site	D4	D5	D6	References
Marine (pelagic)	Fish: red barracuda (*Sphyraena pinguis*), silver croaker (*Pennahia argentata*), chub mackerel (*Scomber japonicus*), Japanese sea bass (*Lateolabrax japonicus*), Japanese sardinella (*Sardinella zunasi*), gizzard shad (*Konosirus punctatus*), Japanese anchovy (*Engraulis japonicus*)	Tokyo Bay, Japan	1.3	1.0	0.8	[91]
Marine (demersal/ pelagic)	Fish: Atlantic cod (*Gadus morhua*), N. Atlantic pollock (*Pollachius pollachius*), poor cod (*Trisopterus minutus*), Vahl's eelpout (*Lycodes vahlii*), European whiting (*Merlangius merlangus*), long rough dab (*Hippoglossoides platessoides*), haddock (*Melanogrammus aeglefinus*), European hake (*Merluccius merluccius*), Norway pout (*Trisopterus esmarkii*), coalfish (*Pollachius virens*), European plaice (*Pleuronectes platessa*), Atlantic herring (*Clupea harengus*) Crustacean: Northern shrimp (*Pandalus borealis*) Molluscs: Mussel sp. A, Mussel sp. B, blue mussel (*Mytilus edulis*) Net plankton Jellyfish Worms	Inner Oslofjord, Norway	0.5/ 0.7	0.3/ 0.4	0.6/ 0.7	[92]
Marine (demersal/ pelagic)	Fish: Atlantic cod (*G. morhua*), haddock (*M. aeglefinus*), long rough dab (*H. platessoides*), coalfish (*P. virens*), Norway pout (*T. esmarkii*), starry skate (*Amblyraja radiata*), common sole (*Solea vulgaris*), European plaice (*P. platessa*), Molluscs: Mussel sp. A, Mussel sp. B Net plankton Jellyfish Worms Sea urchins (*Brissopsis lyrifera*)	Outer Oslofjord, Norway	0.5/ 0.6	0.3/ 0.8	0.3/ 0.9	[92]

(continued)

Table 3 (continued)

Type of food web	Species	Site	D4	D5	D6	References
Marine (mixed)	Fish: Pacific hearing (*Clupea pallasii*), mackerel (*Pneumatophorus japonicus*), greenling (*Hexagrammos otakii*), Schlegel's black rock-fish (*Sebastes schlegelii*), sea catfish (*Synechogobius hasta*) Crustacean: mud crab (*Scylla serrata*) Mollusc: *Mactra quadrangularis* (*Mactra veneriformis*), short-necked clam (*Ruditapes philippinarum*), mussel (*Mytilus galloprovincialis*), Arthritic Neptune (*Neptunea cumingi*), black fovea snail (*Omphalus rustica*) Algae: sea lettuce	Dalian Bay, China	1.2	1.8	1.0	[93]
Freshwater (demersal)	Fish: river carpsucker (*Carpiodes carpio*), quillback (*Carpiodes cyprinus*), white sucker (*Catostomus commersonii*), silver redhorse (*Moxostoma anisurum*), shorthead redhorse (*Moxostoma macrolepidotum*), bluegill sunfish (*Lepomis macrochirus*), smallmouth bass (*Micropterus dolomieu*), largemouth bass (*Micropterus salmoides*), black crappie (*Pomoxis nigromaculatus*), gizzard shad (*Dorosoma cepedianum*), common carp (*Cyprinus carpio*), emerald shiner (*Notropis atherinoides*), white bass (*Morone chrysops*), walleye (*Stizostedion vitreum*), freshwater drum (*Aplodinotus grunniens*) Insects: midge (*Chironomus* sp.), burrowing mayfly (*Hexagenia* sp.)	Lake Pepin, USA	0.2	0.1	0.1	[85]
Freshwater (pelagic)	Fish: fish vendace (*Coregonus albula*), smelt (*Osmerus eperlanus*), brown trout (*Salmo trutta*)	Lake Mjøsa, Norway		2.3		[17]

(continued)

Table 3 (continued)

Type of food web	Species	Site	D4	D5	D6	References
	Zooplankton: water fleas (*Daphnia galeata*), copepods (*Limnicalanus macrurus*) Crustacean: *Mysis relicta*					
Freshwater (pelagic)	Fish: fish vendace (*C. albula*), smelt (*O. eperlanus*), brown trout (*S. trutta*) Crustacean: *M. relicta* Zooplankton: water fleas (*D. galeata*), *Bosmina longispina*, copepods (*L. macrurus*)	Lake Mjøsa, Norway	0.8	3.1	2.7	[18]
Freshwater (pelagic)	Fish: whitefish, smelt (*O. eperlanus*), brown trout (*S. trutta*) Zooplankton: water fleas (*D. galeata*), copepods (*L. macrurus, Eudiaptomus gracilis*), *Heterocope appendiculata*	Lake Randsfjorden, Norway	0.6	2.1	1.5	[18]

analyzed chemicals, such as PCB 180, which indicates that D5 biomagnification was likely governed by species-specific properties such as biotransformation rate or tissue distribution that differed from those of other chemicals.

Two years later, the same authors repeated the study and derived TMF values higher than 1.0 for D5 and D6 (3.1 and 2.7, respectively), confirming the trophic magnification [18]. Moreover, they also found trophic magnification for D5 and D6 (2.1 and 1.5, respectively) in the food web of another Norwegian lake (Lake Randsfjorden), which is under less human impact than Lake Mjøsa [18] (Table 3). Both pelagic food webs included zooplankton, planktivorous fish, and piscivorous fish with brown trout as a top predator. TMF for D4 was below 1.0 in both cases. One explanation for the different biomagnification behaviors of D4 compared to D5 and D6 could be a more rapid metabolism of D4.

These results contrast with other studies reporting trophic dilution for cVMS. Powell et al. [91] studied the bioaccumulation and trophic transfer of cVMS, specifically D4, D5, and D6 in the pelagic marine food web of Tokyo Bay, Japan. The authors analyzed these compounds in sediments and in seven fish species, from high to low trophic levels: red barracuda, silver croaker, chub mackerel, Japanese sea bass, Japanese sardinella, gizzard shad, and Japanese anchovy. The authors found that TMF for cVMS across the sampled food web generally were not statistically different from 1.0 [91], supporting the theory of trophic dilution (Table 3).

Powell et al. [92], recalculated the TMF for a food web in the Oslofjord, Norway, based on field data collected in 2008 [94]. Marine ecosystems such as the Oslofjord typically have very complex food webs that are defined by multiple and interconnected food chains that are confounded by a great diversity in prey organisms and feeding relationships. The studied food web consisted of 22 species and included zooplankton, benthic macroinvertebrates, shellfish, and finfish. The lowest trophic positions were occupied by zooplankton, benthic macroinvertebrates (worms and urchins), and blue mussel (*Mytilus edulis*). Shrimp (*Pandalus borealis*), mussels, and finfish, such as the pleuronectids (flounders), clupeids (herrings), and most gadids (cods), comprised the middle positions. The top of the food web was occupied by cod (*Gadus morhua*). Due to the overlapping of food chains and omnivorous feeding, TMF were calculated for the demersal (composed of fish living close to the bottom and organisms associated with a substrate) and pelagic components of the food web (Table 3). The reported TMF for cVMS were all less than 1.0 across the pelagic (ranging from 0.4 to 0.9) and demersal (ranging from 0.3 to 0.6) components of the sampled food web, suggesting trophic dilution of these compounds. In the same study [92], lipid-normalized concentrations of cVMS in biota were found to be greatest in the lowest trophic level species (i.e., the benthic macroinvertebrates and zooplankton) and to decrease with the increase of the trophic level position.

Trophic dilution was also reported for D4, D5, and D6 in the food web of Lake Pepin, Upper Mississippi River [85]. The food web that was evaluated included surface sediments, 2 benthic macroinvertebrate species (2 genera, 2 families), and 15 fish species (14 genera, 9 families). Benthic detritivores (i.e., *Chironomus* sp. and *Hexagenia* sp.) were in the lowest trophic level and pelagic piscivores (i.e., largemouth bass and walleye) in the highest. Lipid-normalized concentrations of D4, D5, and D6 were highest in the lower trophic levels and significantly decreased up the food web. TMF values for D4, D5, and D6 were all lower than 1.0 (0.8, 0.4, and 0.3, respectively), indicating trophic dilution across the aquatic food web (Table 3).

Trophic magnification factors are preferable to other measures for evaluating bioaccumulation within an ecosystem that has a well-defined food web [95–97] or between multiple ecosystems [98]. However, there is still a high degree of uncertainty in relation to these studies. One of the first sources of uncertainty is the use of isotopic fractioning to assess food webs. These isotopes may be affected by local sources of nitrogen such as agriculture, which will change the ratio of isotopes. Second, contaminants in a food web may originate from multiple sources, and the trophic transfer of a contaminant across a complex food web may be obscured by the overlap and convergence of multiple food chains that comprise the food web. Omnivorous feeding across food chains that have different or multiple exposures to a contaminant may have a strong influence on calculation of TMF, thus making it especially difficult to interpret results [99, 100]. Third, there is still a lack of knowledge on the distribution of VMS in biota tissue, which is essential to select which organs should be analyzed during the study.

Nevertheless, the discrepancy between the studies may reflect the variability in TMF values between different food webs, depending on season, species composition, and location.

4 Occurrence in Biota

Volatile methylsiloxanes have been found in different biota samples from different regions. Most studies were conducted in the aquatic environment, both marine and freshwater, and only recently on terrestrial ecosystems.

In 2013, Wang et al. [1] published a review on the occurrence of VMS in the environment, including in biota samples. In it, the studies of Kaj et al. [11, 12], Brooke et al. [23–25], Warner et al. [83], and Kierkegaard et al. [101–103], reporting VMS concentrations in marine and freshwater fish (liver and muscle), seals, pilot whale, porpoises, seabird eggs, dolphins, and zooplankton, were discussed. Seabird eggs, dolphins, and porpoises all had cVMS below the limits of detection (LOD) [12]. In general, detectable cVMS concentrations in fish from the Arctic and Northern Europe varied from 0.7 to 900 ng g^{-1} (ww, wet weight) for D4, from 2.0 to 344 ng g^{-1} ww for D5, and from 0.7 to 100 ng g^{-1} ww for D6 [1]. Regarding mammals, seal and pilot whale had concentrations varying from <10 to 12 ng g^{-1} ww for D4 (only seals), from <5.0 to 24 ng g^{-1} ww for D5, and from 4.0 to 8.0 ng g^{-1} ww for D6 (only seals) [1]. The lower concentrations of cVMS in mammals compared with fish are likely to be due to their higher elimination rates through respiration and metabolism (see Sect. 2).

Since the review of Wang et al., new studies have been published reporting the occurrence of VMS in a wide variety of biota samples. Because sampling and analytical methodologies differed among them, an overview of each study will be chronologically presented. Two additional studies published before 2013 are also discussed, as they were not included in the study of Wang et al.

In 2009, Powell et al. [85] analyzed different fish species and insects from Lake Pepin, USA (see Table 3 for information on the species). Detectable concentrations of cVMS in fish varied from 1.7 to 59 ng g^{-1} ww for D4, from 7.5 to 447 ng g^{-1} ww for D5, and from 1.7 to 27 ng g^{-1} ww for D6. In insects, concentrations varied from 7.3 to 7.8 ng g^{-1} ww for D4, from 49 to 154 ng g^{-1} ww for D5, and from 3.7 to 10.5 ng g^{-1} ww for D6.

In the study of Borgå et al. [17] published in 2012, concentrations in fish ranged from <0.55 to 4.5 ng g^{-1} ww for D4, from 6.4 to 230 for D5, and from <0.46 to 7.2 for D6. Zooplankton had cVMS concentrations <1.4 ng g^{-1} ww for D4 and ranging from <2.8 to 50 ng g^{-1} ww for D5 and from <2.0 to 4.4 ng g^{-1} ww for D6. Two years later [18], the same authors found concentrations in fish varying from <LOD to 1.1 ng g^{-1} ww for D4, from 32 to 311 ng g^{-1} ww for D5, and from 1.5 to 15 ng g^{-1} ww for D6. In zooplankton, concentrations varied from <0.2 to 2.4 ng g^{-1} ww for D4, from 2.3 to 156 ng g^{-1} ww for D5, and from 0.3 to 2.8 ng g^{-1} ww for D6. In crustaceans, concentrations ranged from <0.4 to 2.0 ng g^{-1} ww for D4, from 9.6 to 50 ng g^{-1} ww for D5, and from <0.9 to 1.6 ng g^{-1} ww for D6. In lake Randsfjorden, Norway (see Table 3 for information on the species), the concentrations varied from <0.1 to 0.9 ng g^{-1} ww for D4, from 0.2 to 115 ng g^{-1} ww for D5, and from <0.2 to 3.0 for D6 in fish, whereas in zooplankton, cVMS varied from <0.1 to 1.2 ng g^{-1} ww for D4, from 1.6 to 53 ng g^{-1} ww for D5, and from <0.1 to 3.0 ng g^{-1} ww for D6.

In 2014, Hong et al. [13] measured VMS in bottom fish (*Hexagrammos otakii*) samples collected from a marine environment in Northeast China. The concentration of the sum of cVMS (D4-D7) and lVMS (L8-L11) ranged from 3.50 to 7.36 ng g^{-1} ww. D6 had the highest concentration, with a mean value of 1.23 ± 0.51 ng g^{-1} ww, followed by L11 (0.925 ± 0.691 ng g^{-1} ww) and D5 (0.786 ± 0.255 ng g^{-1} ww). In the same study, the authors compared VMS in different organs (fish body, muscle, gill, intestine, brain, eye, and sexual gland) and found the highest levels in muscles, followed by gills.

In a study from 2015, Huber et al. [104] analyzed VMS in eggs of three bird species, namely, common eider (*Somateria mollissima*), European shag (*Phalacrocorax aristotelis*), and European herring gull (*Larus argentatus*), from remote islands off the coast of Norway. Cycle VMS concentrations detected in bird eggs were dominated by D5 (<LOD-3.6 ng g^{-1} ww), followed by D6 (<LOD-0.8 ng g^{-1} ww). D4 was not detected in any eggs analyzed, which was explained as being a result of its more rapid hydrolysis rate compared with D5 and D6. The higher concentrations of D5 in relation to D6 were attributable to the higher production and use of D5.

In the same year, Jia et al. [93] analyzed cVMS (D4-D7) and lVMS (L9-L11) in different fish species, crustacean, molluscs, and algae (sea lettuce) from Dalian Bay, China (see Table 3 for information on the species). Linear VMS were only detected in a reduced number of samples, but cVMS were detected in most samples. In fish, detectable cVMS concentrations varied from 6.03 to 38.0 ng g^{-1} ww for D4, from 5.83 to 120 ng g^{-1} ww for D5, from 3.71 to 70.6 ng g^{-1} ww for D6, and from 2.01 to 12.9 ng g^{-1} ww for D7. In molluscs, cVMS concentrations ranged from 5.69 to 13.7 ng g^{-1} ww for D4, from 3.94 to 21.5 ng g^{-1} ww for D5, from 3.74 to 42.1 ng g^{-1} ww for D6, and from 2.05 to 8.14 ng g^{-1} ww for D7. In crustacean, cVMS concentrations varied from 7.56 to 16.6 ng g^{-1} ww for D4, from 14.6 to 25.5 ng g^{-1} ww for D5, from 10.5 to 25.9 ng g^{-1} ww for D6, and from 2.27 to 10.3 ng g^{-1} ww for D7. In sea lettuce, cVMS concentrations ranged from 5.73 to 9.76 ng g^{-1} ww for D4, from 3.99 to 11.3 ng g^{-1} ww for D5, from 3.62 to 37.1 ng g^{-1} ww for D6, and from 2.03 to 6.03 ng g^{-1} ww for D7 [93].

Sanchís et al. [21], also in 2015, measured cVMS and lVMS in samples of phytoplankton, krill (*Euphausia superba*), lichens, mosses, and hair grass (*Deschampsia antarctica*) from the Antarctic Peninsula region. cVMS were detected almost in all samples, while lVMS were only detected in phytoplankton. In terrestrial samples, the highest concentrations were found in hair grass and moss ($\sum$cVMS mean values of 55.1 ng g^{-1} dw and 44.3 ng g^{-1} dw, respectively), followed by lichens ($\sum$cVMS average value of 30.5 ng g^{-1} dw). In the aquatic media, $\sum$cVMS concentrations in krill varied between 4.5 and 145 ng g^{-1} dw and in phytoplankton $\sum$cVMS varied between <LOD and 27 ng g^{-1} dw. The sum of linear VMS in phytoplankton ranged from <LOD to 0.13 ng g^{-1} dw. The authors found a negative correlation between the salinity and the concentrations of VMS in phytoplankton, which was interpreted as a surrogate of contamination from melting ice/snow. In the same region, the authors also found high concentrations of VMS in soil samples, which were explained as being a consequence of the feces of penguins feeding on

krill. Some authors have contested the quality of this study, suggesting the results were a consequence of poor sampling and processing techniques and/or contamination of samples [46, 105]. However, the authors of the Antarctic study have already provided the scientific community with data showing the credibility of the results [106].

Studies reporting VMS in vegetation are almost inexistent. Besides the study of Sanchís et al. [21], only one more study was found. In 2016, Ratola et al. [20] measured VMS in pine needles (*Pinus pinaster* and *Pinus pinea*) from Portugal in areas covering different anthropogenic pressures. The levels of VMS ranged from 2 to 118 ng g^{-1} dw, with D5 and D6 being the predominant compounds. Urban and industrial areas had the highest incidence, suggesting a strong anthropogenic fingerprint.

In 2016, Lucia et al. [51] measured in the Arctic VMS in eggs of black-legged kittiwakes (*Rissa tridactyla*) and of glaucous gull (*Larus hyperboreus*) and in Arctic chars (*Salvelinus alpinus*) (a freshwater fish). The results were below LOD except for D5 in eggs of glaucous gull, whose concentrations ranged from 3.1 to 40.1 ng g^{-1} ww. Unlike black-legged kittiwakes, which feed on invertebrates and small fish, glaucous gull is a generalist and opportunistic predator that feeds on a variety of fish, molluscs, echinoderms, crustaceans, eggs, chicks, and adults of other seabird species, insects, carrion, refuse, and offal. Being diet a potential source of exposure of these species to VMS, the higher concentrations found in glaucous gull eggs in relation to the black-legged kittiwakes ones are understandable.

Wang et al. [107], in 2017, measured linear and cyclic VMS in the blood plasma of common snapping turtles (*Chelydra s. serpentina*), double-crested cormorants (*Phalacrocorax auritus*), and Northwest Atlantic Harbor seals (*Phoca vitulina concolor*) collected from Canadian freshwater and marine ecosystems. The authors selected these species to represent high trophic level piscivores from reptilian, avian, and mammalian taxonomic groups. Among the cVMS, D3, D4, and D6 were found in the three species at <0.010 to 1.4 ng g^{-1} ww, <0.018 to 0.590 ng g^{-1} ww, and <0.035 to 1.31 ng g^{-1} ww, respectively. Snapping turtles, cormorants, and seals from contaminated sites all exhibit high D5 concentrations in comparison with reference sites. In average, D5 varied from 0.143 ± 0.149 (mean ± SD) to 3.59 ± 3.06 ng g^{-1} ww in turtles, from 1.12 ± 0.81 to 7.39 ± 2.25 ng g^{-1} ww in cormorants, and from 0.335 ± 0.239 to 1.20 ± 0.33 ng g^{-1} ww in seals. The environmental contamination was reported to affect the levels of D5 in the different species [107]. VMS found in blood plasma were probably a consequence of a higher rate of exposure than of clearance, as explained in Sect. 2.

Powell et al. [91], also in 2017, while studying the marine food web of Tokyo Bay, Japan (see Table 3 for information on the species), found cVMS concentrations in fish varying from 8.4 to 24.0 ng g^{-1} ww for D4, from 142 to 334 ng g^{-1} ww for D5, and from 3.1 to 12.0 ng g^{-1} ww for D6.

Analyzing different fish species, crustaceans, molluscs, jellyfish, worms, sea urchins, and plankton from the Inner and Outer Oslofjord, Norway (see Table 3 for information on the species), Powell et al. [92], in 2018, found detectable cVMS concentrations in fish, which varied from 0.2 to 40 ng g^{-1} ww for D4, from 1.2 to

3,272 ng g^{-1} ww for D5, and from 0.7 to 43 ng g^{-1} ww for D6. In crustaceans, the levels varied from 0.24 to 2.86 ng g^{-1} ww for D4, from 5.08 to 147 ng g^{-1} ww for D5, and from 1.08 to 4.33 ng g^{-1} ww for D6. In molluscs, concentrations ranged from 0.21 to 3.77 ng g^{-1} ww for D4, from 4.36 to 252 ng g^{-1} ww for D5, and from 1.35 to 8.69 ng g^{-1} ww for D6. Worms had cVMS concentrations that varied between 0.43 to 8.60 ng g^{-1} ww for D4, between 4.01 to 553 ng g^{-1} ww for D5, and between 1.30 to 20.1 ng g^{-1} ww for D6. Jellyfish presented cVMS concentrations below LOD. In net plankton, concentrations varied from 0.61 to 2.81 ng g^{-1} ww for D4, from 10.10 to 368 ng g^{-1} ww for D5, and from 0.60 to 2.94 ng g^{-1} ww for D6. In sea urchins, concentrations ranged from 0.42 to 0.60 ng g^{-1} ww for D4, from 9.19 to 17.6 ng g^{-1} ww for D5, and from 7.12 to 14.3 ng g^{-1} ww for D6. The highest cVMS concentrations were found in the Inner Oslofjord, where anthropogenic pressures are greatest [91]. Regardless of the organism analyzed, D5 was the predominant cVMS.

In a recent study of 2018, Heimstad et al. [108] analyzed cVMS in earthworms (Lumbricidae), fieldfare eggs (*Turdus pilaris*), sparrowhawk eggs (*Accipiter nisus*), brown rat liver (*Rattus norvegicus*), tawny owl addled eggs (*Strix aluco*), red fox liver (*Vulpes vulpes*), and European badger liver (*Meles meles*) from the terrestrial urban environment of Oslo, in Norway. The highest concentrations of cVMS were found in brown rat liver, whose concentrations varied from 4.1 to 50.3 ng g^{-1} ww for D4, from <LOD to 42.0 ng g^{-1} ww for D5, and from <LOD to 22.1 ng g^{-1} for D6. In the remaining species, the levels of cVMS ranged between <LOD and 1.77 ng g^{-1} ww for D4 (the greatest value detected in red fox liver), between <LOD and 5.70 ng g^{-1} ww for D5 (the highest value detected in sparrowhawk eggs), and between <LOD and 4.01 ng g^{-1} for D6 (highest value in fieldfare eggs).

Finally, already in 2019, Zhi et al. [109] analyzed cVMS (D4-D6) and lVMS (L5–L16) in molluscs from culturing rafts in seven cities along the Bohai Sea, China. Species included mussel (*Mytilus galloprovincialis*), venus clam (*Cyclina sinensis*), and oyster (*Crassostrea talienwhanensis*). Concentrations of cVMS ranged from <LOD (1.8) to 47.6 ng g^{-1} ww for D4, from <LOD (2.7) to 77.3 ng g^{-1} ww for D5, and from <LOD (2.4) to 90.4 ng g^{-1} ww for D6. Linear VMS varied between <LOD (2.7) and 16.3 ng g^{-1} ww for L8, between <LOD (2.1) and 17.5 ng g^{-1} ww for L9, between <LOD (1.9) and 12.6 ng g^{-1} ww for L10, between <LOD (1.8) and 15.8 ng g^{-1} ww for L11, between <LOD (1.9) and 17.3 ng g^{-1} ww for L12, between <LOD (2.6) and 13.6 ng g^{-1} ww for L13, between <LOD (2.7) and 15.4 ng g^{-1} ww for L14, between <LOD (3.1) and 13.4 ng g^{-1} ww for L15, and between <LOD (3.0) and 14.7 ng g^{-1} ww for L16. L5–L7 were not found in any mollusc samples.

The studies described in this section clearly show the presence of VMS in biota (including fish, plankton, molluscs, crustacean, birds and seabirds, reptiles, mammals, vegetation) and a predominance of D5 in almost all samples, independently of their geographic origin or type of ecosystem (marine, freshwater, terrestrial). Despite the factors discussed in Sect. 2 related to the ability of organisms for clearance and elimination of VMS, still these compounds were measured in a variety of organisms, which reveals higher exposure rates in relation to clearance. Currently, the observed concentrations do not seem alarmingly high. However, the use of VMS is extensive and rising, and it is possible that the continued use will lead to increased environmental levels, eventually reaching concentrations with clear harmful effect.

5   Conclusions

Volatile methylsiloxanes have been detected in a variety of biota samples across the globe. Their concentrations were found higher in aquatic and terrestrial biota from densely populated areas, though some studies also detected VMS in biota from remote regions. Among the organisms studied so far, the highest levels of VMS were found in aquatic non-air-breathing animals (notably, fish) and the lowest in mammals, probably due to their higher elimination rate. Diet seems to be the main contamination route for fish, especially demersal fish that live in close association with sediments.

The research on the bioaccumulation of VMS has yielded apparently contradictory results, with high laboratory bioconcentration factors on one hand and low field trophic magnification on the other (with some studies supporting trophic magnification and other trophic dilution). The discrepancy of results may be due to different food webs, different species compositions, or different study areas.

However, some authors question the quality of the studies. The analytical determination of VMS in biotic tissues is complicated due to the high volatility, high hydrophobicity, and high background contamination of VMS. The high volatility can contribute to loss of material during sample preparation and extraction. The high hydrophobicity causes VMS to be absorbed by lipids in biota samples, making extraction difficult. The high background contamination from the sampling and extraction procedures, instrumental analysis, and laboratory equipment may mask the low concentrations of VMS typical of biotic tissue samples. Besides these methodological constraints, other uncertainty source is related to the lack of knowledge regarding where VMS are metabolized in different organisms and therefore which organs or tissues should be analyzed for a reliable comparison between organisms.

Acknowledgments The author acknowledges the support from FCT-MCTES (SFRH/BPD/109382/2015) and the project 032084 from 02/SAICT/2017 (LANSILOT).

References

1. Wang D-G, Norwood W, Alaee M et al (2013) Review of recent advances in research on the toxicity, detection, occurrence and fate of cyclic volatile methyl siloxanes in the environment. Chemospere 93:711–725
2. Hamelink JL (1992) Silicones. In: Hutzinger EO (ed) Anthropogenic compounds: detergents. The handbook of environmental chemistry, vol 3, part F. Springer, Berlin, pp 383–394

3. Hobson JF, Atkinson R, Carter WPL (1997) Volatile methylsiloxanes. In: Chandra G (ed) Organosilicon materials. The handbook of environmental chemistry, vol 3, part H. Springer, New York, pp 137–179

4. Mojsiewicz-Pienkowska K, Jamrógiewicz M, Szymkowska K et al (2016) Direct human contact with siloxanes (silicones) – safety or risk part 1. Characteristics of siloxanes (silicones). Front Pharmacol 7:132. https://doi.org/10.3389/fphar.2016.00132

5. Varaprath S, Stutts DH, Kozerski GE (2006) A primer on the analytical aspects of silicones at trace levels - challenges and artifacts - a review. Silicon Chem 3:79–102

6. Genualdi S, Harner T, Cheng Y et al (2013) Global distribution of linear and cyclic volatile methyl siloxanes in air. Environ Sci Technol 45:3349–3354

7. USEPA (2007) High production volume (HPV) challenge program. https://www.echemportal.org/echemportal/participant/participantinfo.action?participantId=9. Accessed 13 Sept 2018

8. McLachlan MS, Kierkegaard A, Hansen KM et al (2010) Concentrations and fate of decamethylcyclopentasiloxane (D5) in the atmosphere. Environ Sci Technol 44:5365–5370

9. Krogseth IS, Kierkegaard A, McLachlan MS et al (2012) Occurrence and seasonality of cyclic volatile methyl siloxanes in Arctic air. Environ Sci Technol 47:502–509

10. Watanabe N, Nakamura T, Watanabe E (1984) Distribution of organosiloxanes (silicones) in water, sediments and fish from the Nagara watershed, Japan. Sci Total Environ 35:91–97

11. Kaj L, Andersson J, Palm CA et al (2005) Results from the Swedish national screening programme 2004 - subreport 4: siloxanes. IVL report B 1643. http://www.imm.ki.se/Datavard/PDF/B1645_adipater.pdf. Accessed 13 Sept 2018

12. Kaj L, Schlabach M, Andersson J et al (2005) Siloxanes in the Nordic environment. TemaNord 593. Nordic Council of Ministers, Copenhagen. http://nordicscreening.org/index.php?module=Pagesetter&type=file&func=get&tid=5&fid=reportfile&pid=4. Accessed 13 Sept 2018

13. Hong W, Jia H, Liu C et al (2014) Distribution, source, fate and bioaccumulation of methyl siloxanes in marine environment. Environ Pollut 191:175–181

14. Sparham C, Egmond RV, Hastie C et al (2011) Determination of decamethylcyclopentasiloxane in river and estuarine sediments in the UK. J Chromatogr A 1218:817–823

15. Kierkegaard A, van Egmond R, McLachlan MS (2011) Cyclic volatile Methylsiloxane bioaccumulation in flounder and Ragworm in the Humber Estuary. Environ Sci Technol 45:5936–5942

16. Sanchís J, Martínez E, Ginebreda A et al (2013) Occurrence of linear and cyclic volatile methylsiloxanes in wastewater, surface water and sediments from Catalonia. Sci Total Environ 443:530–538

17. Borgå K, Fjeld E, Kierkegaard A et al (2012) Food web accumulation of cyclic siloxanes in Lake Mjøsa, Norway. Environ Sci Technol 46:6347–6354

18. Borgå K, Fjeld E, Kierkegaard A et al (2013) Consistency in trophic magnification factors of cyclic methyl siloxanes in pelagic freshwater food webs leading to brown trout. Environ Sci Technol 47:14394–14402

19. McGoldrick D, Chan C, Drouillard K et al (2014) Concentrations and trophic magnification of cyclic siloxanes in aquatic biota from the Western Basin of Lake Erie, Canada. Environ Pollut 186:141–148

20. Ratola N, Ramos S, Homem V et al (2016) Using air, soil and vegetation to assess the environmental behaviour of siloxanes. Environ Sci Pollut Res 23(4):3273–3284

21. Sanchís J, Cabrerizo A, Galbán-Malagón C et al (2015) Unexpected occurrence of volatile Dimethylsiloxanes in Antarctic soils, vegetation, phytoplankton, and krill. Environ Sci Technol 49:4415–4424

22. OECD (2007) Manual for investigation of HPV chemicals. http://www.oecd.org/document/7/0,3343,en_2649_34379_1947463_1_1_1_1,00.html. Accessed 13 Sept 2018

23. Environment Agency (2009) Environmental risk assessment report: decamethylcyclopentasiloxane. Environment Agency, Almondsbury

24. Environment Agency (2009) Environmental risk assessment report: dodecamethylcyclo-hexasiloxane. Environment Agency, Almondsbury
25. Environment Agency (2009) Environmental risk assessment report: octamethylcyclo-tetrasiloxane. Environment Agency, Almondsbury
26. EC & HC (2008) Screening assessment for the challenge: decamethylcyclopentasiloxane (D5) chemical abstracts service registry number 541-02-6. Technical report, Health Canada and Environment Canada, Ottawa
27. EC & HC (2008) Screening assessment for the challenge: dodecamethylcyclohexasiloxane (D6) chemical abstracts service registry number 540-97-6. Technical report, Health Canada and Environment Canada, Ottawa
28. EC & HC (2008) Screening assessment for the challenge: octamethylcyclotetrasiloxane (D4) chemical abstracts service registry number 556-67-2. Technical report. Health Canada and Environment Canada, Ottawa
29. Ministerråd N (2005) Siloxanes in the Nordic environment. Technical report 593, TemaNord, Copenhagen
30. Giesy JP, Solomon KR, Kacew S et al (2016) The case for establishing a board of review for resolving environmental issues: the science court in Canada. Integr Environ Assess Manag 12(3):572–579. https://doi.org/10.1002/ieam.1729
31. Siloxane D5 Board of Review (2011) Report of the Board of Review for deca-methylcyclopentasiloxane (D5), technical. Environment Canada, Ottawa
32. Environment Agency (2014) D4 PBT/vPvB evaluation, technical. Environment Agency, Bristol
33. Environment Agency (2014) D5 PBT/vPvB evaluation, technical. Environment Agency, Bristol
34. Hayden JF, Barlow SA (1972) Structure-activity relationships of organosiloxanes and the female reproductive system. Toxicol Appl Pharmacol 21:68–79
35. Quinn AL, Regan JM, Tobin JM et al (2007) In vitro and in vivo evaluation of the estrogenic, androgenic, and progestagenic potential of two cyclic siloxanes. Toxicol Sci 96:145–153
36. Granchi D, Cavedagna D, Ciapetti G et al (1995) Silicone breast implants: the role of immune-system on capsular contracture formation. J Biomed Mater Res 29:197–202
37. Hea B, Rhodes-Brower S, Miller MR et al (2003) Octamethylcyclotetrasiloxane exhibits estrogenic activity in mice via ER alpha. Toxicol Appl Pharmacol 192:254–261
38. Lieberman MW, Lykissa ED, Barrios R et al (1999) Cyclosiloxanes produce fatal liver and lung damage in mice. Environ Health Perspect 107:161–165
39. Lu Y, Yuan T, Wang W et al (2011) Concentrations and assessment of exposure to siloxanes and synthetic musks in personal care products from China. Environ Pollut 159(12):3522–3528
40. Xu L, Shi Y, Cai Y (2013) Occurrence and fate of volatile siloxanes in a municipal wastewater treatment plant of Beijing, China. Water Res 47:715–724
41. Schweigkofler M, Niessner R (1999) Determination of siloxanes and VOC in landfill gas and sewage gas by canister sampling and GCMS/AES analysis. Environ Sci Technol 33(20):3680–3685
42. Bletsou AA, Asimakopoulos AG, Stasinakis AS et al (2013) Mass loading and fate of linear and cyclic siloxanes in a wastewater treatment Plant in Greece. Environ Sci Technol 47(4):1824–1832
43. Capela D, Ratola N, Alves A (2017) Volatile methylsiloxanes through wastewater treatment plants – a review of levels and implications. Environ Int 102:9–29
44. Xu S, Kropscott B (2012) Method for simultaneous determination of partition coefficients for cyclic volatile methylsiloxanes and dimethylsilanediol. Anal Chem 84(4):1948–1955
45. Huckins JN, Petty JD, Thomas J (1997) Bioaccumulation: how chemicals move from the water into fish and other aquatic organisms. American Petroleum Institute publication number 4656, Midwest Science Center, Columbia
46. Bridges J, Solomon KR (2016) Quantitative weight-of-evidence analysis of the persistence, bioaccumulation, toxicity, and potential for long-range transport of the cyclic volatile methyl siloxanes. J Toxicol Environ Health B 19(8):345–379. https://doi.org/10.1080/10937404.2016.1200505

47. Xu S, Kropscott B (2013) Octanol/air partition coefficients of volatile methylsiloxanes and their temperature dependence. J Chem Eng Data 58:136–142. https://doi.org/10.1021/je301005b

48. Kelly BC, Ikonomou MG, Blair JD et al (2007) Food web-specific biomagnification of persistent organic pollutants. Science 317(5835):236–239

49. Andersen ME, Reddy MB, Plotzke KP (2008) Are highly lipophilic volatile compounds expected to bioaccumulate with repeated exposures? Toxicol Lett 179:85–92. https://doi.org/10.1016/j.toxlet.2008.04.007

50. Seston RM, Powell DE, Woodburn KB et al (2014) Importance of lipid analysis and implications for bioaccumulation metrics. Integr Environ Assess Manag 10:142–144. https://doi.org/10.1002/ieam.1495

51. Lucia M, Gabrielsen GW, Herzke D et al (2016) Screening of UV chemicals bisphenols and siloxanes in the Arctic. Norsk Polarinstitutt, Norwegian Polar Institute, Fram Centre, Tromso

52. Whelan MJ, Estrada E, van Egmond R (2004) A modelling assessment of the atmospheric fate of volatile methyl siloxanes and their reaction products. Chemosphere 57(10):1427–1437

53. Atkinson R (1991) Kinetics of the gas-phase reactions of a series of organosilicon compounds with OH and NO3 radicals and O3 at 297 ± 2 K. Environ Sci Technol 25:863–866. https://doi.org/10.1021/es00017a005

54. SEHSC (2007) Long range transport potential of cyclic methylsiloxanes estimated using a global average chemical fate model: the OECD tool, technical. Silicones Environment Health and Safety Council of North America, Herndon

55. Xu S, Wania F (2013) Chemical fate, latitudinal distribution and long-range transport of cyclic volatile methylsiloxanes in the global environment: a modeling assessment. Chemosphere 93:835–843. https://doi.org/10.1016/j.chemosphere.2012.10.056

56. Navea JG, Xu SH, Stanier CO et al (2009) Heterogeneous uptake of octamethylcyclotetrasiloxane (D4) and decamethylcyclopentasiloxane (D5) onto mineral dust aerosol under variable RH conditions. Atmos Environ 43:4060–4069

57. Durham J (2005) Hydrolysis of octamethylcyclotetrasiloxane (D4). Silicones environment, health and safety council. Study number 10000–102 (cited from the report of the assessment for D4 by Environment Canada and Health Canada)

58. Durham J (2006) Hydrolysis of octamethylcyclotetrasiloxane (D5) silicones environment, health and safety council. Study number 10040–102 (cited from the report of the assessment for D5 by Environment Canada and Health Canada)

59. Xu S (1999) Fate of cyclic methylsiloxanes in soils. 1: the degradation pathway. Environ Sci Technol 33:603–608

60. Xu S, Chandra G (1999) Fate of cyclic methylsiloxanes in soils. 2. Rates of degradation and volatilization. Environ Sci Technol 33:4034–4039

61. Xu S (2007) Estimation of degradation rates of cVMS in soils. HES study no. 10787–102. Health and environmental sciences, Dow Corning Corporation, Auburg, MI (cited from the report of the assessment for cVSM by Environment Agency of England and Wales)

62. Xu S, Miller JA (2008) Aerobic transformation of octamethylcyclotetrasiloxane (D4) in water/sediment system. Centre Européen des silicones (CES), interim report (cited from the report of the assessment for D4 by environment Canada and Health Canada)

63. Xu S (2010) Aerobic and anaerobic transformation of 14CDecamethylcyclopentasiloxane (14C-D5) in the aquatic sediment systems. Centre Européen des Silicones (CES). Dow Corning Corporation

64. Springer T (2007) Decamethylcyclopentasiloxane (D5): a 96-hour study of the elimination and metabolism of orally gavaged 14C-D5 in rainbow trout (Oncorhynchus mykiss). HES study number 10218–101. Centre Europeen des Silicones (CES), Brussels

65. Jovanovic ML, McNett DA, Regan JM et al (2003) Disposition of 14C-decamethylcyclopentasiloxane (D5), in Fischer 344 rats when delivered in various carriers following administration of a single oral dose. Report number 2003-I0000–52391. Dow Corning, Midland

66. Varaprath S, McMahon JM, Plotzke KP (2003) Metabolites of hexamethyldisiloxane and decamethylcyclopentasiloxane in Fischer 344 rat urine: a comparison of a linear and a cyclic siloxane. Drug Metabol Dispos 31:206–214
67. Burkhard LP, Borgå K, Powell DE et al (2013) Improving the quality and scientific understanding of trophic magnification factors (TMFs). Environ Sci Technol 47:1186–1187
68. Gobas FAPC, Powell DE, Woodburn KB et al (2015) Bioaccumulation of decamethylpentacyclosiloxane (D5): a review. Environ Toxicol Chem 34(12):2703–2714
69. Government of Canada (1999) Canadian Environmental Protection Act, Canada Gazette, Part III, vol 22. Public Works and Government Services, Ottawa
70. US Congress (1976) Toxic substances control act, Pub. L. No. 94-469, Washington
71. European Commission (2006) Regulation (EC) 1907/2006 of the European Parliament and of the council of 18 December 2006 concerning the registration, evaluation, authorisation and restriction of chemicals (REACH), establishing a European Chemicals Agency, amending directive 1999/45/EC and repealing council regulation (EEC) 793/93 and commission regulation (EC) no. 1488/94 as well as council directive 76/769/EEC and commission directives 91/155/EEC, 93/67/EEC, 93/105/EC and 2000/21/EC. Off J Eur Union L396:374–375
72. Opperhuizen A, Damen HWJ, Asyee GM (1987) Uptake and elimination by fish of polydimethylsiloxanes (silicones) after dietary and aqueous exposure. Toxicol Environ Chem 13:265–285
73. OECD (2008) SIAR for SIAM 26: hexamethylcyclotrisiloxane. Organisation for Economic Cooperation and Development, Paris
74. Annelin RB, Frye CL (1989) The piscine bioconcentration characteristics of cyclic and linear oligomeric permethylsiloxanes. Sci Total Environ 83:1–11
75. Fackler PH, Dionne E, Hartley DA et al (1995) Bioconcentration by fish of a highly volatile silicone compound in a totally enclosed aquatic exposure system. Environ Toxicol Chem 14:1649–1656
76. Parrott J, Alaee M, Wang D et al (2013) Fathead minnow (Pimephales promelas) egg-to-juvenile exposure to decamethylcyclopentasiloxane (D5). Chemosphere 93:813–818
77. Drottar KR (2005) 14C-Decamethylcyclopentasiloxane (14C-D5): bioconcentration in the fathead minnow (Pimephales promelas) under flow-through test conditions. Dow Corning Corporation, Silicones Environment, Health and Safety Council (SEHSC) (cited from the report of the assessment for D5 by Environment Canada and Health Canada)
78. Drottar KR (2005) 14C-Dodecamethylcyclohexasiloxane (14C–D6): bioconcentration in the fathead minnow (Pimephales promelas) under flow-through test conditions. Dow Corning Corporation, Silicones Environment, Health and Safety Council (SEHSC) (cited from the report of the assessment for D6 by Environment Canada and Health Canada)
79. OECD (2009) SIAR for SIAM 29: dodecamethylcyclohexasiloxane (D6). Organization for Economic Cooperation and Development, Paris. https://hpvchemicals.oecd.org/ui/Default. aspx. Accessed 13 Sept 2018
80. Kent DJ, McNamara PC, Putt AE et al (1994) Octamethylcyclotetrasiloxane in aquatic sediments: toxicity and risk assessment. Ecotoxicol Environ Saf 29:372–389
81. Springborn Smithers Laboratories (2003) Decamethylcyclopentasiloxane (D5)-the full life-cycle toxicity to midge (Chironomus riparius) under static conditions. Silicones environmental, health and safety council (SEHSC) (cited from the report of the assessment for D5 by Environment Canada and Health Canada)
82. Norwood WP, Alaee M, Sverko E et al (2013) Decamethylcyclopentasiloxane (D5) spiked sediment: bioaccumulation and toxicity to the benthic invertebrate Hyalella azteca. Chemosphere 93:805–812
83. Warner NA, Evenset A, Christensen G et al (2010) Volatile siloxanes in the European Arctic: assessment of sources and spatial distribution. Environ Sci Technol 44:7705–7710
84. Krueger HO, Thomas ST, Kendall TZ (2008) D5: a bioaccumulation test with Lumbriculus variegatus using spiked sediment. Wildlife International, Easton

85. Powell DE, Woodburn KB, Drotar KD et al (2009) Trophic dilution of cyclic volatile Methylsiloxane (cVMS) materials in a temperate freshwater Lake. Internal report conducted for Centre Européen des Sililones. Report Number 2009-I0000-60988

86. Woodburn KB, Seston RM, Kim J et al (2018) Benthic invertebrate exposure and chronic toxicity risk analysis for cyclic volatile methylsiloxanes: comparison of hazard quotient and probabilistic risk assessment approaches. Chemosphere 192:337–347

87. Canada (1999) Canadian Environmental Protection Act. http://laws-lois.justice.gc.ca/PDF/C-15.31.pdf. Accessed 13 Sept 2018

88. European Commission (2010) Amending regulation (EC) No 1907/2006 of the European Parliament and of the Council on the Registration, Evaluation, Authorisation and Restriction of Chemicals (REACH) as Regards Annex XIII. http://register.consilium.europa.eu/pdf/en/10/st14/st14860.en10.pdf. Accessed 13 Sept 2018

89. McLachlan MS, Czub G, MacLeod M et al (2011) Bioaccumulation of organic contaminants in humans: a multimedia perspective and the importance of biotransformation. Environ Sci Technol 45:197–202. https://doi.org/10.1021/es101000w

90. Woodburn K, Drottar K, Domoradzki JY et al (2013) Determination of the dietary biomagnification of octamethylcyclotetrasiloxane and decamethylcyclopentasiloxane with the rainbow trout (Oncorhynchus mykiss). Chemosphere 93(5):779–788

91. Powell DE, Suganuma N, Kobayashi K et al (2017) Trophic dilution of cyclic volatile methylsiloxanes (cVMS) in the pelagic marine food web of Tokyo Bay, Japan. Sci Total Environ 578:366–382

92. Powell DE, Schøyen M, Øxnevad S et al (2018) Bioaccumulation and trophic transfer of cyclic volatile methylsiloxanes (cVMS) in the aquatic marine food webs of the Oslofjord, Norway. Sci Total Environ 622–623:127–139

93. Jia H, Zhang Z, Wang C et al (2015) Trophic transfer of methyl siloxanes in the marine food web from coastal area of northern China. Environ Sci Technol 49:2833–2840

94. Powell DE, Durham JA, Huff DW et al (2010) Bioaccumulation and trophic transfer of cyclic volatile Methylsiloxanes (cVMS) materials in the aquatic marine food webs of inner and outer Oslo fjord, Norway. Dow Corning Corporation, Midland. https://www.regulations.gov/contentStreamer?documentId=EPA-HQ-OPPT-2011-0516-0025&attachmentNumber=32&disposition=attachment&contentType=pdf. Accessed 13 Sept 2018

95. Borgå K, Kidd KA, Muir DCG et al (2012) Trophic magnification factors: considerations of ecology, ecosystems, and study design. Integr Environ Assess Manag 8:64–84

96. Law K, Halldorson T, Danell R et al (2006) Bioaccumulation and trophic transfer of some brominated flame retardants in a Lake Winnipeg (Canada) food web. Environ Toxicol Chem 25:2177–2186

97. Muir DCG, Whittle MD, Vault DS et al (2004) Bioaccumulation of toxaphene congeners in the Lake Superior food web. J Great Lakes Res 30:316–340

98. Houde M, Muir DCG, Kidd KA et al (2008) Influence of lake characteristics on the biomagnification of persistent organic pollutants in lake trout food webs. Environ Toxicol Chem 27:2169–2178

99. Kim J, Gobas FAPC, Arnot JA et al (2016) Evaluating the roles of biotransformation, spatial concentration differences, organism home range, and field sampling design on trophic magnification factors. Sci Total Environ 551–552:438–451

100. McLeod AM, Arnot JA, Borgå K et al (2015) Quantifying uncertainty in the trophic magnification factor related to spatial movements of organisms in a food web. Integr Environ Assess Manag 11:306–318

101. Kierkegaard A, Adolfsson-Erici M, McLachlan MS (2010) Determination of cycle volatile methylsiloxanes in biota with a purge and trap method. Anal Chem 82:9573–9578

102. Kierkegaard A, Bignert A, McLachlan MS (2013) Bioaccumulation of decamethyl-cyclopentasiloxane in perch in Swedish lakes. Chemosphere 93(5):789–793

103. Kierkegaard A, Bignert A, McLachlan MS (2013) Cycle volatile methylsiloxanes in fish from the Baltic Sea. Chemosphere 93(5):774–778

104. Huber S, Warner NA, Nygard T et al (2015) A broad cocktail of environmental pollutants found in eggs of three seabird species from remote colonies in Norway. Environ Toxicol Chem 34(6):1296–1308
105. Mackay D, Gobas F, Solomon K et al (2015) Comment on "unexpected occurrence of volatile Dimethylsiloxanes in Antarctic soils, vegetation, phytoplankton, and krill". Environ Sci Technol 49(12):7507–7509
106. Sanchís J, Cabrerizo A, Galbán-Malagón C et al (2015) Response to comments on "unexpected occurrence of volatile Dimethylsiloxanes in Antarctic soils, vegetation, phytoplankton, and krill". Environ Sci Technol 49(12):7510–7512
107. Wang D, de Solla SR, Lebeuf M et al (2017) Determination of linear and cycle volatile methylsiloxanes in blood of turtles, cormorants, and seals from Canada. Sci Total Environ 574:1254–1260
108. Heimstad ES, Nygard T, Herzke D et al (2018) Environmental pollutants in the terrestrial and urban environment 2017. Report M-1076/2018, Norwegian Institute for Air Research, 234 pp
109. Zhi L, Xu L, He X et al (2019) Distribution of methylsiloxanes in benthic mollusks from the Chinese Bohai Sea. J Environ Sci 76:199–207

Volatile Methyl Siloxanes in Polar Regions

Ingjerd S. Krogseth and Nicholas A. Warner

Contents

Abstract This chapter reviews volatile methyl siloxanes (VMS) in polar regions (i.e., at latitudes above the polar circles), including their sources, measured concentrations, and the effect of polar environmental conditions on behavior of VMS. Knowledge about VMS in polar regions has been centered on cyclic VMS (cVMS) due to their widespread use and presence in the environment. Due to their high volatility, cVMS are mainly emitted to and remain in the atmosphere, where they eventually degrade. cVMS are present in Arctic air due to both long-range atmospheric transport and local sources within the Arctic. There is no evidence that cVMS deposit to surface media to a significant extent, not even under polar environmental conditions. However, cVMS are emitted via wastewater, and many Arctic communities have limited wastewater treatment where low removal efficiency of cVMS from wastewater can result in high emissions. cVMS concentrations in sediments and aquatic biota close to wastewater outlets in the Norwegian Arctic are comparable to those at temperate latitudes. Sporadic detections of cVMS in biota and surface media in remote Arctic and Antarctic regions need further investigation to be confirmed. Very few measurements are reported for linear VMS (lVMS) in polar regions, with a majority of studies reporting findings below detection limits.

I. S. Krogseth (✉) and N. A. Warner
NILU – Norwegian Institute for Air Research, Fram Centre, Tromsø, Norway
e-mail: isk@nilu.no; nw@nilu.no

V. Homem and N. Ratola (eds.), *Volatile Methylsiloxanes in the Environment*,
Hdb Env Chem (2020) 89: 279–314, DOI 10.1007/698_2019_388,
© Springer Nature Switzerland AG 2019, Published online: 7 September 2019

The understanding of how Arctic conditions, including low temperatures and strong seasonality, affect the environmental behavior and bioaccumulation of VMS has been expanded through modelling studies. However, important knowledge gaps remain regarding temperature dependence of partitioning behavior in aquatic environments, biotransformation rates in polar biota, and the influence of physiological and behavioral adaptations of polar biota on bioaccumulation of VMS.

Keywords Arctic, Environmental behavior, Measurements, Modelling, Polar, Volatile methyl siloxanes

1 Sources of Volatile Methyl Siloxanes to Polar Regions

1.1 Long-Range Environmental Transport

Due to the high volatility of VMS, the majority of VMS emissions to the environment are to the atmosphere [1–3]. In urban and rural regions, atmospheric concentrations of the cVMS normally dominate over the lVMS, assumedly due to higher emissions [4–6]. Once in the atmosphere, VMS can undergo degradation by hydroxyl radicals to water-soluble silanols, which are removed from the atmosphere through wet deposition [7, 8]. The atmospheric half-lives ($\tau_{1/2}$) of VMS due to reaction with hydroxyl radicals range from 2 to 20 days based on both experimental and modelled estimates [7, 9–11]. This meets the long-range atmospheric transport (LRAT) criteria set out under the Stockholm Convention ($\tau_{1/2} > 2$ days) for chemicals displaying sufficient atmospheric persistence to reach to remote areas, including the Arctic [12]. The first measurements of VMS in Arctic air were as part of the Global Atmospheric Passive Air Sampling (GAPS) network in 2009 [4]. Here, the cVMS hexamethylcyclotrisiloxane (D3), octamethylcyclotetrasiloxane (D4), decamethylcyclopentasiloxane (D5), and dodecamethylcyclohexasiloxane (D6) and the lVMS octamethyltrisiloxane (L3), decamethyltetrasiloxane (L4), and dodecamethylpentasiloxane (L5) were measured at two remote Arctic locations in North America (Alert, Canada) and Europe (the Zeppelin Observatory, Ny-Ålesund, Svalbard). All cyclic VMS (cVMS) were detected, whereas linear VMS (lVMS) were below detection limits. These measurements confirmed that cVMS undergo LRAT to the Arctic, supporting model predictions (Sect. 3.1.1) [4]. At the Arctic sites, measured concentrations of D3 and D4 were higher than concentrations of D5 and D6, in contrast to the urban sites in the GAPS study [4] and elsewhere [5, 13–15] where D5 normally dominates. However, D3 and D4 are both more volatile [16] and have longer atmospheric half-lives [7] than D5 and D6, making them more prone to undergo LRAT.

In 2011, a more detailed study of LRAT of cVMS to the Arctic was carried out at the Zeppelin Observatory. Concentrations of D3, D4, D5, and D6 in air were measured with a newly developed active air sampling method [14] allowing for a temporal resolution down to 24 h [17]. Here, D5 was the dominant congener, with a clear seasonal pattern displaying higher concentrations in winter than in summer. This finding is in agreement with model predictions [17, 18] and is attributed to reduced atmospheric degradation due to lower atmospheric concentration of hydroxyl radicals during the dark polar night in winter. However, some uncertainty is associated with these results due to sampling artifacts caused by degradation and formation of cVMS on the air sampling sorbent used (hydroxyl-substituted polystyrene-divinylbenzene (ENV+)) [5, 17]. cVMS have been regularly monitored at the Zeppelin Observatory since 2013 using the same method in which correction for storage degradation has been applied [19–23]. All measured concentrations of VMS in Arctic air are more thoroughly discussed in Sect. 2.1.1.

Despite continuous input of cVMS to Arctic air through LRAT, deposition to Arctic surface media is unlikely due to their high volatility. This has been corroborated by several modelling studies (Sect. 3.1.1), and concentrations of cVMS in remote Arctic areas (except in air) are very low and often under detection limits (Sect. 2). In 2015, occurrence of VMS in Antarctic soil, vegetation, phytoplankton, and krill was reported and claimed to be present due to LRAT followed by deposition and water runoff to marine waters [24]. However, no concentrations were measured in air or snow to support this hypothesis, and the data quality of the study was questioned [25–27].

Long-range transport of VMS to polar regions via oceanic currents or riverine input has not been investigated, but is considered unlikely due to the low water solubility, high volatility, and high hydrophobicity of VMS as well as their propensity to undergo hydrolysis in water. Long-range transport of VMS to polar regions with biota has also not been investigated, but cannot be ruled out. In fact, some detection of VMS in Arctic birds (Sect. 2.2.3) may be explained by the migratory behavior of these species, where VMS exposure may have occurred in temperate regions prior to migrating to the Arctic.

1.2 *Local Sources*

In addition to undergoing LRAT, VMS are present in the Arctic environment due to local emissions from settlements within the Arctic. VMS are used in a range of industrial and consumer products, and considerable environmental emissions result from the use of personal care products, either by volatilizing to air or being emitted via wastewater effluent [1–3]. This means that use of VMS-containing products within the polar regions can be a significant local source of VMS to both air and water.

VMS can be efficiently removed during wastewater treatment, with removal efficiencies of 90% or greater [28]. However, small Arctic communities often have very limited or no wastewater treatment. Hence, resultant per capita emissions in Arctic communities can be higher than in more populated areas. The impact of local cVMS emissions through wastewater in Arctic Norwegian communities has been investigated in Tromsø (70°N, population (pop.) 70,000), Hammerfest (71°N, pop.

7,000), Longyearbyen (78°N, pop. 2000), and Ny-Ålesund (79°N, pop. 40–150). Of these, Tromsø has mechanical treatment of the wastewater, while the other communities discharge the untreated wastewater directly to receiving waters. In Tromsø, both D4, D5, and D6 have been detected in wastewater, and D5 dominated with concentrations up to 20,000 ng/L (Table 1) [29, 30]. In Hammerfest, concentrations in wastewater were lower (D5 concentrations up to 1,415 ng/L), but all VMS were still detected (Table 1) [31]. In Longyearbyen, VMS concentrations in sediment decreased with increasing distance to the wastewater outlet in the fjord, while VMS concentrations in sediment were below detection limits outside Ny-Ålesund [32, 33]. These results clearly illustrate how population size impacts local emissions of cVMS in Arctic communities. The resultant concentrations in the receiving environments (i.e., sediment and biota) of Tromsøysund (Tromsø), Lake Storvannet and Hammerfest harbor (Hammerfest), Adventfjorden (Longyearbyen), and Kongsfjorden (Ny-Ålesund) are discussed in Sect. 2. Wastewater from Arctic communities has not yet been analyzed for lVMS. Wastewater from Tromsø in 2017 were screened for three siloxane compounds of varying substitution: trimethyl-tris(trifluoropropyl)-cyclotrisiloxane (D3F), tetraethenyl-tetramethyl-cyclotetrasiloxane (D4Vn), and heptamethylphenylcyclotetrasiloxane (D4Ph), but they were not found above detection limits [30].

Local emissions in Antarctica will be much lower than in the Arctic due to a smaller human presence. However, anthropogenic activities have been shown to be exposure sources for various chemicals (i.e., heavy metals, polycyclic aromatic hydrocarbons (PAHs), polychlorinated biphenyls (PCBs), and polybrominated diphenyl ethers (PBDEs)) in Antarctica [34] and may act as local sources of VMS as well.

2 Concentrations of Volatile Methylsiloxanes in Polar Regions

Environmental concentrations of VMS in the Arctic have recently been reviewed in an assessment report from the Arctic Monitoring and Assessment Program (AMAP) [29]. This section builds and expands upon that report, with an overview of measured concentrations in Tables 1–4.

2.1 The Physical Environment

2.1.1 Atmosphere

Some of the key measurements of VMS in Arctic air have already been discussed (Sect. 1.1). The two most substantial datasets are the measured concentrations in air at Alert and Zeppelin as part of the GAPS study (2009, 2013, 2015) [4, 35] and the

Table 1 Measured concentrations of cVMS in the physical environment of polar regions

Matrix	Location	U/R	Year	n	Unit	D3	D4	D5	D6	Reference
Air	Alert	R	2009	1	ng/m³	10	12	0.6	0.3	Genualdi et al. [4]
Air	Alert	R	2013, 2015	4	ng/m³	0.5–13	1.6–72	2.0–26	nd–39	Rauert et al. [35]
Air	Zeppelin	R	2009	1	ng/m³	17	16	4.0	0.5	Genualdi et al. [4]
Air	Zeppelin	R	2013, 2015	3	ng/m³	4.5–9.4	18–67	6.4–25	1.3–3.8	Rauert et al. [35]
Air	Zeppelin	R	2011	24	ng/m³	nd–3.0	nd–2.1	0.2–3.6	0.1–0.8	Krogseth et al. [17]
Air	Zeppelin	R	2013–2016	46	ng/m³		nd–5.7	0.6–7.3	0.1–2.6	Bohlin-Nizzetto et al. [19–22]
Air	Zeppelin	R	2017	52	ng/m³			0.02–4.5	0.003–0.2	Bohlin-Nizzetto et al. [23]
Air	Tromsø urban sites	U	2017	3	ng/m³		7–20	26–103	4–13	Schlabach et al. [30]
Air	Tromsø remote site	R	2017	1	ng/m³		15	2	nd	Schlabach et al. [30]
Soil	Antarctica	R	2009	11	ng/g dw	nd–25	nd–24	nd–110	nd–42	Sanchis et al. [24]
Vegetation	Antarctica	R	2009	17	ng/g dw	nd–5.7	nd–21	nd–55	0.9–88	Sanchis et al. [24]
Wastewater	Tromsø	U	2014	9	ng/L		100–330	1,000–20,000	180–900	Warner et al. [29]
Wastewater	Tromsø	U	2017	6	ng/L		128–408	165–1998	97–1,240	Schlabach et al. [30]
Wastewater	Hammerfest	U	2014	8	ng/L		nd–52	23–1,415	nd–100	Krogseth et al. [31]
River water	Hammerfest	U	2014	6	ng/L		nd	nd	nd	Krogseth et al. [31]
Lake water	Storvannet (Hammerfest)	U	2014	8	ng/L		nd	nd	nd	Krogseth et al. [31]
Marine water	Tromsøysund (Tromsø)	U	2014	4	ng/L		nd	nd	nd	Warner et al. unpublished
Marine sediment	Tromsøysund (Tromsø)	U	2010–2011	6	ng/g ww		nd	3.5–12	nd–4.5	Warner et al. [39]

(continued)

Table 1 (continued)

Matrix	Location	U/R	Year	n	Unit	D3	D4	D5	D6	Reference
Marine sediment	Nipøya (Tromsø)	S	2010–2011	–	ng/g ww		nd	nd	nd	Warner et al. [39]
Marine sediment	Tromsøysund (Tromsø)	U	2014	4	ng/g dw		nd	20–60	4.5–11	Warner et al. unpublished
Freshwater sediment	Storvannet (Hammerfest)	U	2014	8	ng/g dw		4–16	73–328	17–75	Krogseth et al. [31]
Marine sediment	Hammerfest harbor	U	2014	1	ng/g dw		10	351	32	Krogseth et al. unpublished
Marine sediment	Adventfjorden (Longyearbyen)	U	2009	5	ng/g dw		nd	0.7–2.1	nd	Warner et al. [32]
Marine sediment	Kongsfjorden (Ny-Ålesund)	S	2009	5	ng/g dw		nd	nd	nd	Warner et al. [32]
Marine sediment	Kongsfjorden, Liefdefjorden	S	2008	4	ng/g dw		nd	nd	nd	Evenset et al. [42]
Marine sediment	Barents Sea	R	2006–2007	11	ng/g dw		nd–40	nd–13	nd	Bakke et al. [43]

U urban (local point source), *S* semi-urban, *R* remote (no local source). *ww* wet weight, *dw* dry weight, *nd* not detected

Table 2 Measured concentrations of IVMS in the physical environment and biota in polar regions

Matrix	Tissue	Location	U/R	Year	n	Unit	L2	L3	L4	L5	L6	Reference
Air		Alert	R	2009	1	ng/m³		nd	nd	nd		Genualdi et al. [4]
Air		Alert	R	2013, 2015	4	ng/m³		nd	nd	nd		Rauert et al. [35]
Air		Zeppelin	R	2009	1	ng/m³		nd	nd	nd		Genualdi et al. [4]
Air		Zeppelin	R	2013, 2015	3	ng/m³		nd	nd	nd–0.04		Rauert et al. [35]
Soil		Antarctica	R	2009	11	ng/g dw		0.007–0.573	nd–0.602	nd–0.606	nd–0.313	Sanchís et al. [24]
Vegetation		Antarctica	R	2009	17	ng/g dw		nd	nd	nd	nd	Sanchís et al. [24]
Phytoplankton	WB	Antarctica	R	2009	11	ng/g dw		0.006–0.088	nd–0.017	nd–0.015	nd–0.12	Sanchís et al. [24]
Krill (*Euphausia superba*)	WB	Antarctica	R	2009	11	ng/g dw		nd–0.082	nd	nd	nd	Sanchís et al. [24]
Polar cod (*Boreogadus saida*)	L	Liefdefjorden	R	2008	2	ng/g ww	nd	nd	nd	nd		Evenset et al. [42]
Polar cod (*B. saida*)	L	Billefjorden	R	2008	4	ng/g ww	nd	nd–0.17	nd	nd		Evenset et al. [42]
Polar cod (*B. saida*)	L	Moffen	R	2008	5	ng/g ww	nd	nd	nd	nd		Evenset et al. [42]
Atlantic cod (*Gadus morhua*)	L	Kongsfjorden (Ny-Ålesund)	S	2008	5	ng/g ww	nd	nd–0.33	nd	nd		Evenset et al. [42]
Kittiwake (*Rissa tridactyla*)	L	Kongsfjorden (Ny-Ålesund)	S	2008	5	ng/g ww	nd	nd	nd	nd		Evenset et al. [42]
Kittiwake (*R. tridactyla*)	L	Liefdefjorden	R	2008	4	ng/g ww	nd	nd	nd	nd		Evenset et al. [42]
Common eider (*Somateria mollissima*)	L	Kongsfjorden (Ny-Ålesund)	S	2008	5	ng/g ww	nd	nd	nd	nd		Evenset et al. [42]

U urban (local point source), *S* semi-urban, *R* remote (no local source). *WB* whole body, *L* liver. *ww* wet weight, *dw* dry weight, *nd* not detected

Table 3 Measured concentrations of cVMS in aquatic biota of polar regions

Species	Tissue	Location	U/R	Year	n	Unit	D3	D4	D5	D6	Reference
Freshwater											
Pea clams (*Pisidium* sp.)	WB	Storvannet (Hammerfest)	U	2014	2	ng/g ww		4.7 ± 0.4	107 ± 4.5	12 ± 1.2	Krogseth et al. [44]
Chironomid larvae (Chironomidae sp.)	WB	Storvannet (Hammerfest)	U	2014	2	ng/g ww		9.9 ± 0.3	60 ± 1.2	9.3 ± 0.1	Krogseth et al. [44]
Three-spined stickleback (*Gasterosteus aculeatus*)	WB	Storvannet (Hammerfest)	U	2014	5	ng/g lw		101–687	681–6,942	63–293	Krogseth et al. [44]
Three-spined stickleback (*G. aculeatus*)	WB	Storvannet (Hammerfest)	U	2015	5	ng/g lw		89–129	623–1,028	71–332	Warner et al. unpublished
Brown trout (*Salmo trutta*) (stationary)	M	Storvannet (Hammerfest)	U	2014	13	ng/g lw		nd	383–2,434	nd	Krogseth et al. [44]
Brown trout (*S. trutta*) (stationary)	L	Storvannet (Hammerfest)	U	2014	13	ng/g lw		nd	560–3,647	nd–464	Krogseth et al. [44]
Brown trout (*S. trutta*) (stationary)	M	Storvannet (Hammerfest)	U	2015	5	ng/g lw		nd	680–1,394	239–309	Warner et al. unpublished
Brown trout (*S. trutta*) (anadromous)	M	Storvannet (Hammerfest)	U	2014	2	ng/g lw		nd	332–757	nd	Krogseth et al. unpublished
Brown trout (*S. trutta*) (anadromous)	L	Storvannet (Hammerfest)	U	2014	2	ng/g lw		nd	nd–1831	nd–460	Krogseth et al. unpublished
Arctic char (*Salvelinus alpinus*) (stationary)	M	Storvannet (Hammerfest)	U	2014	11	ng/g lw		nd–1,277	757–8,264	nd–809	Krogseth et al. [44]
Arctic char (*S. alpinus*) (stationary)	L	Storvannet (Hammerfest)	U	2014	10	ng/g lw		nd–2,586	986–24,375	182–952	Krogseth et al. [44]
Arctic char (*S. alpinus*) (stationary)	M	Storvannet (Hammerfest)	U	2015	5	ng/g lw		273–624	1,616–8,439	164–3,165	Warner et al. unpublished
Arctic char (*S. alpinus*) (anadromous)	M	Storvannet (Hammerfest)	U	2014	4	ng/g lw		nd–321	520–2,238	nd	Krogseth et al. unpublished

					n						
Arctic char (*S. alpinus*) (anadromous)	L	Storvannet (Hammerfest)	U	2014	4	ng/g lw		nd–519	710–3,168	nd–238	Krogseth et al. unpublished
Arctic char (*S. alpinus*) (stationary)	M	Lake Ellasjøen (Bear Island)	R	2015	5	ng/g ww		nd	nd	nd	Lucia et al. [54]
Arctic char (*S. alpinus*) (stationary)	M	Erlingvatnet (Svalbard)	R	2015	5	ng/g ww		nd	nd	nd	Lucia et al. [54]
Marine											
Phytoplankton	WB	Antarctica	R	2009	11	ng/g dw	nd–10	0.3–3.5	0.3–27	0.1–8.8	Sanchís et al. [24]
Krill (*Euphausia superba*)	WB	Antarctica	R	2009	11	ng/g dw	4.5–154	12–117	21–63	12–73	Sanchís et al. [24]
Zooplankton	WB	Kongsfjorden (Ny-Ålesund)	S	2009	3	–		nd	nd	nd	Warner et al. [32]
Zooplankton	WB	Liefdefjorden	R	2009	3	–		nd	nd	nd	Warner et al. [32]
Polar cod (*Boreogadus saida*)	L	Liefdefjorden	R	2008	2	ng/g ww	5.8–10	6.3–9.2	18.6–19.1	10.7	Evenset et al. [42]
Polar cod (*B. saida*)	L	Billefjorden	R	2008	4	ng/g ww	nd	nd–3.9	6.9–12	nd	Evenset et al. [42]
Polar cod (*B. saida*)	L	Moffen	R	2008	5	ng/g ww	3.6–9.9	3.6–7.8	nd–5.1	2.2–3.8	Evenset et al. [42]
Atlantic cod (*Gadus morhua*)	L	Tromsøysund (Tromsø)	U	2010–2011	12	ng/g lw		16–111	338–2,530	29–139	Warner et al. [39]
Atlantic cod (*G. morhua*)	L	Nipøya (Tromsø)	S	2010–2011	8	ng/g lw		5.6–15	30–1,260	4.6–146	Warner et al. [39]
Atlantic cod (*G. morhua*)	L	Tromsøysund (Tromsø)	U	2017	14	ng/g ww		nd–60	55–718	13–149	Green et al. [58]
Atlantic cod (*G. morhua*)	M	Tromsøysund (Tromsø)	U	2014	10	ng/g lw		41–313	72–903	nd–187	Warner et al. unpublished

(continued)

Table 3 (continued)

Species	Tissue	Location	U/R	Year	n	Unit	D3	D4	D5	D6	Reference
Atlantic cod (*G. morhua*)	M	Hammerfest harbor	U	2014	5	ng/g lw		nd–3,114	857–4,033	nd–1,089	Krogseth et al. unpublished
Atlantic cod (*G. morhua*)	L	Hammerfest harbor	U	2014	5	ng/g lw		69–5,821	467–14,530	24–2,970	Krogseth et al. unpublished
Atlantic cod (*G. morhua*)	L	Isfjorden	S	2017	12	ng/g ww		nd–27.4	6.9–23	4.9–18	Green et al. [58]
Atlantic cod (*G. morhua*)	L	Adventfjorden (Longyearbyen)	U	2009	5	ng/g lw		nd	45–358	5.3–31	Warner et al. [32]
Atlantic cod (*G. morhua*)	L	Kongsfjorden (Ny-Ålesund)	S	2009	5	ng/g lw		nd	13–29	11–53	Warner et al. [32]
Atlantic cod (*G. morhua*)	L	Kongsfjorden (Ny-Ålesund)	S	2008	5	ng/g ww	4.3–7.1	2.9–3.9	2.7–4.6	nd	Evenset et al. [42]
Shorthorn sculpin (*Myoxocephalus scorpius*)	L	Adventfjorden (Longyearbyen)	U	2009	5	ng/g lw		nd	54–2,150	nd–31	Warner et al. [32]
Shorthorn sculpin (*M. scorpius*)	L	Liefdefjorden	R	2009	5	ng/g lw		nd	nd–11	nd–10	Warner et al. [32]

U urban (local point source), *S* semi-urban, *R* remote (no local source). *WB* whole body, *L* liver, *M* muscle. *lw* lipid weight, *ww* wet weight, *dw* dry weight, *nd* not detected

Table 4 Measured concentrations of cVMS in air-breathing biota of polar regions

Species	Tissue	Location	U/R	Year	n	Unit	D3	D4	D5	D6	Reference
Kittiwake (*Rissa tridactyla*)	L	Kongsfjorden (Ny-Ålesund)	S	2008	5	ng/g ww	nd–3.8	nd–3.5	nd–1.5	nd	Evenset et al. [42]
Kittiwake (*R. tridactyla*)	L	Liefdefjorden	R	2008	4	ng/g ww	nd	nd	nd	nd	Evenset et al. [42]
Common eider (*Somateria mollissima*)	L	Kongsfjorden (Ny-Ålesund)	S	2008	5	ng/g ww	nd	nd	nd	nd	Evenset et al. [42]
Glaucous gull (*Larus hyperboreus*)	L	Bear Island, Norway	R	2003–2005	10	ng/g ww			32–69	nd	Knudsen et al. [59]
Kittiwake (*R. tridactyla*)	E	Kongsfjorden (Ny-Ålesund)	S	2013	5	ng/g ww		nd	nd	nd	Lucia et al. [54]
Glaucous gull (*L. hyperboreus*)	E	Kongsfjorden (Ny-Ålesund)	S	2014	5	ng/g ww		nd–5.8	nd–40	nd	Lucia et al. [54]
Common eider (*S. mollissima*)	E	Røst, Norway	S	2012	4	ng/g ww		nd	nd	nd	Huber et al. [60]
European shag (*Phalacrocorax aristotelis*)	E	Røst, Norway	S	2012	4	ng/g ww		nd	nd–1.4	nd	Huber et al. [60]
Herring gull (*Larus argentatus*)	E	Røst, Norway	S	2012	4	ng/g ww		nd	nd–1.5	nd	Huber et al. [60]
Common eider (*S. mollissima*)	E	Kongsfjorden (Ny-Ålesund)	S	2017	5	ng/g		nd	nd	nd	Schlabach et al. [30]
European shag (*P. aristotelis*)	E	Røst, Norway	S	2017	5	ng/g		nd	nd	nd	Schlabach et al. [30]
Glaucous gull (*L. hyperboreus*)	E	Kongsfjorden (Ny-Ålesund)	S	2017	5	ng/g		nd	nd	nd	Schlabach et al. [30]
Kittiwake (*R. tridactyla*)	E	Kongsfjorden (Ny-Ålesund)	S	2017	5	ng/g		nd	nd	nd	Schlabach et al. [30]

(continued)

Table 4 (continued)

Species	Tissue	Location	U/R	Year	n	Unit	D3	D4	D5	D6	Reference
Common gull (*Larus canus*)	E	Grindøya, Tromsø	U	2017	5	ng/g		nd	nd	nd	Schlabach et al. [30]
Bearded seals (*Erignathus barbatus*)	B	Kongsfjorden (Ny-Ålesund)	S	2009	5	ng/g lw		nd	nd	0.8–1.1	Warner et al. [32]
Mink (*Neovison vison*)	L	Sommarøy (Tromsø)	S	2013–2014	5	ng/g		nd	nd	nd	Schlabach et al. [30]
Polar bear (*Ursus maritimus*)	P	Svalbard	R	2017	10	ng/g		nd	nd	nd	Schlabach et al. [30]
Humans (*Homo sapiens*)	P	Northern Norway (MISA)	U	2009	17	ng/mL		nd–2.7	nd	nd	Hanssen et al. [66]
Humans (*H. sapiens*)	P	Norway (NOWAC)	U	2005	94	ng/mL		nd–13	nd–3.9	nd–3.2	Hanssen et al. [66]

U urban (local point source), *S* semi-urban, *R* remote (no local source). *L* liver, *E* egg, *P* plasma, *B* blubber. *lw* lipid weight, *ww* wet weight, *nd* not detected annual monitoring of cVMS at Zeppelin carried out by NILU – Norwegian Institute for Air Research (2011, 2013–2017) [17, 19–22]. These two datasets are not directly comparable as concentrations are measured with two different techniques. In GAPS, sorbent-impregnated polyurethane foam passive air samplers (SIP-PAS) are used to measure semiquantitative time-integrated concentrations over a 3-month period [4, 35]. In the annual monitoring at Zeppelin, ENV+ solid-phase active air samplers (SPE-AAS) were used to collect samples in a summer campaign and a winter campaign with a temporal resolution of 1–5 days [22]. From 2017, the monitoring frequency has been increased to weekly sampling of cVMS at Zeppelin [23]. There is some uncertainty in these results due to sampling artifacts caused by degradation and formation of cVMS on the ENV+ sorbent during sampling and storage [5, 17]. Comparison of measured cVMS concentrations in air at Zeppelin obtained with the two different methods for all years shows that concentrations of D3 and D4 are higher when measured with the SIP-PAS than with SPE-AAS (Table 1). However, due to the current methodological problems with the ENV+ sorbent [17], improved methods are needed to evaluate concentrations of these cVMS in Arctic air [36]. The SPE-AAS method is currently being improved to address these issues, and results using a different sorbent material look promising [23, 36].

No significant time trends are apparent in the Arctic datasets, except for a slight decrease in D5 and D6 concentrations at Zeppelin from 2015 to 2017 [23]. Whether this is year-to-year variation, or a consistent decrease with time that will continue, needs to be confirmed by continued annual monitoring [23]. Consistent seasonal trends in D5 concentrations have been observed at Zeppelin, with higher concentrations in winter than in summer due to reduced atmospheric degradation during the polar night period due to lower concentrations of hydroxyl radicals. Seasonal trends for D6 show the same tendency, but are less consistent [17, 22]. The seasonal pattern for D5 was not reproduced with the SIP-PAS, but this could be due to the long deployment times of the samplers resulting in average time-integrated concentrations [35].

Another type of passive air sampler based on a polystyrene-divinylbenzene copolymeric resin (XAD-PAS) was deployed in the city of Tromsø and at a remote site outside of Tromsø in 2017 [30]. The measured concentrations are presented on an ng per cubic meter basis using a passive sampling rate determined for VMS in the city of Toronto, Canada, for the same sampler design [6]. However, results should be interpreted with care as passive sampling rates will be influenced by environmental conditions including temperature and wind speeds. Despite uncertainties in the concentrations, the results clearly indicate that cVMS concentrations are higher in Tromsø than outside of the city (Table 1), indicating Tromsø represents a local source of VMS to air (Sect. 1.2). These samples, as well as selected SPE-AAS samples from Zeppelin in 2017, were also screened for D3F, D4Vn, and D4Ph. None of them were detected in Zeppelin samples, but D4Ph was detected at low concentrations in samples from Tromsø [30]. Linear VMS have not yet been detected in Arctic air, except for one detection of L5 in a SIP-PAS from Zeppelin (Table 2) [35].

2.1.2 Soil and Vegetation

The only reported cVMS and lVMS concentrations in soil and vegetation in polar regions are from samples collected in the Antarctica in 2009 [24]. However, concentrations reported in Antarctic soil exceed those found in urban and agricultural areas in Europe, including biosolid-amended soil [37, 38]. Findings from Antarctica should be interpreted with care as data quality within this study has been questioned [25–27].

2.1.3 Water, Snow, and Ice

VMS are difficult to measure in environmental water samples due to their low water solubility. When present in water, they are removed relatively quickly by hydrolysis, volatilization, sediment burial, and/or advection pathways (Sect. 3.1.2). cVMS were not found above limits of quantification in river water and lake water (Lake Storvannet) from the Hammerfest region in Northern Norway [31] and also not in marine water from Tromsøysund (Tromsø) (Warner et al., unpublished data) (Table 1). Concentrations of VMS in wastewater were discussed in Sect. 1.2. No measurements have been reported for VMS in snow or ice.

2.1.4 Sediments

Most measurements of cVMS in Arctic sediments (Table 1) have been done in areas impacted by local wastewater emissions. cVMS were detected in sediments directly outside Tromsø, but not in sediments outside of Nipøya, only 30 km northeast of Tromsø [39]. In a follow-up study, concentrations of D5 and D6 in sediment near Tromsø ranged from 1.0 to 4.0 µg/g organic carbon (OC) (20–60 ng/g dry weight (dw)) and 0.3–0.6 µg/g OC (4.5–11 ng/g dw), respectively (Warner et al., unpublished data) (Fig. 1, Table 1). D4 was not measured above detection limits. Highest D5 and D6 concentrations were found at the sampling site (site 3) between the two main wastewater outlets (Fig. 1), and concentrations were significantly lower at the three remaining sites.

Concentrations of cVMS in freshwater sediments in Lake Storvannet (Hammerfest) in 2014 were 207 ± 30, $3,775 \pm 973$, and 848 ± 211 ng/g OC for D4, D5, and D6, respectively [31]. While Storvannet receives unintentional wastewater emissions through leaking pipes and overflow events from the combined sewer system, the untreated wastewater is emitted to the marine harbor. cVMS concentrations in one sediment sample from the marine harbor (close to the wastewater outlet) were comparable to Lake Storvannet on a dry weight basis (Table 1), but much higher when normalized to OC content (1.0, 35, and 3.2 µg/g OC for D4, D5, and D6, respectively) (Krogseth et al., unpublished data) due to a lower OC content in the marine (1.0%) than in the freshwater (5.0%) sediments.

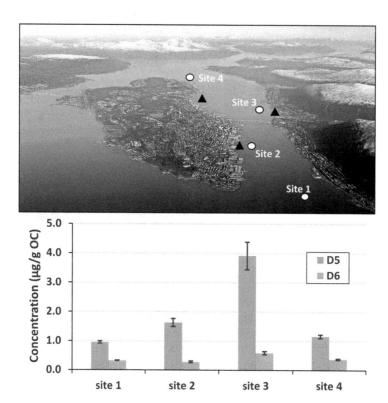

Fig. 1 Concentrations of cVMS in sediment (µg/g OC) at selected sites (○) outside of Tromsø in relation to wastewater effluent outflows (△). Arial photograph of Tromsø provided by Jon Terje Eiterå

A 15 cm sediment core was also collected from the deepest part of Lake Storvannet in 2014, sliced into 1 cm layers, and analyzed for cVMS (Krogseth et al., unpublished data). Age of the individual core layers was determined using Pb-210 isotopes. D4, D5, and D6 were detected in the upper 3 to 5 cm of the core, with decreasing concentrations observed with sediment core depth (Fig. 2). Lake Storvannet is an oligotrophic lake with a low sediment deposition rate, and the upper 5 cm of the core was dated to correspond to the last 35 years (i.e., the early 1980s to 2014). After the first 5 cm of the core, there was a large portion of sediment with higher dry weight content and lower OC content that had all been deposited in the late 1970s (possibly due to construction work around the lake during that time period). cVMS could no longer be detected in this part of the core (Fig. 2). The results are in reasonable agreement with detection of cVMS from the early 1970s in sediment cores from Lake Pepin (USA) [40] and Ulsan Bay (Korea) [41], reflecting global trends in production and use of cVMS [40, 41].

In the Svalbard archipelago, D5 was the only cVMS measured above detection limits in sediments from Adventfjorden [32], whereas all cVMS were below

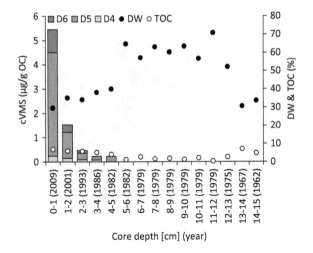

Fig. 2 Concentrations of cVMS, dry weight (DW), and total organic carbon (TOC) content in a dated sediment core from Lake Storvannet, 2014. The year represents the mean year of each slice. Only concentrations above limits of quantification are included

detection limits in sediment from Kongsfjorden and the remote Liefdefjorden [32, 33, 42]. These findings can be explained by the lower population in Ny-Ålesund (Kongsfjorden) compared to Longyearbyen (Adventfjorden) (Sect. 1.2). D4 and D5 were detected in one and two sediment samples, respectively, from remote areas in the Barents Sea [43]. However, concentrations were close to detection limits, and these findings should be corroborated as they were measured prior to proper quality control routines were established for VMS [33].

2.2   Organisms, Food Webs, and Humans

2.2.1   Freshwater Biota

The most extensive study of VMS in freshwater biota in polar regions is a food web study carried out in Lake Storvannet, Hammerfest [31, 44]. Concentrations of D4, D5, and D6 were measured in benthic invertebrates (the pea clams *Pisidium* sp. and chironomid larvae) and in the three dominating fish species in the lake (three-spined sticklebacks (*Gasterosteus aculeatus*), brown trout (*Salmo trutta*), and Arctic char (*Salvelinus alpinus*)) (Table 3). For the two top-predator species (char and trout), cVMS concentrations were measured in both muscle and liver tissue. Measured concentrations were highest for D5, followed by D4 and D6. This study was carried out just prior to renovations to the combined sewer system aiming to reduce unintentional wastewater emissions to the lake. In 2015, a follow-up study was carried out to investigate whether cVMS concentrations in the lake had changed after these renovations. Concentrations in fish were comparable (Warner et al., unpublished data, Table 3) to those found in 2014 (prior to renovation) [44]. This could be due to continued overflow events to the lake, in spite of the renovations, and/or persistence of cVMS in the system. These results need to be followed up to evaluate the effect of reduced wastewater emissions to the lake.

In comparison to other findings, concentrations in char from Storvannet were higher than in char from the wastewater-impacted Lake Vättern in Southern Sweden [45], while concentrations in trout were lower than concentrations in trout from the wastewater-impacted Lake Mjøsa in Southern Norway [46, 47]. None of the cVMS displayed trophic magnification in Storvannet. This agrees with findings in Lake Pepin [48], the Oslofjord [49, 50], and Tokyo Bay, but contrasts with findings in Lake Mjøsa and Randsfjorden [46, 47]. Lipid-normalized cVMS concentrations were significantly higher in the liver than in muscle for both trout and char, suggesting that cVMS tissue partitioning is not only controlled by lipids [51]. Dynamic model simulations were used to elucidate the behavior of cVMS both in the physical environment and in the organisms and food web of Lake Storvannet (Sect. 3.2).

Lake Storvannet is a lake subjected to strong seasonality with ice cover for about 6 months of the year and only a short plankton bloom [31]. The char and trout that were included in the food web study were stationary and remain in the lake all year round. A substantial fraction of the char and trout in Storvannet are however anadromous, i.e., they migrate to sea in summer to feed on the vast marine food supplies before returning to the lake for overwintering. A few anadromous trout ($n = 2$) and char ($n = 4$) were also sampled in 2014 (Krogseth et al., unpublished data). They were not included in the food web study as their exposures to cVMS are from both Lake Storvannet and the marine environment. The anadromous char has a low feeding rate when overwintering in the lake and has a mainly pelagic diet in the marine areas in summer [52]. The cVMS concentrations in the anadromous char and trout were significantly lower than in their stationary relatives (Table 3). This indicates that even if the anadromous fish migrate to marine areas impacted by anthropogenic activities (including wastewater effluents) [53], they are less exposed to cVMS than the resident fish staying in Lake Storvannet all year round. This is likely a result of their pelagic, rather than benthic, diet in the marine areas. However, Atlantic cod caught close to the wastewater outlet in Hammerfest harbor displayed the highest cVMS concentrations measured in fish in the Arctic (Krogseth et al., unpublished data) (Sect. 2.2.2). The cod may both eat pelagic, benthic, and piscivorous prey and scavenge on anthropogenic waste in the harbor area.

Arctic char from two remote Arctic lakes that are not impacted by any local sources (Lake Ellasjøen (Bear Island) and Erlingvatn (Svalbard)) have also been analyzed for cVMS and tris(trimethylsiloxy)phenylsilane (M3T(Ph)), but all compounds were below detection limits [54]. D5 was detected in char and trout from two remote lakes in Northern Norway, but standard quality control procedures for siloxanes were not followed in this analysis [55].

2.2.2 Marine Biota

The highest concentrations in Arctic marine biota have been measured in Atlantic cod caught close to the wastewater outlet in Hammerfest harbor (Krogseth et al., unpublished data) (Table 3). This is in agreement with high cVMS concentrations in sediment from Hammerfest harbor (Sect. 2.1.4) and illustrates the impact of local emissions in relatively small Arctic communities that do not have any wastewater treatment (Sect. 1.2). Similar to the freshwater fish in Storvannet [44], the Atlantic cod also displayed higher lipid-normalized cVMS concentrations in liver than in muscle (Table 3), although not statistically significant. Investigations of cVMS in Atlantic cod in the area around Tromsø in 2010–2011 showed how cVMS concentrations in cod liver decreased with increasing distance from point sources in urban areas [39]. This study also highlighted decreasing concentrations of D4 and D6 in cod liver with increasing length and weight of the fish, highlighting that other factors (i.e., growth dilution, non-lipid interactions, metabolic or dietary differences with age) play a role in cVMS accumulation. In a follow-up study in 2014, cVMS were measured in cod muscle from Tromsøysund (Warner et al., unpublished data, Table 3), but no clear correlations with length and weight were observed. D5 was detected in all cod muscle samples ($n = 10$), while detection frequency was lower for D4 (60%) and D6 (10%). This agrees with findings in char from Storvannet [44] as well as in flounder from the Humber Estuary [56] where detection frequency was higher for D4 than D6. This is in contrast to higher detections of D6 than D4 in sediments of the same areas (Table 1) [31, 56]. A possible explanation is slow mass transfer kinetics of D6 across the gastrointestinal tract due to its high molecular mass, resulting in a lower bioaccumulation potential of D6 than D4 [56, 57].

Measured concentrations of cVMS in Atlantic cod in Adventfjorden, Isfjorden (which connects to Adventfjorden), and Kongsfjorden on Svalbard are lower than in Tromsø and Hammerfest, indicating a lower cVMS exposure due to the smaller populations of Longyearbyen and Ny-Ålesund, respectively [32, 42, 58]. The linear siloxane L3 was detected in Atlantic cod in Kongsfjorden in 2008 [42], but this has not been further investigated. cVMS were also measured in shorthorn sculpin, a bottom-dwelling fish, in Adventfjorden and Liefdefjorden [42]. Measured concentrations in Adventfjorden can be explained by the local wastewater emissions, while measured concentrations of D5 and D6 above detection limits in Liefdefjorden were somewhat surprising as there are no local sources in this fjord, except possible emissions from cruise ships [32, 33].

Several studies have investigated cVMS in marine aquatic biota in remote polar regions. Both cVMS and lVMS have been detected in phytoplankton and krill from Antarctica [24], but the data quality of this study has been questioned [25–27]. cVMS could not be measured above detection limits in zooplankton from Kongsfjorden and Liefdefjorden from the Svalbard archipelago [32]. Detection of both cVMS and L3 has been reported in polar cod caught in areas on the west coast of Svalbard not impacted by local sources [42]. However, the measured concentrations were sporadic and close to detection limits. Further research is needed to corroborate these measurements. However, based on the limited data from the Arctic, concentrations of VMS in marine biota from areas impacted by local wastewater emissions are significantly higher compared to biota in remote locations (Table 3).

2.2.3 Birds and Mammals

cVMS have been analyzed in a range of seabirds from Svalbard and Northern Norway (Table 4). The first reports of cVMS in Arctic seabirds were concentrations of 32–69 ng/g wet weight (ww) D5 in the liver of dead or dying glaucous gulls (*Larus hyperboreus*) collected on Bear Island in 2003–2005 [59]. However, these concentrations were measured prior to proper quality control protocols for cVMS were established [33]. Besides these samples, the highest concentrations have been measured in eggs of glaucous gulls from Kongsfjorden [54]. The glaucous gull is a scavenging bird, which may feed upon anthropogenic waste. cVMS have also been measured sporadically above detection limits in the liver of kittiwake (*Rissa Tridactyla*) from Kongsfjorden [42] and in eggs of European shag (*Phalacrocorax aristotelis*) and herring gull (*Larus argentatus*) from Røst [60] (Table 4). However, the vast majority of measurements in birds are below detection limits (Table 4). Similarly, lVMS have not been detected in livers from kittiwakes and common eiders from Kongsfjorden and Liefdefjorden [42]. M3T(Ph) were not detected in glaucous gull eggs [54], and D4Ph, D3F, and D4Vn were not detected in a range of bird eggs sampled in 2017 [30]. D6 has been detected at low concentrations in blubber from bearded seals from Kongsfjorden [32], but no cVMS, D4Ph, D3F, or D4Vn were detected in liver from mink outside Tromsø or in plasma from polar bears on Svalbard [30] (Table 4).

The measured cVMS concentrations in air-breathing birds and mammals (Table 4) are much lower than in water-respiring biota from the same areas (Table 3). This is in agreement with similar findings from the Baltic Sea where lipid-normalized D5 concentrations were much lower in gray seals than in herring [61]. This can mainly be explained by the high volatility of cVMS, resulting in rapid elimination of cVMS from air-breathing biota through exhalation [62, 63]. In addition, cVMS are biotransformed in mammals [64]. Further investigation is needed to assess cVMS in seabirds and maternal transfer of cVMS from birds to eggs [65]. However, considering the efficient elimination mechanisms, cVMS are not expected to be of concern for air-breathing top predators in polar regions [29].

2.2.4 Humans

cVMS have been measured in blood plasma samples from pregnant women (MISA cohort) and postmenopausal women (NOWAC cohort) in Northern Norway and all of Norway, respectively (Table 4) [66]. In the North-Norwegian pregnant women, detection frequency of D4 was only 18% (highest concentration 2.7 ng/mL), while

D5 and D6 were not detected [66]. All cVMS were detected in the postmenopausal women from all of Norway, but D5 and D6 were below detection limits in most of the samples. No significant correlation was observed between concentrations of cVMS and use of personal care products, and the differences between the two cohorts could be explained by sampling year and age of the donors [66].

3   Understanding the Behavior of Volatile Methyl Siloxanes in Polar Regions

While knowledge of lVMS in polar regions, and cVMS in Antarctic areas, is still scarce, cVMS are clearly present in the Arctic (Sect. 2). The polar regions are characterized by a cold and highly seasonal environment. How is the environmental behavior and bioaccumulation of VMS affected by polar environmental conditions? These aspects will be discussed here, based on existing modelling studies and measurement data.

   To understand how VMS are impacted by the cold temperatures in polar regions, it is essential to understand the temperature dependence of their environmental partitioning properties. Hence, these are briefly introduced here and will be discussed in more detail in later sections. The key partitioning properties are the partition coefficients between air and water (K_{AW}), octanol and air (K_{OA}), octanol and water (K_{OW}), and organic carbon and water (K_{OC}) (Fig. 3). Their temperature dependence are controlled by the respective enthalpies of phase change, i.e., ΔU_{AW}, ΔU_{OA}, ΔU_{OW}, and ΔU_{OC}, respectively. VMS are both hydrophobic and volatile. Hence, their K_{OW} and K_{OC} coefficients are comparable to polychlorinated biphenyls (PCBs), while their K_{AW} and K_{OA} coefficients are much higher and lower than for PCBs, respectively (exemplified by PCB28 and PCB153 in Fig. 3). At lower temperatures, VMS become less volatile (i.e., K_{AW} decrease and K_{OA} increase) [16, 67, 68]. The situation for K_{OW} and K_{OC} is more complex. K_{OW} has been measured to decrease at lower temperatures [16, 68, 69]. This implies a decrease in hydrophobicity of VMS at lower temperatures which is in contrast to what is normally observed for other organic contaminants such as PCBs (Fig. 3) [70]. For many years, the temperature dependence of K_{OW} was assumed valid for K_{OC} of VMS in lack of measured values of ΔU_{OC}. However, K_{OC} for VMS has now been measured to increase at lower temperatures [71], opposite as for K_{OW} (Fig. 3). This has significant implications for their predicted environmental behavior in polar aquatic environments (Sects. 3.1.2 and 3.2).

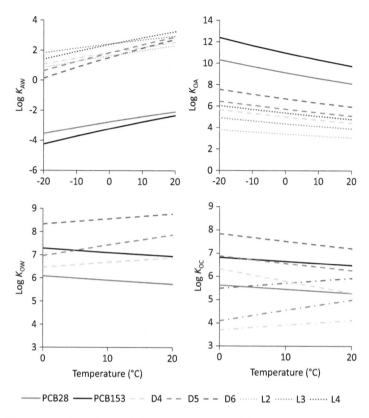

Fig. 3 Temperature dependence of the key partition coefficients for selected VMS [16, 67–69]. PCB28 and PCB153 have been included as reference compounds [70]. Two sets of properties have been included for K_{OC} of cVMS, based on either measurements of K_{OC} and ΔU_{OC} [71, 72] (dashed lines) or measurements of K_{OC} [73] assuming $\Delta U_{OC} = \Delta U_{OW}$ (interrupted lines)

3.1 Environmental Behavior

3.1.1 Atmosphere

Because of the high volatility of VMS, the two environmental characteristics most relevant for the atmospheric behavior of VMS in polar regions are (1) low temperatures influencing both partitioning and atmospheric degradation and (2) seasonally variable daylight length influencing concentrations of hydroxyl radicals and hence atmospheric degradation.

Partitioning between gaseous and particulate phases in the atmosphere, as well as between gaseous air and surface media like soil and vegetation is controlled by K_{OA}. Similarly, partitioning between the gaseous phase of the atmosphere and water is controlled by K_{AW}. At lower temperatures, VMS increasingly partition out of the gaseous phase to organic matter and water (Fig. 3), making them more prone for deposition to surface media. However, even at $-20°C$, K_{AW} of VMS are predicted to be several orders of magnitude higher than K_{AW} of PCBs due to the high volatility and low water solubility of VMS [68, 70]. The difference between K_{OA} of VMS and PCBs is smaller than for K_{AW} due to the lipophilic nature of VMS; at $-20°C$ K_{OA} of D6 is approaching K_{OA} of PCB-18 at $+20°C$ [68, 70]. This implies that even at the low temperatures experienced in the Arctic, VMS will remain mainly within the gas phase and are unlikely to deposit from the atmosphere to surface media to a significant extent. This is supported by modelling studies. Xu and Wania estimated the characteristic travel distance (CTD) and the transfer efficiency (TE) of D4, D5, and D6 using the OECD P_{OV} and LRTP screening tool [74]. The cVMS were estimated to have high CTDs (2,500–5,300 km) (i.e., able to be transported over long distances in the atmosphere), but low estimated TEs (0.004–0.02%), (i.e., a low potential to be deposited). Hence, they fall in the category of "fliers" that do not deposit to surface media even in Arctic conditions [74, 75]. Xu and Wania also used the global multimedia model GloboPOP to estimate the Arctic Contamination Potential (ACP), which is a similar metric to TE as it estimates the potential fraction of a chemical emitted globally that can deposit to Arctic surface media [74, 75]. The calculated ACPs for cVMS were very small and 4–5 orders of magnitude lower than ACP of well-known long-range transported contaminants [74]. These results were later reproduced for cVMS and expanded to D3 and lVMS by Kim et al. [76]. This agrees with a majority of measured concentrations below detection limits in sediments and biota in remote Arctic areas that do not have any local sources of VMS (Sect. 2).

The major atmospheric removal mechanism of VMS is degradation by hydroxyl radicals produced by sunlight, resulting in silanols that are removed from the atmosphere through wet deposition [7, 8]. During winter in polar regions (i.e., polar night), the atmospheric half-lives for VMS will be greatly enhanced due to the lack of sunlight and hence lower concentrations of hydroxyl radicals. Following Webster et al. [77, 78], a minimum concentration of hydroxyl radicals in the polar night atmosphere of 6×10^3 molecules/cm^3, instead of 7.7×10^5 molecules/cm^3 as used by Atkinson et al. in his calculations for VMS [7], would increase the estimated half-life of D5 due to reaction with hydroxyl radicals from 6.7 days [7] at general conditions to 2.4 years during the polar night. In addition, the half-life of VMS will also be increased by the low temperatures in polar regions. Bernard et al. recently showed that the reaction rates of hexamethyldisiloxane (L2), L3, D3, and D4 with hydroxyl radicals decrease with lower temperatures [10]. This is in agreement with previous results for D4, D5, and D6 [11, 79]. Based on the results from Bernard et al. [10], reaction rates for L2, D3, and D4 will be approximately 35% lower at $-20°C$ than at $+20°C$. As a result, model simulations have predicted higher atmospheric concentrations of D5 in winter than in summer at high latitudes which has been confirmed by measurements both in rural Sweden and at the Zeppelin Observatory [17, 18]. Less is known about the seasonal pattern of other VMS, but D6 shows the same tendency as D5 (Sect. 2.1.1).

In addition to hydroxyl radical degradation, aerosol-mediated degradation has recently been proposed by Xu et al. [80]. Atmospheric concentration ratios of D5/D6

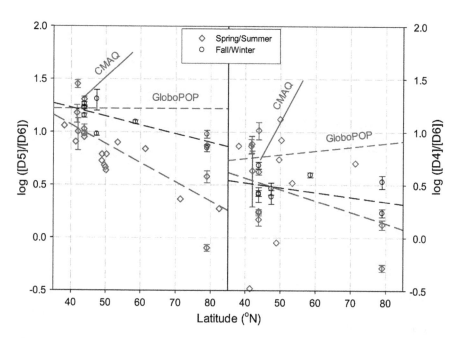

Fig. 4 Correlation of log-normalized concentration ratios of D5/D6 (left) and D4/D6 (right). Dashed lines represent regressions of empirical data from spring/summer (◊, red) and fall/winter (○, black). Estimated yearly trends by the Community Multiscale Air Quality (CMAQ) [84] (blue solid line) and the GloboPOP [80] (blue dashed line) models are labelled. Reprinted from Chemosphere, Vol 228, Xu, S., Warner, N. A., Bohlin-Nizzetto, P. Durham, J., McNett, D. Long-range transport potential and atmospheric persistence of cyclic volatile methylsiloxanes based on global measurements, pages 460–468, Copyright (2019), with permission from Elsevier

were observed to decrease along a south to north transect (40°N to 80°N) based on empirical measurements. This indicates that D5 undergoes faster removal from the atmosphere compared to D6 as they are transported northwards from emission source regions to polar regions. This was contrary to model predictions where D6 ($\tau_{1/2}$ = 5.2 days) was predicted to be removed faster compared to D5 ($\tau_{1/2}$ = 6.9 days) when only hydroxyl radical degradation was assumed to occur (Fig. 4) [80]. No trend was observed in regressions between D4/D6 concentration ratio and latitude due to large variation in the empirical data, which can be attributed to analytical uncertainty surrounding D4 measurements do to degradation on sorbents used in active air sampling methodology. Irreversible sorption of D4 and D5 on mineral and sulfate aerosols has been observed in laboratory experiments [81]. This may be linked to surface reactions of polymerization at high surface loadings and hydrolysis at environmental relevant concentrations [82]. Surface-facilitated degradation has been reported earlier in soils where rates of degradation proceeded in the order of D4 > D5 > D6 [83]. Based on concentration ratios observed, these findings suggest that aerosol-mediated degradation in addition to hydroxyl radical degradation occurs and needs further elucidation.

Concentrations of D5 in Arctic air have been predicted with the environmental multimedia models GloboPOP [74], DEHM [4, 17], and BETR-Global [85]. These concentrations have been based on estimated emissions of D5 to air in the Northern Hemisphere, as well as physicochemical properties and degradation rates of D5. Xu and Wania estimated concentrations of D5 to range from about 0.6 ng/m^3 (summer) to 6 ng/m^3 (winter) in Arctic regions with the GloboPOP model [74], which also agrees with predictions from the BETR-Global model [85]. The DEHM model predicted D5 concentrations to range from 0.1 ng/m^3 (summer) to 10 ng/m^3 (winter) at the Zeppelin Observatory in August–December 2009, accounting for actual meteorological data from the same time period [17]. The predicted concentrations and seasonality are in good agreement with measurements of D5 in Arctic air (Table 1) [4, 17, 74], indicating that we have a relatively good understanding of the atmospheric behavior of this compound. There is still a need for modelling studies to increase the understanding of the other VMS in Arctic air.

To summarize, the atmospheric behavior of VMS in the Arctic is mostly affected by the seasonal degradation by hydroxyl radicals. The low temperatures do not seem to significantly affect their partitioning behavior, as VMS are very volatile even at Arctic temperatures. If VMS deposit to surface media, it is most likely to occur in winter when temperatures are low, and degradation is limited. However, the models predict that this will still be negligible, and no empirical evidence has been observed so far.

3.1.2 Aquatic Environments

To understand how Arctic conditions influence VMS behavior in aquatic environments, it is important to understand the typical behavior of VMS in aquatic systems in terms of their overall partitioning and main removal processes. Due to their low water solubility, VMS that are emitted to water (i.e., with wastewater) are often relatively quickly removed from the water phase through volatilization to air, hydrolysis, sediment burial, and/or advection depending on the specific system characteristics such as lake depth, water residence time, or sedimentation rates [76, 86]. This means that a range of environmental characteristics may influence the behavior of VMS in aquatic environments, also in polar regions. Here, particular focus will be placed upon (1) how lower temperatures influence partitioning and degradation and (2) how environmental changes occurring with the strong seasonality in polar regions (e.g., ice cover, hydrodynamics, organic carbon flows) influence overall environmental behavior of VMS.

In aquatic environments, partitioning between air and water is influenced by K_{AW}, while partitioning between water and organic matter (i.e., dissolved organic matter, particulate organic matter, and sediments) is influenced by K_{OC}. As already discussed (Sect. 3.1.1), log K_{AW} of VMS is high (>0, Fig. 3) even at low temperatures, resulting in net volatilization from water to air in aquatic systems receiving wastewater effluents [86–89]. The temperature effect on K_{OC} of VMS is still poorly understood. The temperature dependence of K_{OW} (ΔU_{OW}) is often assumed to represent the temperature dependence of K_{OC} (ΔU_{OC}) in lack of available data for ΔU_{OC}. However, the only two studies reporting the effect of temperature on K_{OW} and K_{OC} of VMS show contrasting results with lower temperature: decreasing K_{OW} (i.e., positive ΔU_{OW}) [16, 68] and increasing K_{OC} (i.e., negative ΔU_{OC}) [71] (Fig. 3). The implication of assuming one or the other temperature dependence (ΔU_{OW} or ΔU_{OC}) is profound for predicted behavior in polar aquatic systems, where water temperatures can be close to zero degrees Celsius for most of the year. Environmental partitioning behavior of cVMS (D4, D5, and D6) in Lake Storvannet, Hammerfest (Norway), was modelled using the QWASI model for the physical environment [31] and an updated bentho-pelagic ACC-HUMAN model for the food web (Sect. 3.2) [44]. In the QWASI model, ΔU_{OW} was assumed to represent the temperature dependence of K_{OC} (i.e., decreasing with decreasing temperature) as no measured values of ΔU_{OC} were available at that time [31]. This resulted in predictions of lower partitioning of cVMS to suspended particulate matter and sediments at lower temperatures, leading to a higher fraction of VMS in the dissolved water phase. This again resulted in a higher predicted removal from lake water in winter than in summer and hence a lower predicted overall persistence in winter than in summer [31]. K_{OC} and ΔU_{OC} were identified as the main variables of uncertainty in modelling cVMS behavior in Storvannet (including its food web) [31, 44]. However, the recently measured ΔU_{OC} values for VMS produce contrasting results to those reported earlier and predict VMS to behave more in accordance with known behavior for traditional POPs, i.e., higher partitioning to suspended particulate matter and sediments at lower temperatures, slower removal, and higher overall persistence. This means that conclusions on the aquatic behavior of VMS in cold systems will be strongly dependent on the choice of ΔU_{OW} or ΔU_{OC}. This was highlighted by a modelling study of Adventfjorden using the EQC model [90], in agreement with the results for Lake Storvannet. Depending on which values that were used for K_{OC} and ΔU_{OW} or ΔU_{OC}, predicted overall residence times for D5 in Adventfjorden differed with 1,100 days [90]. When using the K_{OC} from Kozerski et al. [73] and ΔU_{OW}, almost 100% of D5 was predicted to disappear from the fjord within 1 year after a hypothetical stop of emissions of D5 to the fjord. When using the measured K_{OC} and ΔU_{OC} from Panagopoulos et al. [71, 72], about 65% of D5 was predicted to still remain in the fjord after 1 year [90]. This illustrates how important it is to address these uncertainties, and more research is critically needed, particularly as there is currently no scientific consensus. In a recent modelling study of predicted persistence and response times of linear and cyclic VMS in various aquatic environments using the dynamic QWASI model, ΔU_{OW} was again used to represent the temperature dependence of K_{OC}, resulting in predicted higher partitioning to water and lower overall persistence at lower temperatures [76]. Hence, this is currently the most critical knowledge gap to understand the behavior of VMS in cold aquatic environments.

One of the reasons uncertainty in partitioning in aquatic environments becomes so important at low temperatures for VMS is due to temperature effects on degradation in water. In temperate areas, VMS undergo relatively rapid hydrolysis in water with half-lives of 4, 70, and 204 days for D4, D5, and D6, respectively, at pH

7 and 25°C [1–3]. These half-lives decrease at both acidic and alkaline conditions, but increase at lower temperatures. Hence, while degradation is predicted to be the main removal mechanism from water for VMS, particularly D4, in some temperate aquatic environments [86], removal through degradation was predicted to be negligible for VMS in Lake Storvannet due to the low water temperatures [31]. Similarly, the main degradation mechanism of VMS in sediments has been assumed to be hydrolysis of VMS present in the sediment pore water [86, 87]. Hence, degradation by hydrolysis in sediment is also expected to slow down at lower temperatures. If sorption to organic matter (K_{OC}) increases at lower temperatures, this will lower the fraction of VMS in the dissolved phase, further slowing down degradation by hydrolysis. As a result, other elimination mechanisms like volatilization, advection, and sediment burial will be more important in cold aquatic environments than in warmer regions [31].

In addition to the low water temperatures, many water bodies in polar regions are ice covered for significant time periods of the year. Ice cover can affect the behavior and persistence of VMS in aquatic systems as ice cover hinders volatilization, which is an important removal mechanism of VMS from water due to their high K_{AW} values. Hence, when volatilization is physically hindered by ice cover, persistence in the water phase is predicted to increase. This has been illustrated in modelling studies both for the Baltic Sea [91], Lake Storvannet [31], and Adventfjorden [90]. Moreover, VMS in aquatic systems can also be affected by the strong seasonality in flows of water and organic carbon in polar regions. Frozen grounds, precipitation in the form of snow, lower runoff, and lower riverine inputs may lead to a lower input of both water and organic carbon to polar aquatic systems in winter compared to summer. In Adventfjorden, residence times of VMS in water were generally predicted to be lower in summer than in winter due to the lack of ice cover and the possibility for VMS to volatilize to air [90]. However, when a higher concentration of suspended particulate matter in the fjord in the summer time was taken into account, predicted residence times increased [90]. In Lake Storvannet, average water residence times are much lower in summer (9 days) than in winter (38 days) [31]. This impacts the removal of VMS from the lake through advection with water, which was predicted to be the main elimination mechanism of VMS from this lake. The reason advection dominates is because (1) the water turnover time is short as it is a small lake at the receiving end of a large drainage basin and (2) that volatilization, hydrolysis, and sediment deposition are slowed down by ice cover, low temperatures, and slow deposition rates, respectively (all properties characteristic of arctic and subarctic oligotrophic lakes) [31]. However, in Lake Storvannet, the high water flows in summer also lead to a higher input of VMS to the system. The wastewater emissions to Lake Storvannet are unintentional, occurring through leaking pipes and combined sewer overflow events [31]. Hence, VMS emissions to the lake peak during high water flow events such as in snow melt season and during heavy rainfalls [31]. Even if this is a special case, it illustrates how important the particular characteristics of an aquatic system are to understand VMS behavior.

Predictions of actual VMS concentrations in water and sediment in polar regions still remain to be conducted. The modelling for Adventfjorden with dynamic QWASI used a hypothetical emission rate and did not look at actual concentrations [90]. For Lake Storvannet, it was not possible to predict concentrations in water and sediment within the lake due to intermittent and unintentional wastewater emissions of unknown magnitude [31]. Thus, an inverse modelling approach was applied to predict the VMS emissions to the lake based on the measured concentrations in sediment. However, predicted wastewater emissions were unrealistically high based on measured VMS concentrations in sewage and knowledge of the system. This was attributed to the uncertainty in the temperature dependence of K_{OC} [31], and more reasonable estimates would likely have been achieved if ΔU_{OC} had been applied instead of ΔU_{OW} [44].

3.2   Bioavailability and Bioaccumulation

In this section, we will discuss the potential effects of polar environmental conditions on VMS bioavailability and bioaccumulation based on available measurements (Sect. 2.2) and modelling studies. Focus is placed on aquatic biota in environments impacted by local emissions of cVMS. No studies modelling VMS bioaccumulation behavior in air-breathing biota or humans in polar regions have been carried out so far. However, as discussed in Sect. 2.2, bioaccumulation risk to air-breathing organisms is considered low due to them possessing additional mechanisms for elimination (via air respiration), which is supported by current findings within the environment.

The low temperatures in polar regions impact both bioavailability and bioaccumulation in aquatic environments through temperature effects on K_{OC} and on the lipid-water partition coefficient (K_{LW}), respectively. Higher K_{OC} implies higher sorption to sediment and suspended solids and less bioavailability for water-respiring organisms. Higher K_{LW} implies higher partitioning out of the water phase to lipids in biota. Due to the lack of measurements for K_{LW} for VMS, K_{OW} is often assumed representative for K_{LW}. However, there is still little evidence for this and even less for membrane lipids than for storage lipids [92, 93]. The temperature dependence of K_{LW} for VMS has never been measured directly. It has been estimated to decrease at lower temperatures for D4, D5, and D6, similar to K_{OW} [94]. If this is the case, this implies a relationship with temperature contrary to known contaminants like the PCBs, which have higher partitioning to lipids at lower temperatures. Also, it implies opposite behavior for K_{OC} and K_{LW} for VMS at lower temperatures and that sorption to organic carbon in sediments and suspended matter increase, but sorption to lipids decrease. This was assumed in the modelling of VMS in the food web of Lake Storvannet using a bentho-pelagic version of the ACC-HUMAN model [44]. As zooplankton is only present for a very short period of the year in Storvannet, benthos is a key dietary item for the fish (Sect. 2.2.1). In Lake Storvannet, the main source of VMS to the fish was predicted to be ingestion of benthos, while ventilation to water was predicted to be the main elimination route of VMS from fish. The results were very sensitive to the K_{OC} and its temperature dependence, as this controls uptake of VMS in benthos [44]. Ventilation would have been less important as an elimination mechanism for fish if K_{OW} had been assumed to increase at lower temperatures (like K_{OC}) instead of decrease. Again, this illustrates how uncertainty in the temperature dependence of K_{OC}, K_{OW}, and K_{LW} critically constrains our understanding of the behavior of VMS in cold aquatic environments.

Low temperatures will most likely affect biotransformation of VMS in aquatic biota, as biotransformation is generally assumed to slow down at lower temperatures [95]. However, there are no studies on biotransformation of VMS in polar species, and temperature-dependent biotransformation of VMS is a key knowledge gap to understand the behavior of VMS in biota. With regard to physical characteristics of polar aquatic environments (e.g., ice cover, seasonal hydrodynamics, etc.), they will affect concentrations in water and sediment (Sect. 3.1.2) and hence exposure to biota. In addition, the seasonal environment leads to seasonal food availability for many polar species. Polar biota have adapted ways to deal with this, including seasonal diets, seasonal migrations, dormitory states in winter, and often highly seasonal lipid contents. Potential effects of these adaptations on VMS concentrations in polar biota (relative to in more temperate areas) have not been thoroughly investigated yet. cVMS concentrations were observed to be lower in anadromous char (migrating to sea in summer) than in stationary char in Storvannet, which can be explained by their different exposure in the different environments (Sect. 2.2.1). This highlights the key importance of local sources of VMS to explain observed VMS concentrations in polar biota (Sect. 2.2), and comparison of different sites and species is not possible without taking the different exposure scenarios into account.

4   Summary and Future Directions

Most of the knowledge of VMS in polar regions result from monitoring and modelling of cVMS at a limited number of sites in the Norwegian Arctic (Tromsø, Hammerfest, Longyearbyen, Ny-Ålesund), notably with a few exceptions. It is quite clear from the existing measurements that cVMS are definitely present within the Arctic. Long-range atmospheric transport of cVMS to the Arctic does occur, but evidence to date suggests that they have a negligible potential to deposit to surface media. Measurements of VMS in Arctic surface media such as snow, soil, or vegetation are still lacking. However, concentrations in these media are expected to be negligible due to the high volatility of VMS even under polar environmental conditions, which is supported by modelling studies. Occurrence of cVMS in sediments and aquatic biota within the Arctic is mainly a result of local wastewater emissions, and concentrations in air-breathing biota are much lower than concentrations in water-respiring biota. Sporadic detections of VMS at low levels in sediments, fish, and seabirds at remote locations in the Arctic, as well as detection of

VMS in Antarctica, need to be further investigated. However, the combined measurement and modelling evidence for cVMS shows that they are not likely to deposit from the atmosphere to surface media, even under polar environmental conditions. Considering the low deposition potential of cVMS, and the even higher volatility of lVMS than cVMS (Fig. 3), lVMS are also not likely to deposit in polar regions to any significant extent.

Our understanding of VMS behavior in polar environmental conditions is increasing, but is still limited. One of the key uncertainties that need to be addressed is the temperature dependence of partitioning of VMS between water, organic carbon, and lipids as it currently restricts our understanding of the environmental behavior and bioaccumulation of VMS in cold aquatic environments. Temperature dependence of biotransformation in polar biota is also a key knowledge gap, as well as the influence of ecological adaptations to the polar climate, such as seasonal behavior, food availability, and lipid dynamics, on VMS bioaccumulation.

Acknowledgments The Research Council of Norway (#222259, #267574) and the Fram Centre Flagship research programme for Hazardous Substances – effects on ecosystem and human health.

References

1. Brooke D, Crookes M, Gray D, Robertson S (2009) Environmental risk assessment report: decamethylcyclopentasiloxane. Environment Agency of England and Wales, Bristol
2. Brooke D, Crookes M, Gray D, Robertson S (2009) Environmental risk assessment report: octamethylcyclotetrasiloxane. Environment Agency of England and Wales, Bristol
3. Brooke D, Crookes M, Gray D, Robertson S (2009) Environmental risk assessment report: dodecamethylcyclohexasiloxane. Environment Agency of England and Wales, Bristol
4. Genualdi SGS, Harner T, Cheng Y, MacLeod M, Hansen KM, van Egmond R, Shoeib M, Lee SC (2011) Global distribution of linear and cyclic volatile methyl siloxanes in air. Environ Sci Technol 45(8):3349–3354. https://doi.org/10.1021/es200301j
5. Kierkegaard A, McLachlan MS (2013) Determination of linear and cyclic volatile methylsiloxanes in air at a regional background site in Sweden. Atmos Environ 80:322–329. https://doi.org/10.1016/j.atmosenv.2013.08.001
6. Krogseth IS, Zhang X, Lei YD, Wania F, Breivik K (2013) Calibration and application of a passive air sampler (XAD-PAS) for volatile methyl siloxanes. Environ Sci Technol 47 (9):4463–4470. https://doi.org/10.1021/es400427h
7. Atkinson R (1991) Kinetics of the gas-phase reactions of a series of organosilicon compounds with OH and NO3 radicals and O3 at 297 +/− 2K. Environ Sci Technol 25(5):863–866. https:// doi.org/10.1021/es00017a005
8. Whelan MJ, Estrada E, van Egmond R (2004) A modelling assessment of the atmospheric fate of volatile methyl siloxanes and their reaction products. Chemosphere 57(10):1427–1437. https://doi.org/10.1016/j.chemosphere.2004.08.100
9. MacLeod M, Kierkegaard A, Genualdi S, Harner T, Scheringer M (2013) Junge relationships in measurement data for cyclic siloxanes in air. Chemosphere 93(5):830–834. https://doi.org/10. 1016/j.chemosphere.2012.10.055
10. Bernard F, Papanastasiou DK, Papadimitriou VC, Burkholder JB (2018) Temperature dependent rate coefficients for the gas-phase reaction of the OH radical with linear (L2, L3) and cyclic

(D3, D4) permethylsiloxanes. J Phys Chem A 122(17):4252–4264. https://doi.org/10.1021/acs. jpca.8b01908

11. Safron A, Strandell M, Kierkegaard A, Macleod M (2015) Rate constants and activation energies for gas-phase reactions of three cyclic volatile methyl siloxanes with the hydroxyl radical. Int J Chem Kinet 47(7):420–428. https://doi.org/10.1002/kin.20919

12. UNEP (2009) Stockholm convention of persistent organic pollutants (POPs) as amended in 2009. Text and Annexes

13. Buser AM, Kierkegaard A, Bogdal C, MacLeod M, Scheringer M, Hungerbühler K (2013) Concentrations in ambient air and emissions of cyclic volatile methylsiloxanes in Zürich, Switzerland. Environ Sci Technol 47(13):7045–7051. https://doi.org/10.1021/es3046586

14. Kierkegaard A, McLachlan MS (2010) Determination of decamethylcyclopentasiloxane in air using commercial solid phase extraction cartridges. J Chromatogr A 1217(21):3557–3560. https://doi.org/10.1016/j.chroma.2010.03.045

15. Yucuis RA, Stanier CO, Hornbuckle KC (2013) Cyclic siloxanes in air, including identification of high levels in Chicago and distinct diurnal variation. Chemosphere 92:905–910. https://doi. org/10.1016/j.chemosphere.2013.02.051

16. Xu S, Kozerski G, Mackay D (2014) Critical review and interpretation of environmental data for volatile methylsiloxanes: partition properties. Environ Sci Technol 48(20):11748–11759. https://doi.org/10.1021/es503465b

17. Krogseth IS, Kierkegaard A, McLachlan MS, Breivik K, Hansen KM, Schlabach M (2013) Occurrence and seasonality of cyclic volatile methyl siloxanes in Arctic air. Environ Sci Technol 47(1):502–509. https://doi.org/10.1021/es3040208

18. McLachlan MS, Kierkegaard A, Hansen KM, van Egmond R, Christensen JH, Skjøth CA (2010) Concentrations and fate of decamethylcyclopentasiloxane (D(5)) in the atmosphere. Environ Sci Technol 44(14):5365–5370. https://doi.org/10.1021/es100411w

19. Bohlin-Nizzetto P, Aas W, Krogseth IS (2014) Monitoring of environmental contaminants in air and precipitation, annual report 2013. Norwegian Environment Agency, M-202/2014. NILU – Norwegian Institute for Air Research, NILU OR 29/2014, Kjeller, Norway

20. Bohlin-Nizzetto P, Aas W, Warner N (2015) Monitoring of environmental contaminants in air and precipitation, annual report 2014. Norwegian Environment Agency, M-368/2015. NILU – Norwegian Institute for Air Research, NILU OR 19/2015, Kjeller, Norway

21. Bohlin-Nizzetto P, Aas W (2016) Monitoring of environmental contaminants in air and precipitation, annual report 2015. Norwegian Environment Agency, M-579/2016. NILU – Norwegian Institute for Air Research, NILU report 14/2016, Kjeller, Norway

22. Bohlin-Nizzetto P, Aas W, Warner N (2017) Monitoring of environmental contaminants in air and precipitation, annual report 2016. Norwegian Environment Agency, M-757/2017. NILU – Norwegian Institute for Air Research, NILU report 17/2017, Kjeller, Norway

23. Bohlin-Nizzetto P, Aas W, Warner NA (2018) Monitoring of environmental contaminants in air and precipitation, annual report 2017. Norwegian Environment Agency, M-1062/2018. NILU – Norwegian Institute for Air Research, NILU report 13/2018, Kjeller, Norway

24. Sanchís J, Cabrerizo A, Galbán-Malagón C, Barceló D, Farré M, Dachs J (2015) Unexpected occurrence of volatile dimethylsiloxanes in Antarctic soils, vegetation, phytoplankton, and krill. Environ Sci Technol 49(7):4415–4424. https://doi.org/10.1021/es503697t

25. Mackay D, Gobas F, Solomon K, Macleod M, McLachlan M, Powell DE, Xu S (2015) Comment on "Unexpected occurrence of volatile dimethylsiloxanes in Antarctic soils, vegetation, phytoplankton, and krill". Environ Sci Technol 49(12):7507–7509. https://doi.org/10. 1021/acs.est.5b01936

26. Warner NA, Krogseth IS, Whelan MJ (2015) Comment on "Unexpected occurrence of volatile dimethylsiloxanes in Antarctic soils, vegetation, phytoplankton, and krill". Environ Sci Technol 49(12):7504–7506. https://doi.org/10.1021/acs.est.5b01612

27. Sanchís J, Cabrerizo A, Galbán-Malagón C, Barceló D, Farré M, Dachs J (2015) Response to comments on "Unexpected occurrence of volatile dimethylsiloxanes in Antarctic soils,

vegetation, phytoplankton and krill". Environ Sci Technol 49(12):7510–7512. https://doi.org/10.1021/acs.est.5b02184

28. Wang D-G, Norwood W, Alaee M, Byer JD, Brimble S (2013) Review of recent advances in research on the toxicity, detection, occurrence and fate of cyclic volatile methyl siloxanes in the environment. Chemosphere 93(5):711–725. https://doi.org/10.1016/j.chemosphere.2012.10.041

29. Warner NA (2017) Siloxanes. AMAP assessment 2016: chemicals of emerging Arctic concern. Arctic Monitoring and Assessment Programme (AMAP), Oslo, pp 131–139

30. Schlabach M, Bavel Bv, Lomba JAB, Borgen A, Fjeld E, Gabrielsen GW, Götsch A, Halse AK, Hanssen L, Krogseth IS, Nikiforov V, Nygård T, Bohlin-Nizzetto P, Reid M, Rostkowski P (2018) Screening programme 2017 – AMAP assessment compounds. Norwegian Environment Agency, M-1080/2018. NILU – Norwegian Institute for Air Research, NILU report 21/2018, Kjeller, Norway

31. Krogseth IS, Whelan MJ, Christensen GN, Breivik K, Evenset A, Warner NA (2017) Understanding of cyclic volatile methyl siloxane fate in a high latitude lake is constrained by uncertainty in organic carbon–water partitioning. Environ Sci Technol 51(1):401–409. https://doi.org/10.1021/acs.est.6b04828

32. Warner NA, Evenset A, Christensen G, Gabrielsen GW, Borgå K, Leknes H (2010) Volatile siloxanes in the European Arctic: assessment of sources and spatial distribution. Environ Sci Technol 44(19):7705–7710. https://doi.org/10.1021/es101617k

33. Warner NA, Kozerski G, Durham J, Koerner M, Gerhards R, Campbell R, McNett DA (2013) Positive vs. false detection: a comparison of analytical methods and performance for analysis of cyclic volatile methylsiloxanes (cVMS) in environmental samples from remote regions. Chemosphere 93:749–756. https://doi.org/10.1016/j.chemosphere.2012.10.045

34. Tin T, Fleming ZL, Hughes KA, Ainley DG, Convey P, Moreno CA, Pfeiffer S, Scott J, Snape I (2008) Impacts of local human activities on the Antarctic environment. Antarct Sci 21(1):3–33. https://doi.org/10.1017/S0954102009001722

35. Rauert C, Shoeib M, Schuster JK, Eng A, Harner T (2018) Atmospheric concentrations and trends of poly- and perfluoroalkyl substances (PFAS) and volatile methyl siloxanes (VMS) over 7 years of sampling in the Global Atmospheric Passive Sampling (GAPS) network. Environ Pollut 238:94–102. https://doi.org/10.1016/j.envpol.2018.03.017

36. Warner N, Nikiforov V, Krogseth IS, Kierkegaard A, Bohlin-Nizzetto P (2018) Reducing sampling artifacts in air measurements: improvement of active air sampling methodologies for accurate measurements of cyclic volatile methylsiloxanes in remote regions. Organohalogen Compd 80:465–468

37. Companioni-Damas EY, Santos FJ, Galceran MT (2012) Analysis of linear and cyclic methylsiloxanes in sewage sludges and urban soils by concurrent solvent recondensation – large volume injection – gas chromatography-mass spectrometry. J Chromatogr A 1268:150–156. https://doi.org/10.1016/j.chroma.2012.10.043

38. Sánchez-Brunete C, Miguel E, Albero B, Tadeo JL (2010) Determination of cyclic and linear siloxanes in soil samples by ultrasonic-assisted extraction and gas chromatography–mass spectrometry. J Chromatogr A 1217(45):7024–7030. https://doi.org/10.1016/j.chroma.2010.09.031

39. Warner NA, Nøst TH, Andrade H, Christensen G (2014) Allometric relationships to liver tissue concentrations of cyclic volatile methyl siloxanes in Atlantic cod. Environ Pollut 190:109–114. https://doi.org/10.1016/j.envpol.2014.03.031

40. Powell DE, Durham J, Darren WH, Kozerski GE, Böhmer T, Gerhards R (2010) Deposition of cyclic volatile methylsiloxane (cVMS) materials to sediment in a temperate freshwater lake: a historical perspective. Dow Corning Corporation, HES study no. 10725-108. Auburn, MI

41. Lee S-Y, Lee S, Choi M, Kannan K, Moon H-B (2018) An optimized method for the analysis of cyclic and linear siloxanes and their distribution in surface and core sediments from industrialized bays in Korea. Environ Pollut 236:111–118. https://doi.org/10.1016/j.envpol.2018.01.051

42. Evenset A, Leknes H, Christensen GN, Warner NA, Remberger M, Gabrielsen GW (2009) Screening of new contaminants in samples from the Norwegian Arctic. Norwegian Pollution Control Authority, TA-2510/2009. Akvaplan-niva, Report 4351-1, Tromsø, Norway

43. Bakke T, Boitsov S, Brevik EM, Gabrielsen GW, Green N, Helgason LB, Klungsøyr J, Leknes H, Miljeteig C, Måge A, Rolfsnes BE, Savinova T, Schlabach M, Skaare BB, Valdersnes S (2008) Mapping selected organic contaminants in the Barents Sea 2007. Norwegian Pollution Control Authority, TA-2400/2008. Norwegian Institute for Water Research, NIVA-report 5589-2008, Oslo, Norway

44. Krogseth IS, Undeman EM, Evenset A, Christensen GN, Whelan MJ, Breivik K, Warner NA (2017) Elucidating the behavior of cyclic volatile methylsiloxanes in a subarctic freshwater food web: a modeled and measured approach. Environ Sci Technol 51(21):12489–12497. https://doi.org/10.1021/acs.est.7b03083

45. Kierkegaard A, Adolfsson-Erici M, McLachlan MS (2010) Determination of cyclic volatile methylsiloxanes in biota with a purge and trap method. Anal Chem 82(22):9573–9578. https://doi.org/10.1021/ac102406a

46. Borgå K, Fjeld E, Kierkegaard A, McLachlan MS (2012) Food web accumulation of cyclic siloxanes in Lake Mjøsa, Norway. Environ Sci Technol 46(11):6347–6354. https://doi.org/10.1021/es300875d

47. Borgå K, Fjeld E, Kierkegaard A, McLachlan MS (2013) Consistency in trophic magnification factors of cyclic methyl siloxanes in pelagic freshwater food webs leading to brown trout. Environ Sci Technol 47(24):14394–14402. https://doi.org/10.1021/es404374j

48. Powell DE, Woodburn KB, Drotar KD, Durham J, Huff DW (2009) Trophic dilution of cyclic volatile methylsiloxane (cVMS) materials in a temperate freshwater lake. Dow Corning Corporation, HES study no. 10771-108. Auburn, MI

49. Powell DE, Durham J, Huff DW, Böhmer T, Gerhards R, Koerner M (2010) Bioaccumulation and trophic transfer of cyclic volatile methylsiloxane (cVMS) materials in the aquatic marine food webs of the inner and outer Oslofjord, Norway. Dow Corning Corporation, HES study no. 11060-108. Auburn, MI

50. Powell DE, Schøyen M, Øxnevad S, Gerhards R, Böhmer T, Koerner M, Durham J, Huff DW (2018) Bioaccumulation and trophic transfer of cyclic volatile methylsiloxanes (cVMS) in the aquatic marine food webs of the Oslofjord, Norway. Sci Total Environ 622:127–139. https://doi.org/10.1016/j.scitotenv.2017.11.237

51. McGoldrick DJ, Chan C, Drouillard KG, Keir MJ, Clark MG, Backus SM (2014) Concentrations and trophic magnification of cyclic siloxanes in aquatic biota from the Western Basin of Lake Erie, Canada. Environ Pollut 186:141–148. https://doi.org/10.1016/j.envpol.2013.12.003

52. Rikardsen AH, Amundsen PA, Bodin PJ (2003) Growth and diet of anadromous Arctic charr after their return to freshwater. Ecol Freshw Fish 12(1):74–80. https://doi.org/10.1034/j.1600-0633.2003.00001.x

53. Jensen JLA, Christensen GN, Hawley KH, Rosten CM, Rikardsen AH (2016) Arctic charr exploit restricted urbanized coastal areas during marine migration: could they be in harm's way? Hydrobiologia 783(1):335–345. https://doi.org/10.1007/s10750-016-2787-6

54. Lucia M, Gabrielsen GW, Herzke D, Christensen GN (2016) Screening of UV chemicals, bisphenols and siloxanes in the Arctic. Norwegian Environment Agency, M-598/2016. Norwegian Polar Institute, Brief report no. 039, Tromsø, Norway

55. Jartun M, Fjeld E, Bæk K, Løken KB, Rundberget T, Grung M, Schlabach M, Warner NA, Johansen I, Lyche JL, Berg V, Nøstbakken OJ (2018) Monitoring of environmental contaminants in freshwater ecosystems. Norwegian Environment Agency, M-1106/2018. Norwegian Institute for Water Research and Norwegian University of Life Sciences, Oslo, Norway

56. Kierkegaard A, van Egmond R, McLachlan MS (2011) Cyclic volatile methylsiloxane bioaccumulation in flounder and ragworm in the Humber Estuary. Environ Sci Technol 45 (14):5936–5942. https://doi.org/10.1021/es200707r

57. Gobas FAPC, Muir DCG, Mackay D (1988) Dynamics of dietary bioaccumulation and fecal elimination of hydrophobic organic chemicals in fish. Chemosphere 17(5):943–962

58. Green NW, Schøyen M, Hjermann DØ, Øxnevad S, Ruus A, Lusher A, Beylich B, Lund E, Tveiten L, Håvardstun J, Jenssen MTS, Ribeiro AL, Bæk K (2018) Contaminants in coastal waters of Norway 2017. Norwegian Environment Agency, M-1120/2018. Norwegian Institute for Water Research, NIVA-report 7302-2018, Oslo, Norway

59. Knudsen LB, Sagerup K, Polder A, Schlabach M, Josefsen TD, Strøm H, Skåre JU, Gabrielsen GW (2007) Halogenated organic contaminants and mercury in dead or dying seabirds on Bjørnøya (Svalbard). Norwegian Pollution Control Authority, TA-2222/2007. Norwegian Polar Institute, SPFO-Report 977/2007, Tromsø, Norway

60. Huber S, Warner NA, Nygård T, Remberger M, Harju M, Uggerud HT, Kaj L, Hanssen L (2015) A broad cocktail of environmental pollutants found in eggs of three seabird species from remote colonies in Norway. Environ Toxicol Chem 34(6):1296–1308. https://doi.org/10.1002/etc.2956

61. Kierkegaard A, Bignert A, McLachlan MS (2013) Cyclic volatile methylsiloxanes in fish from the Baltic Sea. Chemosphere 93(5):774–778. https://doi.org/10.1016/j.chemosphere.2012.10.048

62. Tobin JM, McNett DA, Durham JA, Plotzke KP (2008) Disposition of decamethylcyclopentasiloxane in Fischer 344 rats following single or repeated inhalation exposure to 14C-decamethylcyclopentasiloxane (14C-D5). Inhal Toxicol 20(5):513–531. https://doi.org/10.1080/08958370801935075

63. Sarangapani R, Teeguarden J, Andersen ME, Reitz RH, Plotzke KP (2003) Route-specific differences in distribution characteristics of octamethylcyclotetrasiloxane in rats: analysis using PBPK models. Toxicol Sci 71(1):41–52. https://doi.org/10.1093/toxsci/71.1.41

64. Domoradzki JY, Sushynski CM, Sushynski JM, McNett DA, Van Landingham C, Plotzke KP (2017) Metabolism and disposition of [(14)C]-methylcyclosiloxanes in rats. Toxicol Lett 279 (Suppl 1):98–114. https://doi.org/10.1016/j.toxlet.2017.05.002

65. Lu Z, Martin PA, Burgess NM, Champoux L, Elliott JE, Baressi E, De Silva AO, de Solla SR, Letcher RJ (2017) Volatile methylsiloxanes and organophosphate esters in the eggs of European starlings (Sturnus vulgaris) and congeneric gull species from locations across Canada. Environ Sci Technol 51(17):9836–9845. https://doi.org/10.1021/acs.est.7b03192

66. Hanssen L, Warner NA, Braathen T, Odland JO, Lund E, Nieboer E, Sandanger TM (2013) Plasma concentrations of cyclic volatile methylsiloxanes (cVMS) in pregnant and postmenopausal Norwegian women and self-reported use of personal care products (PCPs). Environ Int 51:82–87. https://doi.org/10.1016/j.envint.2012.10.008

67. Xu S, Kropscott B (2013) Octanol/air partition coefficients of volatile methylsiloxanes and their temperature dependence. J Chem Eng Data 58(1):136–142. https://doi.org/10.1021/je301005b

68. Xu S, Kropscott B (2014) Evaluation of the three-phase equilibrium method for measuring temperature dependence of internally consistent partition coefficients (KOW, KOA, and KAW) for volatile methylsiloxanes and trimethylsilanol. Environ Toxicol Chem 33(12):2702–2710. https://doi.org/10.1002/etc.2754

69. Xu S, Kropscott B (2012) Method for simultaneous determination of partition coefficients for cyclic volatile methylsiloxanes and dimethylsilanediol. Anal Chem 84(4):1948–1955. https://doi.org/10.1021/ac202953t

70. Schenker U, MacLeod M, Scheringer M, Hungerbühler K (2005) Improving data quality for environmental fate models: a least-squares adjustment procedure for harmonizing physicochemical properties of organic compounds. Environ Sci Technol 39(21):8434–8441. https://doi.org/10.1021/es0502526

71. Panagopoulos D, Jahnke A, Kierkegaard A, MacLeod M (2017) Temperature dependence of the organic carbon/water partition ratios (KOC) of volatile methylsiloxanes. Environ Sci Technol Lett 4(6):240–245. https://doi.org/10.1021/acs.estlett.7b00138

72. Panagopoulos D, Jahnke A, Kierkegaard A, MacLeod M (2015) Organic carbon/water and dissolved organic carbon/water partitioning of cyclic volatile methylsiloxanes: measurements and polyparameter linear free energy relationships. Environ Sci Technol 49(20):12161–12168. https://doi.org/10.1021/acs.est.5b02483

73. Kozerski GE, Xu S, Miller J, Durham J (2014) Determination of soil-water sorption coefficients of volatile methylsiloxanes. Environ Toxicol Chem 33(9):1937–1945. https://doi.org/10.1002/etc.2640

74. Xu S, Wania F (2013) Chemical fate, latitudinal distribution and long-range transport of cyclic volatile methylsiloxanes in the global environment: a modeling assessment. Chemosphere 93 (5):835–843. https://doi.org/10.1016/j.chemosphere.2012.10.056

75. Wania F (2006) Potential of degradable organic chemicals for absolute and relative enrichment in the Arctic. Environ Sci Technol 40(2):569–577. https://doi.org/10.1021/es051406k

76. Kim J, Mackay D, Whelan MJ (2018) Predicted persistence and response times of linear and cyclic volatile methylsiloxanes in global and local environments. Chemosphere 195:325–335. https://doi.org/10.1016/j.chemosphere.2017.12.071

77. Webster E, Mackay D, Wania F (1998) Evaluating environmental persistence. Environ Toxicol Chem 17(11):2148–2158. https://doi.org/10.1002/etc.5620171104

78. Altshuller AP (1989) Ambient air hydroxyl radical concentrations – measurements and model predictions. J Air Waste Manage Assoc 39(5):704–708. https://doi.org/10.1080/08940630.1989.10466556

79. Xiao R, Zammit I, Wei Z, Hu W-P, MacLeod M, Spinney R (2015) Kinetics and mechanism of the oxidation of cyclic methylsiloxanes by hydroxyl radical in the gas phase: an experimental and theoretical study. Environ Sci Technol 49(22):13322–13330. https://doi.org/10.1021/acs.est.5b03744

80. Xu S, Warner N, Bohlin-Nizzetto P, Durham J, McNett D (2019) Long-range transport potential and atmospheric persistence of cyclic volatile methylsiloxanes based on global measurements. Chemosphere 228:460–468. https://doi.org/10.1016/j.chemosphere.2019.04.130

81. Kim J, Xu S (2016) Sorption and desorption kinetics and isotherms of volatile methylsiloxanes with atmospheric aerosols. Chemosphere 144:555–563. https://doi.org/10.1016/j.chemosphere.2015.09.033

82. Kim J, Xu S, Varaprath S (2009) Removal of trace-level D4 in air by mineral aerosol. SETAC Europe 19th Annual Meeting, Gothenburg, Sweden

83. Xu S, Chandra G (1999) Fate of cyclic methylsiloxanes in soils. 2. Rates of degradation and volatilization. Environ Sci Technol 33(22):4034–4039. https://doi.org/10.1021/es990099d

84. Janechek NJ, Hansen KM, Stanier CO (2017) Comprehensive atmospheric modeling of reactive cyclic siloxanes and their oxidation products. Atmos Chem Phys 17(13):8357. https://doi.org/10.5194/acp-17-8357-2017

85. MacLeod M, von Waldow H, Tay P, Armitage JM, Wöhrnschimmel H, Riley WJ, McKone TE, Hungerbühler K (2011) BETR global – a geographically-explicit global-scale multimedia contaminant fate model. Environ Pollut 159(5):1442–1445. https://doi.org/10.1016/j.envpol.2011.01.038

86. Whelan MJ (2013) Evaluating the fate and behaviour of cyclic volatile methyl siloxanes in two contrasting North American lakes using a multi-media model. Chemosphere 91 (11):1566–1576. https://doi.org/10.1016/j.chemosphere.2012.12.048

87. Whelan MJ, Breivik K (2013) Dynamic modelling of aquatic exposure and pelagic food chain transfer of cyclic volatile methyl siloxanes in the Inner Oslofjord. Chemosphere 93(5):794–804. https://doi.org/10.1016/j.chemosphere.2012.10.051

88. Hughes L, Mackay D, Powell DE, Kim J (2012) An updated state of the science EQC model for evaluating chemical fate in the environment: application to D5 (decamethylcyclopentasiloxane). Chemosphere 87(2):118–124. https://doi.org/10.1016/j.chemosphere.2011.11.072

89. Mackay D, Hughes L, Powell DE, Kim J (2014) An updated quantitative water air sediment interaction (QWASI) model for evaluating chemical fate and input parameter sensitivities in aquatic systems: application to D5 (decamethylcyclopentasiloxane) and PCB-180 in two lakes. Chemosphere 111:359–365. https://doi.org/10.1016/j.chemosphere.2014.04.033

90. Panagopoulos D, MacLeod M (2018) A critical assessment of the environmental fate of linear and cyclic volatile methylsiloxanes using multimedia fugacity models. Environ Sci Process Impacts 20(1):183–194. https://doi.org/10.1039/c7em00524e

91. Undeman E, Gustafsson B, Humborg C, McLachlan M (2015) Application of a novel modeling tool with multistressor functionality to support management of organic contaminants in the Baltic Sea. Ambio 44(3):498–506. https://doi.org/10.1007/s13280-015-0668-2

92. Seston RM, Powell DE, Woodburn KB, Kozerski GE, Bradley PW, Zwiernik MJ (2014) Importance of lipid analysis and implications for bioaccumulation metrics. Integr Environ Assess Manag 10(1):142–144. https://doi.org/10.1002/ieam.1495

93. Jahnke A, Holmbäck J, Andersson RA, Kierkegaard A, Mayer P, MacLeod M (2015) Differences between lipids extracted from five species are not sufficient to explain biomagnification of nonpolar organic chemicals. Environ Sci Technol Lett 2(7):193–197. https://doi.org/10.1021/acs.estlett.5b00145

94. Kozerski GE, McNett D (2015) Determination of storage lipid-to-air partition coefficients and their temperature dependence for Octamethylcyclotetrasiloxane (D4; CAS 556-67-2), Decamethylcyclopentasiloxane (D5; CAS 541-02-6) and Dodecamethylcyclohexasiloxane (D6; CAS 540-97-6). Dow Corning Corporation, HES Study Number 17240-108, Auburn, MI

95. Arnot JA, Mackay D, Parkerton TF, Bonnell M (2008) A database of fish biotransformation rates for organic chemicals. Environ Toxicol Chem 27(11):2263–2270. https://doi.org/10.1897/08-058.1

Concluding Remarks and Future Perspectives

Vera Homem and Nuno Ratola

Contents

Abstract This final chapter intends to convey to the readers the main lessons learned so far, the most important challenges found and an overview of the possible research trends yet to explore in the study of volatile methylsiloxanes (VMSs) in the environment. From all the excellent contributions to this book, besides a comprehensive and thorough display of properties, levels, trends and advances of the current state of the art in several environmental matrices, a number of gaps were detected and discussed and new solutions suggested. Therefore, it is expected that in the near future, the development of new management strategies; new findings regarding the occurrence, behaviour and toxicity of volatile methylsiloxanes; the improvement of risk assessment studies; and the inclusion of chronic exposure assays of mixtures of these substances are the pathways to enhance the expertise of scientists and stakeholders on the science of siloxanes.

Keywords Environment, Future research, Knowledge gaps, Limitations, Volatile methylsiloxanes

V. Homem (✉) and N. Ratola (✉)
LEPABE – Laboratory for Process Engineering, Environment, Biotechnology and Energy, Faculty of Engineering, University of Porto, Porto, Portugal
e-mail: vhomem@fe.up.pt; nrneto@fe.up.pt

V. Homem and N. Ratola (eds.), *Volatile Methylsiloxanes in the Environment*,
Hdb Env Chem (2020) 89: 315–320, DOI 10.1007/698_2019_411,
© Springer Nature Switzerland AG 2019, Published online: 26 November 2019

1 Findings and Challenges

Although the number of studies of volatile methylsiloxanes (VMSs) in the environment has been increasing, crucial information is still missing. VMSs have been detected in larger quantities in wastewater treatment plant (WWTP) influents and sludge, biogas, sediment and air samples. The most common explanation for this predominant presence in such matrices was related to their unique physicochemical properties (high octanol-water and soil organic carbon-water partitioning coefficients, low water solubility and low vapour pressure), as well as to the increase in emission rates and sources (e.g. industrial activities, consumer products as personal care and household products) [1, 2]. However, the current insight about partition behaviour of VMSs in the environment is not enough to explain all results satisfactorily. More tests need to be performed to assess accurately some physicochemical properties and their dependence on environmental factors (e.g. temperature). This gap restricts our understanding of the environmental behaviour and bioaccumulation potential of VMSs. It also hampers the development of reliable environmental fate models. The assessment of VMSs biodegradation capabilities, decay mechanisms, as well as their potential for trophic biomagnification is also necessary. The few tests found in the literature have shown some controversy. In some of the studies, the authors found evidences that VMSs do not accumulate and, consequently, do not undergo biomagnification, while in others an opposite trend is demonstrated [3]. Although VMSs have been detected in biota samples, especially aquatic non-air-breathing animals, tests should be conducted to clarify possible mechanism of accumulation, metabolic pathways in different organisms and trophic transfer. Moreover, there are still many uncertainties regarding VMS assessment of PBT (*persistence, bioaccumulation, toxicity*) properties. In addition, questions have emerged about the persistence of VMSs in the environment and their potential toxicity. Long-range atmospheric transport (LRAT) potential has been discussed and is another matter that lacks consensus. The limited data on the presence of VMSs in remote regions (polar regions) is ambiguous as some authors suggest the presence of VMSs as a result of LRAT, while others explain their presence due to the choice of sampling locations near human settlings [4, 5]. Regarding toxicity, an extremely scarce number of studies is available, which is one of the most pointed out knowledge deficits. Without this information, the data to conduct environmental risk assessment studies is insufficient, and the preliminary reports published so far yield some ambiguous conclusions.

In short, it is essential to increase our knowledge of VMS properties, but also to extend research towards the establishment of long-term environmental monitoring plans with repeated sampling in key sites, that may be defined with the help of advanced chemistry transport modelling. This will be fundamental to establish an accurate environmental fingerprint of these compounds, in order to better understand the extent of their presence in numerous matrices, concentrations changes over time and possible effects and impacts on the environment.

There are reasons that may justify the lack of these field studies. VMSs are ubiquitous compounds, being present in numerous consumer products that are part of our daily routines. In fact, not only do we bring these organic chemicals with us into the lab, using personal care products, but they are also essential components in lab material and equipment (e.g. syringes, lubricants, flasks, septa, etc.) [6]. Therefore, extreme care must be taken to avoid inadvertent contamination, which may lead to erroneous conclusions (false positives). This task requires demanding efforts and measures, such as the handling of samples in cleanrooms or clean air cabinets, employing siloxane-free materials, and banning the use of certain personal care products among laboratory personnel, extra care in sampling, etc.

One of the major challenges of researchers is the development of analytical protocols for VMSs, overcoming background contamination and possible analytical artefacts. For this kind of trace analysis, in which gas chromatography-mass spectrometry (GC-MS) is commonly used as the instrumental method, the internal standard calibration approach is highly recommended. However, the choice of suitable internal standards for VMSs analysis is limited. The compound predominantly used is tetrakis(trimethylsilyloxy)silane (a.k.a. M4Q). Although with a behaviour similar to that of VMSs and availability at an attractive price, its presence in environmental matrices has been reported in a recent study [7], and, therefore, if it cannot be avoided, extra care must be taken when using M4Q as an internal standard. Mass-labelled VMS compounds with deuterium (D) or carbon-13 isotopes (^{13}C) may be used alternatively. However, this is currently a very expensive option, which makes it still inaccessible to a significant number of research laboratories.

Another important gap found in the literature is that other siloxane-based compounds (e.g. chlorinated or fluorinated VMSs) remain largely unknown in the environment, despite being generated daily as by-products of industrial processes. The study of their occurrence and behaviour is essential to understand the overall environmental picture of this family of emerging pollutants.

2 Future Perspectives

An interesting topic addressed in this book (cf. chapter "Presence of Siloxanes in Sewage Biogas and Their Impact on Its Energetic Valorization") is the energetic recovery of biogas and the consequences that the presence of VMSs has in the process. Indeed, renewable energy policies coupled with economic, environmental and climate benefits have encouraged biogas production, with a positive trend over the last few years that meets the principles of circular economy. In Europe, biogas has been mainly used for the production electricity through energy conversion systems (e.g. cogeneration engines) [8]. In addition, some WWTPs also produce biogas to meet their energy needs and eventually sell the surplus to power grids whenever allowed by national legislation. But VMSs can pose a risk to the performance of these energy conversion systems, due to the formation of silicon dioxide deposits during biogas combustion, causing damages to equipment that diminish the quality and energetic potential of biogas, and ultimately require early part replacements. Nowadays, most biogas-producing WWTPs have purification systems (typically based on adsorption processes), but the investment and maintenance costs are high [8].

However, being a renewable fuel beneficial to the economy and the environment, it is expected that biogas production may be increased in the next years, and thus, new technological solutions to reduce or eliminate VMSs and other undesired chemicals may emerge. A recent line of research on this issue is the thermal pre-treatment of sludge. This solution can be advantageous as it not only prevents siloxanes in biogas acting before the anaerobic digestion but also removes them (and eventually other volatile or semi-volatile contaminants) from the sludge. Also, some studies have tried to incorporate certain strains of microorganisms in order to degrade VMSs, and this could be a way for WWTPs to select the most indicated strains for that purpose without compromising the normal biological treatment activity. In both these cases, with an appropriate conditioning, the sludge in question may be further digested and used as a safe fertilizer.

Another future trend that is, for instance, in line with the societal challenges of EU framework programmes focused on sustainable cities, and mobility is the purification of biogas to higher-quality biomethane for use as vehicle fuel or for injection into regional or national natural gas grids. In fact, Europe is the world's prominent biomethane producer destined to those two ends, with Germany leading by country, followed by Sweden [8]. However, not all countries have this process within their legislation (particularly the southern and eastern nations), which leaves room for an expansion of the net production and the geographical coverage. Once again, the development of new green removal technologies will likely walk hand in hand with this technological development. And whenever VMSs are present in these processes, their removal will be mandatory to achieve the desired sustainability.

Still, the major concerns of the scientific community regarding VMSs are the sources and their rampant production and use in consumer goods. A clear example of this is the publication of an amendment to Annex XVII to Regulation (EC) No. 1907/2006 of REACH (*Registration, Evaluation, Authorisation and Restriction of Chemicals*) in January 2018, restricting the concentration of D4 and D5 to less than 0.1% w/w in rinse-off cosmetic products. This measure enters into force after 31 January 2020 and was adopted due to the fact that "a risk to the environment arises from the presence of D4 and D5 in certain cosmetic products that are washed off with water after application, because of their hazard properties as a PBT and a vPvB substance in the case of D4 and a vPvB substance in the case of D5" [9]. Moreover, the European Chemicals Agency (ECHA) proposed in mid-January 2019 new restrictions under the REACH protocols. Leave-on personal care products and other consumer/professional products (e.g. dry cleaning, waxes and polishes, washing and cleaning products) containing D4/D5/D6 in concentrations higher than 0.1% shall not be placed on the market as well as wash-off cosmetic products containing D6 [10]. On the other hand, CES – Silicones Europe (a non-profit trade organisation representing all major producers of silicones in Europe) declared that ongoing regulatory activities are disproportionate and do not guarantee environmental protection since the use of silicone polymers reduces the carbon footprint [11, 12]. The same organisation also expressed concerns about the potential impacts of this regulation on the use (and potential replacement) of silicone polymers in existing sustainable solutions and future technologies. CES expects to be able to continue using siloxanes, as in its opinion the socioeconomic benefit outweighs the environmental advantage of the potential reduction that would be achieved in siloxane emissions [11, 12]. In fact, if these restrictions were approved, the industry will have to find new alternatives.

In addition to these actions, it is also necessary to develop more appropriate risk management strategies. For that, it is expected that in the near future, the regulatory agencies can include these compounds in long-term environmental monitoring plans, identify trends and improve our knowledge on human and environmental exposure, as well as assess the risk in realistic scenarios. It is also important to highlight the importance of including new toxicity studies on chronic effects of the presence of VMS mixtures in the environment, which are unknown to this day.

These are only a few indications for the future of research in siloxanes. We believe the ground for evolution in this field is still considerable and that many other approaches will contribute to the continuous clarification of the presence, behaviour and impacts of these ubiquitous chemicals.

References

1. Global Silicones Council (2016) Socio-economic evaluation of the global silicones industry. Final Report, London. https://sehsc.americanchemistry.com/Socio-Economic-Evaluation-of-the-Global-Silicones-Industry-Final-Report.pdf. Accessed Jan 2018
2. Mojsiewicz-Pieńkowska K, Krenczkowska D (2018) Evolution of consciousness of exposure to siloxanes – review of publications. Chemosphere 191:204–217
3. Homem V, Capela D, Silva JS, Cincinelli A, Santos L, Alves A, Ratola N (2017) An approach to the environmental prioritisation of volatile methylsiloxanes in several matrices. Sci Total Environ 579:506–513
4. Sanchís J, Cabrerizo A, Galbán-Malagón C, Barceló D, Farré M, Dachs J (2015) Unexpected occurrence of volatile dimethylsiloxanes in antarctic soils, vegetation, phytoplankton, and krill. Environ Sci Technol 49:4415–4424
5. Warner NA, Evenset A, Christensen G, Gabrielsen GW, Borgå K, Leknes H (2010) Volatile siloxanes in the European arctic: assessment of sources and spatial distribution. Environ Sci Technol 44:7705–7710
6. Chainet F, Lienemann C-P, Courtiade M, Ponthusa J, François O, Donard X (2011) Silicon speciation by hyphenated techniques for environmental, biological and industrial issues: a review. J Anal At Spectrom 26:30–51
7. Wang D-G, Steer H, Pacepavicius G, Smyth SA, Kinsman L, Alaee M (2013) Seasonal variation and temperature-dependent removal efficiencies of cyclic volatile methylsiloxanes in fifteen wastewater treatment plants. Organohalogen Compd 75:1286–1290
8. Scarlat N, Dallemand J-F, Fahl F (2018) Biogas: developments and perspectives in Europe. Renew Energy 129(Part A):457–472
9. European Commission (2018) Commission Regulation (EU) 2018/35 of 10 January 2018 amending Annex XVII to Regulation (EC) No 1907/2006 of the European Parliament and of the Council concerning the Registration, Evaluation, Authorisation and Restriction of Chemicals (REACH) as regards octamethylcyclotetrasiloxane ('D4') and decamethylcyclopentasiloxane ('D5'). Off J Eur Union L 6:45–47

10. European Chemical Agency (ECHA) (2018) Annex XV restriction report – proposal for a restriction – D4, D5 and D6. Finland. https://echa.europa.eu/registry-of-restriction-intentions/-/dislist/details/0b0236e181a55ade. Accessed June 2019
11. CES – Silicones Europe (2018) Media statement: the addition of D4, D5 and D6 to the candidate list under REACH is disproportionate and endangers critical beneficial uses. https://www.silicones.eu/wp-content/uploads/2018/06/The-addition-of-D4-D5-and-D6-to-the-Candidate-list-under-REACH-is-disproportionate-and-endangers-critical-beneficial-uses-1.pdf. Accessed July 2019
12. Oziel C (2019) Echa's new siloxanes restriction plan faces industry opposition. Chemical Watch – Global Risk & Regulation News. https://chemicalwatch.com/74170/echas-new-siloxanes-restriction-plan-faces-industry-opposition. Accessed July 2019

CPSIA information can be obtained
at www.ICGtesting.com
Printed in the USA
LVHW050412060721
691900LV00001B/63

9 783030 501372